Springer Texts in Statistics

Springer Texts in Statistics

Alfred	Elements of Statistics for the Life and Social Sciences
Blom	Probability and Statistics: Theory and Applications
Chow and Teicher	Probability Theory: Independence, Interchangeability, Martingales. Second Edition
Christensen	Plane Answers to Complex Questions: The Theory of Linear Models
du Toit, Steyn and Strumpf	Graphical Exploratory Data Analysis
Kalbfleisch	Probability and Statistical Inference: Volume 1: Probability. Second Edition
Kalbfleisch	Probability and Statistical Inference: Volume 2: Statistical Inference. Second Edition
Keyfitz	Applied Mathematical Demography. Second Edition
Kiefer	Introduction to Statistical Inference
Kokoska and Nevison	Statistical Tables and Formulae
Madansky	Prescriptions for Working Statisticians
McPherson	Statistics in Scientific Investigation: Basis, Application and Interpretation
Nguyen and Rogers	Fundamentals of Mathematical Statistics: Volume I: Probability for Statistics
Nguyen and Rogers	Fundamentals of Mathematical Statistics: Volume II: Statistical Inference

Thomas J. Santner Diane E. Duffy

The Statistical Analysis of Discrete Data

With 30 Illustrations

Springer-Verlag
New York Berlin Heidelberg
London Paris Tokyo Hong Kong

Thomas J. Santner
School of Operations Research
and Industrial Engineering
Cornell University
Ithaca, NY 14853
USA

Diane E. Duffy
Bell Communications Research
Morristown, NJ 07960
USA

Mathematics Subject Classification (1980): 62-07, 62H17, 62H12, 62H15, 62J99, 62F15

Library of Congress Cataloging-in-Publication Data
Santner, Thomas J.
The statistical analysis of discrete data / Thomas J. Santner and
Diane E. Duffy.
p. cm. —(Springer texts in statistics)
Bibliography: p.
Includes indexes.
ISBN 0-387-97018-5
1. Multivariate analysis. I. Duffy, Diane E. II. Title.
III. Series.
QA278.S26 1989
519.5'35—dc20 89-34062

Printed on acid-free paper.

Camera-ready copy prepared using LaTEX.
Printed and bound by R. R. Donnelley & Sons, Harrisonburg, Virginia.
Printed in the United States of America.

9 8 7 6 5 4 3 2 1

ISBN 0-387-97018-5 Springer-Verlag New York Berlin Heidelberg
ISBN 3-540-97018-5 Springer-Verlag Berlin Heidelberg New York

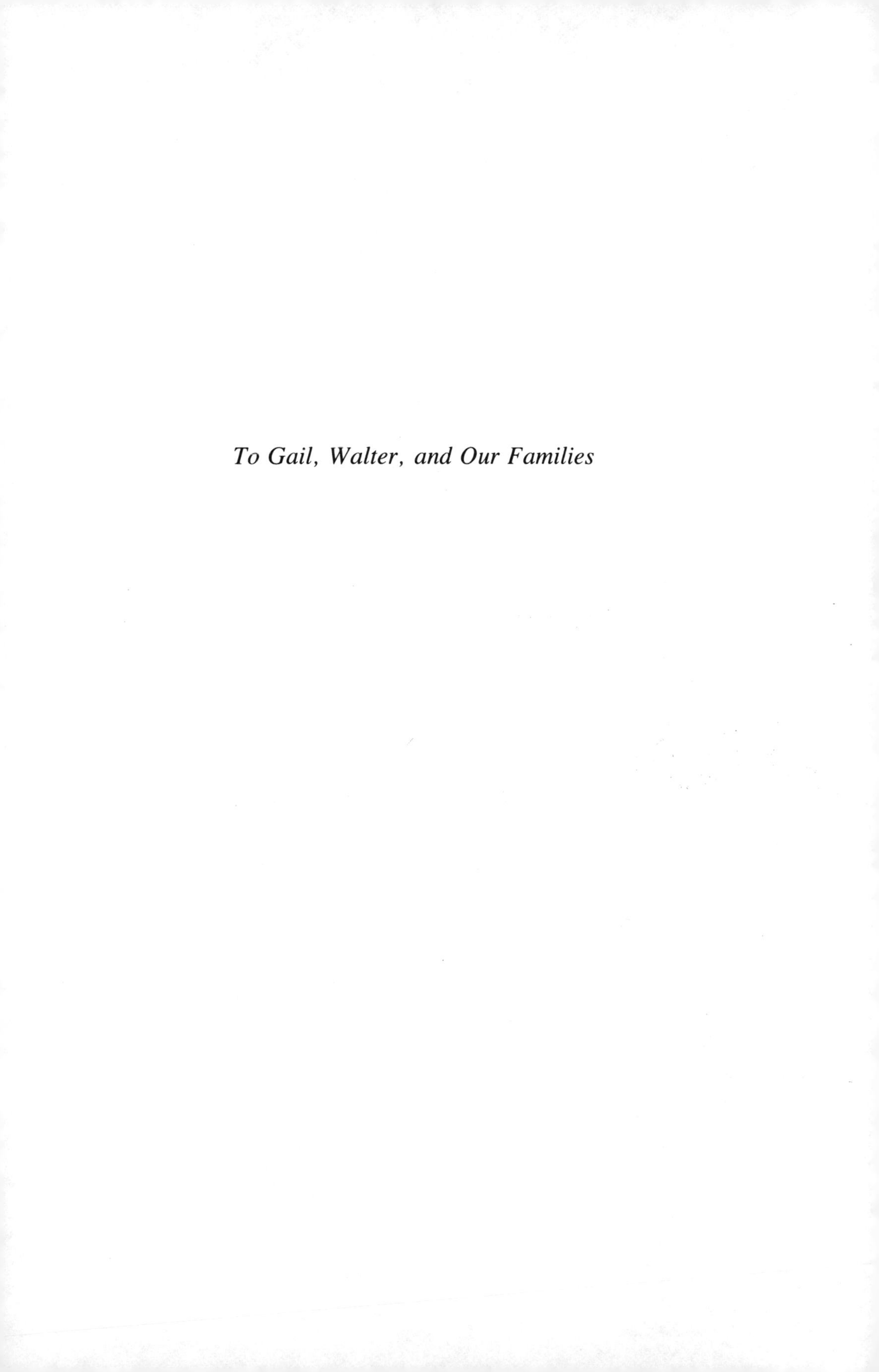

To Gail, Walter, and Our Families

Contents

Preface

The Statistical Analysis of Discrete Data provides an introduction to current statistical methods for analyzing discrete response data. The book can be used as a course text for graduate students and as a reference for researchers who analyze discrete data. The book's mathematical prerequisites are linear algebra and elementary advanced calculus. It assumes a basic statistics course which includes some decision theory, and knowledge of classical linear model theory for continuous response data. Problems are provided at the end of each chapter to give the reader an opportunity to apply the methods in the text, to explore extensions of the material covered, and to analyze data with discrete responses. In the text examples, and in the problems, we have sought to include interesting data sets from a wide variety of fields including political science, medicine, nuclear engineering, sociology, ecology, cancer research, library science, and biology.

Although there are several texts available on discrete data analysis, we felt there was a need for a book which incorporated some of the myriad recent research advances. Our motivation was to introduce the subject by emphasizing its ties to the well-known theories of linear models, experimental design, and regression diagnostics, as well as to describe alternative methodologies (Bayesian, smoothing, etc.); the latter are based on the premise that external information is available. These overriding goals, together with our own experiences and biases, have governed our choice of topics.

The text covers both single sample problems (Chapter 2) and problems with structured means which can be studied via loglinear and logistic models (Chapters 3 through 5). Classical maximum likelihood estimators, as well as estimators based on Bayesian, smoothing, shrinkage, and ridge approaches, are described for estimating parameters in structured and unstructured problems. Maximum likelihood estimation theory for loglinear models is developed via the notion of linear projection to highlight the similarities with maximum likelihood estimation for normal linear models. In addition to the standard testing and estimation formulations, problems of simultaneous interval estimation, multiple comparisons, and ranking and selection are considered with references to the appropriate literature. We describe and compare various small sample methods, especially for common confidence interval problems, as these are widely used in applications. Descriptions of recent research on graphical models for contingency tables, and diagnostic tools for loglinear models and logistic regression are included.

This book is not intended to cover every aspect of the statistical analysis of discrete data. Many important topics, such as measures of association, models for measuring change, the analysis of ordinal data, incomplete and missing data, and the analysis of panel and repeated measurement data are not included. Some of these subjects are sufficiently complicated to warrant book-length treatments, and many are well covered either by existing texts or by comprehensive survey papers.

We are grateful to many members of the statistics community, especially colleagues at Cornell University and at Bellcore for their support, ideas, examples, and references which added considerably to the text. Sid Dalal and Ed Fowlkes introduced us to the space shuttle example (Example 1.2.9), Chuck McCulloch suggested the example on pirating eagles (Example 1.2.8), and Joe Gastwirth pointed out the data on police exams used in Problem 5.15. We would like to thank those Cornell graduate students, especially Andy Forbes, who criticized early versions of the course on discrete data from which this book originated. We deeply appreciate the efforts of Jon Kettenring who gave the completed manuscript a critical reading which improved the final version. Of course, any errors or omissions in the final book are the sole responsibility of the authors. We would like to acknowledge Kathy King's skillful typing of portions of the manuscript, and the help of the Springer-Verlag staff. The first author's research was partially supported by the U.S. Army Research Office through the Mathematical Sciences Institute at Cornell University. Finally, we would like to thank Gail Santner and Walter Willinger for help with proofreading and for patience and support during the long period in which we kept assuring them that the book was "almost" done.

Thomas J. Santner
Diane E. Duffy

Introduction

1.1 Classes of Statistical Problems

Statistical problems can be classified according to the types of variables observed. Two different criteria for distinguishing variables are important in this book. First, it is convenient to differentiate between (i) responses and (ii) explanatory variables (which affect the responses). In a given problem how one makes the distinction depends on the study design and the scientific goals of the investigation. Second, variables can be distinguished according to their scale of measurement. Four measurement scales are described below.

Nominal Scale

A nominal scale categorizes the data into distinct groups. Examples of variables measured on nominal scales are: sex (male/female) and race (Black/Caucasian/Native American/other).

Ordinal Scale

An ordinal scale both categorizes into groups and orders the groups. Examples of variables measured using ordinal scales are: pain (none/moderate/severe) and socio-economic status (low/middle/high).

Interval Scale

Interval scales categorize, order, and quantify comparisons between *pairs* of measurements. An example is temperature measured in °F for which it is clear that the difference between 40°F and 20°F is equal to that between 100°F and 80°F; i.e., 40°F − 20°F = 100°F − 80°F. Similarly, this difference is less than that between 50°F and 35°F. However, it is not true that 40°F is twice as hot as 20°F (think of changing these measurements to the centigrade scale). Comparisons between individual measurements cannot be performed with an interval scale. Interval scales require a unit of measurement and an *arbitrary* origin.

Ratio Scale

A ratio scale categorizes, orders, quantifies comparisons between pairs of measurements, and quantifies comparisons between individual measurements. One example is temperature measured on the Kelvin scale; it is true that 40°K is twice as hot as 20°K. A second example is length. A ratio scale requires a unit of measurement and an *absolute* origin.

These measurement scales are used to classify variables in the following way. Variables measured on either nominal or ordinal scales are called *qualitative* while those measured on either interval or ratio scales are called *quantitative.* For example, the explanatory variable, "patient sex" would be qualitative while "baking temperature" of a kiln in a manufacturing process would be quantitative.

Variables are called *discrete* if they can assume either a finite or countable number of values; they are called *continuous* if they can assume any value in some interval. For example, the response "number of accidents at an intersection over a fixed period of time" is discrete while the response "output voltage from an electrical circuit" is continuous. Discrete variables can be either quantitative or qualitative; continuous variables are quantitative. The discrete response "pain" measured on the ordinal scale (none/moderate/severe) is qualitative. The discrete response "number of accidents at an intersection" is measured on a ratio scale and hence is quantitative (six accidents are twice as many as three accidents).

This book studies problems for which the response variables are discrete. The next section provides some motivating examples of these problems. The final section of Chapter 1 reviews discrete distributions. Chapter 2 describes both maximum likelihood and alternative analyses for data from a single binomial distribution (Section 2.1), single unstructured multinomial distribution (Section 2.2), and several Poisson distributions with unstructured means (Section 2.3). The reader should review the material in Appendix 4 on large sample theory in preparation for reading Chapter 2. Sections 3.1–3.3 study the classical theory of maximum likelihood estimation of a vector of means from either Poisson, multinomial, or product multinomial random variables which follow a loglinear model. Section 3.4 considers alternative methods for such data. The reader should be familiar with Appendices 1 and 2 on linear algebra and the maximization of concave functions, respectively, before reading this material. Chapter 4 considers cross-classified data in detail. Discussions of variable selection, residual analysis, and collapsing are included. The final chapter studies problems with a single response variable and one or more qualitative or quantitative explanatory variables. Section 5.2 considers the problems of comparing $T \geq 2$ binomial populations. Section 5.3 covers the likelihood analysis of binary regression data. More specialized topics in logistic regression, including recently developed methods of inference and graphical assessments of fit and influence, are discussed in Section 5.4. Section 5.5 introduces problems of stratified or matched data which lead to models with many strata-level nuisance parameters.

1.2 Examples

The examples discussed below are divided into two groups: (i) discrete univariate or multivariate problems meaning that (all) the variable(s) is (are) qualitative and (ii) discrete response regression problems consisting of a mixture of discrete and continuous variables of which one or more discrete variables are responses and the remainder are explanatory variables affecting the distribution of the responses. Discrete multivariate data are ordinarily displayed as a contingency table. One or more of the variables are responses and the remainder (if any) are explanatory variables; the extremes range from the case of all the variables being responses to the case of a single response with all the other variables being explanatory. The examples below illustrate all the possibilities above.

Example 1.2.1. The data in Table 1.2.1 arose in an engineering application described in Drinkwater and Hastings (1967). The counts in the table are the number of times in one year that each of 550 army vehicles was sent for repair. The data consist of a univariate discrete response for each of the 550 vehicles. The Poisson distribution is a possible probability model for these data since there is a fixed period of time over which the study was conducted. The general question of assessing goodness-of-fit to the Poisson distribution will be explored in Section 2.3. These data are analyzed in Problem 2.22.

Table 1.2.1. Frequency of Repair for 550 Vehicles (Reprinted with permission from Drinkwater and Hastings: "An Economic Replacement Model," *Operations Research Quarterly* 18, 1967.)

	Number of Repairs						
	0	1	2	3	4	5	6+
Number of Vehicles	295	190	53	5	5	2	0

Example 1.2.2. Table 1.2.2 from Carp and Rowland (1983) concerns judicial decisions made between 1933 and early 1977 by judges appointed by Presidents Johnson and Nixon to serve in the Federal District Courts. The decisions are classified by the type of case and whether they are liberal or conservative opinions. These data were collected from the Federal Supplement, a major publisher of trial court opinions, and consist of a trivariate

discrete response for each relevant case. The data are displayed in a multinomial contingency table with 76 = 2 (Johnson or Nixon appointee) × 2 (liberal or conservative opinion) × 19 (type of case) cells. The goals of the study are to analyze whether Nixon appointees hand down relatively fewer liberal decisions than Johnson appointees, and whether the type of case effects the comparison between the liberal-conservative behavior of Johnson appointees versus Nixon appointees. These questions are addressed in Problem 5.36.

Table 1.2.2. Judicial Decisions in Federal District Courts from 1933 to 1977 with Liberal (L) and Conservative (C) Opinions (Reprinted with permission from *Policymaking and Politics in the Federal District Courts* by R.A. Carp and C.K. Rowland. Univ. of Tennessee Press, 1983.)

	Nixon Appointees		Johnson Appointees	
Type of Case	L	C	L	C
Race Discrimination	101	172	279	144
14th Amendment	234	434	513	387
Criminal Court Motions	134	447	352	598
Fair Labor Standards Act	45	57	134	58
Local Econ. Regulation	41	37	112	23
Freedom of Expression	74	97	234	120
Women's Rights	44	55	64	30
Union Members vs. Union	17	46	40	32
Environmental Protection	73	82	129	63
Freedom of Religion	37	42	111	63
U.S. Habeas Corpus Pleas	41	97	146	219
Criminal Conviction	29	55	45	52
U.S. Commercial Reg.	126	84	292	92
St. Habeas Corpus Pleas	102	289	282	627
Indian Rights and Law	10	15	18	20
Union vs. Company	74	95	102	94
Employee vs. Employer	46	85	45	67
Alien Petitions	24	25	44	37
Voting Rights Cases	36	34	36	35

Example 1.2.3. The data in this example originally appeared in Madsen (1976). They come from a survey taken in Copenhagen, Denmark which studied satisfaction with housing conditions. The study was conducted in twelve areas of the city with similar social status and consisting of rental units built between 1960 and 1968. A total of 1681 persons surveyed were classified with respect to the following four attributes:

(1) Type of housing (4 levels—tower blocks, apartment houses with less than 5 stories, atrium houses, and terraced houses),

(2) Satisfaction with housing conditions (3 levels—low, medium, and high),

(3) Degree of contact with other residents (2 levels—low and high), and

(4) Feeling of influence on apartment management (3 levels—low, medium, and high).

The resulting $4 \times 3 \times 2 \times 3$ contingency table of data is displayed in Table 1.2.3. The goal of the investigation is to study the relationships between the type of housing and the other three variables. Thus one might view housing type as an explanatory variable and the other three as discrete responses. Additional details about these data will be given in a case study in Section 4.5.

Table 1.2.3. One Thousand Six Hundred and Eighty-One Persons Classified According to Satisfaction, Contact, Influence, and Type of Housing (Reprinted with permission from M. Madsen: "Statistical Analysis of Multiple Contingency Tables: Two Examples," *Scandinavian Journal of Statistics,* 1976. The Almquist & Wiksell Periodical Company.)

Contact		Low			High		
Satisfaction		Low	Medium	High	Low	Medium	High
Housing	Influence						
Tower blocks	Low	21	21	28	14	19	37
	Medium	34	22	36	17	23	40
	High	10	11	36	3	5	23
Apartments	Low	61	23	17	78	46	43
	Medium	43	35	40	48	45	86
	High	26	18	54	15	25	62
Atrium houses	Low	13	9	10	20	23	20
	Medium	8	8	12	10	22	24
	High	6	7	9	7	10	21
Terraced houses	Low	18	6	7	57	23	13
	Medium	15	13	13	31	21	13
	High	7	5	11	5	6	13

The appropriate sampling model for cross-classified discrete multivariate responses (i.e., Examples 1.2.2 and 1.2.3) is multinomial with cells formed by the cross-classification. Most of the questions formulated in such cases can be phrased in terms of the joint distribution of the classification variables. Five major approaches have been proposed for performing inference on this joint distribution:

(i) likelihood methods,

(ii) weighted least squares,

(iii) information theoretic approaches,

(iv) Bayesian methods, and

(v) smoothing techniques.

The first three are frequentist in nature while (iv) and (v) assume additional information is available about the problem. The analysis of unstructered multinomial data is studied in Chapter 2 while Chapter 4 discusses cross-classified data.

Example 1.2.4. Schneider et al. (1979) report the data in Table 1.2.4 on the outcome of a clinical trial investigating the effects of vitamin C therapy on a genetic metabolic renal disorder called nephropathic cystosis. The trial contained two groups, those receiving and those not receiving vitamin C. The response was clinical improvement (Y/N). The data are bivariate with one binary response variable and one binary explanatory variable.

A stochastic model for the number of patients in the two groups experiencing clinical improvement is that of two independent binomial populations with possibly different probabilities of clinical improvement. The study is prospective because individuals are selected at random to enter the two groups (vitamin C versus control) and then are followed forward in time to determine whether clinical improvement occurs. The research question is whether vitamin C increases the probability of clinical improvement and, if so, to quantify the increase. These data are discussed further in Problems 2.8 and 5.9.

A retrospective study to address the research question posed in Example 1.2.4 could be conducted by examining two groups of nephropathic cystosis patients, one of which exhibited improvement and the other of which did not. Each patient would be classified according to whether they had taken vitamin C (in the past) or not. The following paragraphs introduce retrospective studies and contrast them with prospective investigations.

Table 1.2.4. Results of a Two-Group Clinical Trial Measuring the Effect of Vitamin C on Nephropathic Cystosis (Reprinted with permission from J. Schneider, J. Schlesselman, S. Mendoza, S. Orloff, J. Thoene: "Ineffectiveness of Ascorbic Acid Therapy in Nephropathic Cystinosis," *New England Journal of Medicine,* vol. 300, pg. 756, 1979. Massachusetts Medical Society.)

	Clinical Improvement	
	Y	N
vitamin C	24	8
no vitamin C	29	3

Suppose m_1 subjects, called cases (D), having a certain disease are studied together with m_2 disease-free subjects, called controls ($\sim D$). The problem is to determine whether or not exposure to a binary factor E (yes/no) affects the chance a subject becomes diseased. Examples of exposure factors are fluoride use, cigarette smoking, and alcohol consumption. The data for each subject are the disease state ($D \mid \sim D$) and the exposure status ($E \mid \sim E$). Retrospective studies derive their name from the fact that one looks backward in time to determine the exposure status of the individuals. Table 1.2.5 displays the generic form of retrospective data. Note that the numbers of diseased (m_1) and disease-free (m_2) subjects are determined by the study design. One interesting hypothesis is that exposure does not affect the chance of contracting the disease; i.e., $H_0 : P[D \mid E] = P[D \mid \sim E]$. If these probabilities are not the same, then a confidence interval for some measure of discrepancy between $P[D \mid E]$ and $P[D \mid \sim E]$ is of interest. The problem that arises is that only $P[E \mid D]$ and $P[E \mid \sim D]$ (and functions of them) are directly estimable from the data since m_1 and m_2 are fixed by design. These considerations and other issues in the analysis of retrospective studies will be discussed in Sections 5.1 and 5.2.

Table 1.2.5. Generic 2×2 Table of Retrospective Case Control Data

	Cases	Controls
Exposed	Y_1	Y_2
Unexposed	$m_1 - Y_1$	$m_2 - Y_2$
	m_1	m_2

Example 1.2.5. Tuyns et al. (1977) record the data in Table 1.2.6 on the occurrence of esophogeal cancer and alcohol consumption. The study is retrospective as the number of cases and controls are fixed at 200 and 775, respectively. The problem is to determine whether there is a difference in the incidence of esophogeal cancer between individuals who consume

alcohol at the two levels considered in the study and, if so, to quantify the association. These data are considered further in Problem 5.10.

Table 1.2.6. Number of Cases and Controls (Cancer-Free) Classified by Their Alcohol Consumption (Reprinted with permission from *Statistical Methods in Cancer Research, Vol. I: The Analysis of Cases* by N.E. Breslow and N.E. Day. International Agency for Research on Cancer, World Health Organization, Geneva, Switzerland, 1980.)

		Cases	Controls
alcohol	0–79 gr/day	104	666
consumption	80+ gr/day	96	109
		200	775

Example 1.2.6. The data in Table 1.2.7 are from a tumorigenicity experiment reported by Innes et al. (1969). Four strain-by-sex combinations of mice were used in the study with some mice being treated with the fungicide Avadex (Av) and others not (C). The response is the development of a tumor within two years (Y/N). The object was to determine the possible carcinogenic activity of Avadex.

The data have one binary response variable and three explanatory variables. As in Example 1.2.4, this is a prospective study. An interesting question concerning the design of the experiment is why so many control animals were used compared to treated animals. These data are analyzed in Section 5.5.

Table 1.2.7. Numbers of Mice Developing Tumor within Two Years in a Tumorigenicity Experiment

	Sex	Treatment	Tumor Y	Tumor N
Strain X	M	Av	4	12
		C	5	74
	F	Av	2	14
		C	3	84
Strain Y	M	Av	4	14
		C	10	80
	F	Av	1	14
		C	3	79

Example 1.2.7. Table 1.2.8 is data reported in Farewell (1982) from a clinical trial using several combination chemotherapies in the treatment of cancer. The response variable is severity of nausea measured on an ordinal scale (0 := none, 1, 2, 3, 4, 5 := very severe); the explanatory variable is

binary (presence or absence of cisplatinum in the combination chemotherapy). The object of the trial is to determine the effect of cisplatinum on the severity of patient nausea. One possible model for these data is proposed in Problem 5.4.

Table 1.2.8. Severity of Nausea Classified by the Use of Cisplatinum (Reprinted with permission from V. Farewell: "A Note on the Regression Analysis of Ordinal Data," *Biometrika,* vol. 69, pg. 538, 1982. Biometrika Trust.)

	Severity					
	0	1	2	3	4	5
Cis	7	7	3	12	15	14
No Cis	43	39	13	22	15	29

Example 1.2.8. Knight and Skagen (1988) collected the data in Table 1.2.9 during a field study on the foraging behavior of wintering bald eagles. The data concern 160 attempts by a (pirating) bald eagle to steal chum salmon from another (feeding) bald eagle. For each attempt the size (L = large, S = small) and age (A = adult, I = immature) of the pirating eagle, and the size of the feeding eagle were recorded, along with whether or not the attempted theft was successful. The responses are the number of successful thefts out of the total number attempted. The question of interest is to quantify the effects of the three explanatory variables on the probability of a theft is successful, and to determine if there are any interactions between them. Problem 5.24 addresses these questions.

Table 1.2.9. Pirating Attempts by Bald Eagles (Reprinted with permission from R.L. Knight and S.K. Skagen: "Agonistic Asymmetry and the Foraging of Bald Eagles," *Ecology,* 69. Ecological Society of America, 1988.)

Number of Successful Attempts	Number of Attempts	Size of Pirating Eagle	Age of Pirating Eagle	Size of Feeding Eagle
17	24	L	A	L
29	29	L	A	S
17	27	L	I	L
20	20	L	I	S
1	12	S	A	L
15	16	S	A	S
0	28	S	I	L
1	4	S	I	S
100	160			

Example 1.2.9. This example is taken from the work of Dalal, Fowlkes, and Hoadley (1988, 1989). The data in Table 1.2.10, from the "Report

to the President by the Presidential Commission on the Space Ship Challenger Accident," concern space shuttle flights prior to the January 20, 1986 Challenger explosion. Of the 24 missions prior to the Challenger launch, data are available on 23 missions; the hardware for one flight was lost at sea. The first three columns of Table 1.2.10 list the NASA flight designation, flight dates, and orbiter used for each of the remaining 23 missions. The space shuttle uses two booster rockets to help lift it into orbit. Each booster rocket consists of several pieces whose joints are sealed with rubber O-rings. O-rings are designed to prevent the escape of hot gases produced during combustion. Each booster contains 3 primary O-rings which are inspected post-flight for certain types of damage ("blowby" and "erosion"). Table 1.2.10 lists the number of primary field O-rings (out of 6 per mission) showing signs of damage along with the launch temperature (°F). The Challenger exploded after being launched at 31°F. Before each launch, a pressure leak test of the sealing ability of the O-rings was performed and it is possible that the test itself loosened the seal. Column 6 lists the pressure at which this test was conducted.

Dalal, Fowlkes and Hoadley (1988, 1989) provide extensive background discussion and analysis of the data. One problem is to quantify the information which these data provide about the relationship between launch temperature, leak test pressure, and O-ring damage. Problem 5.31 considers this issue.

Examples 1.2.4–1.2.9 consider discrete response regression data of the form $(Y_i, m_i, \mathbf{x}_i)$, $i = 1(1)T$, where $Y_1, \ldots, Y_T$ are mutually independent binomial (or multinomial) random variables with Y_i based on m_i trials and having vector of cell probabilities $\mathbf{p}_i$. The vector $\mathbf{x}_i = (x_{i1}, \ldots, x_{ik})'$ is a k-dimensional covariate whose components of $\mathbf{x}_i$ can be either quantitative or qualitative. The vector of success probabilities $\mathbf{p}_i = \mathbf{p}(\mathbf{x}_i)$ depends on the covariates. Chapter 5 will consider regression models for the binary response case.

Example 1.2.10. Table 1.2.11 from Moore and Beckman (1988) concerns failures for 90 valves from one pressurized nuclear reactor. For each valve the number of failures and the operating time (in 100 hours) were recorded, as well as five factors which may affect the rate (per 100 hours operating time) at which the valves failed. The five explanatory variables are:

System: 1 = containment, 2 = nuclear, 3 = power conversion, 4 = safety, 5 = process auxiliary.

Operator type: 1 = air, 2 = solenoid, 3 = motor driven, 4 = manual.

Table 1.2.10. NASA Designation, Flight Dates, Orbiter,[1] Launch Temperature (°F), Number of Primary O-rings Showing Damage (out of 6 per Flight), and Pressure (psi) of Pre-Launch Test for Each of Twenty-Three Pre-Challenger Flights

NASA Designation	Flight Dates	Orbiter	Temp.	Number Damaged	Pres.
STS-51-C	1/24/85–1/27/85	DI	53	2	200
STS-41-B	2/3/84–2/11/84	CH	57	1	200
STS-61-C	1/12/86–1/18/86	CO	58	1	200
STS-41-C	4/6/84–4/13/84	CH	63	1	200
STS-1	4/12/81–4/14/81	CO	66	0	50
STS-51-A	11/8/84–11/16/84	DI	67	0	200
STS-51-D	4/12/85–4/19/85	DI	67	0	200
STS-6	4/4/83–4/9/83	CH	67	0	50
STS-5	11/11/82–11/16/82	CO	68	0	50
STS-3	3/22/81–3/30/81	CO	69	0	50
STS-9	11/28/83–12/8/83	CO	70	0	200
STS-51-G	6/17/85–6/24/85	DI	70	0	200
STS-41-D	8/30/84–9/5/84	DI	70	1	200
STS-2	11/12/81–11/14/81	CO	70	1	50
STS-7	6/18/83–6/24/83	CH	72	0	50
STS-8	8/30/83–9/6/83	CH	73	0	100
STS-51-B	4/29/85–5/6/85	CH	75	0	200
STS-61-A	10/30/85–11/6/85	CH	75	2	200
STS-51-I	8/27/85–9/3/85	DI	76	0	200
STS-61-B	11/26/85–12/3/85	AT	76	0	200
STS-41-G	10/5/84–10/13/84	CH	78	0	200
STS-51-J	10/3/85–10/10/85	AT	79	0	200
STS-51-F	7/29/85–8/6/85	CH	81	0	200

[1] AT := Atlantis, CH := Challenger, CO := Columbia, DI := Discovery.

Valve type: 1 = ball, 2 = butterfly, 3 = diaphragm, 4 = gate,
5 = globe, 6 = directional control.

Head size: 1 =$\leq$2 inches, 2 = 2–10 inches, 3 = 10–30 inches.

Operation mode: 1 = normally closed, 2 = normally open.

Problem 3.6 explores one approach to these data based on models linear in the natural logarithm of the rates.

Table 1.2.11. Number of Valve Failures for 90 Valves in a Pressurized Nuclear Reactor (Reprinted with permission from L.M. Moore and R.J. Beckman: "Approximate One-Sided Tolerance Bounds on the Number of Failures," *Technometrics,* vol. 30, no. 3, 1988. American Society for Quality Control and American Statistical Association.)

System	Oper. Type	Valve Type	Size	Operating Mode	Failures	Time (100 Hrs)
1	3	4	3	1	2	1,752
1	3	4	3	2	2	1,752
1	3	5	1	1	1	876
2	1	2	2	2	0	876
2	1	3	2	1	0	876
2	1	3	2	2	0	438
2	1	5	1	1	2	1,752
2	1	5	1	2	4	2,628
2	1	5	2	1	1	438
2	1	5	2	2	2	438
2	2	5	2	2	3	876
2	3	4	2	1	0	876
2	3	4	2	2	0	1,752
2	3	4	3	1	0	1,314
2	3	4	3	2	0	438
2	3	5	1	1	1	876
2	3	5	2	2	0	1,752
2	3	5	3	2	0	876
2	4	3	1	2	0	438
2	4	3	2	1	0	438
2	4	4	1	1	2	438
2	4	5	2	1	0	876
3	1	1	2	1	1	15,768
3	1	1	2	2	2	1,752
3	1	1	3	2	0	876
3	1	2	2	1	0	876
3	1	2	3	1	3	3,504
3	1	3	2	1	1	6,570
3	1	3	2	2	0	1,752
3	1	4	1	1	0	438
3	1	4	1	2	0	876
3	1	4	2	1	5	4,818
3	1	4	2	2	23	2,628
3	1	4	3	2	21	1,752
3	1	5	1	1	0	1,752

Table 1.2.11. (cont.)

System	Oper. Type	Valve Type	Size	Operating Mode	Failures	Time (100 Hrs)
3	1	5	1	2	0	1,752
3	1	5	2	1	11	13,578
3	1	5	2	2	3	13,578
3	1	5	3	2	2	438
3	1	6	2	1	1	876
3	1	6	2	2	0	438
3	1	6	3	2	0	438
3	2	6	2	2	1	876
3	3	2	2	1	0	438
3	3	2	3	2	0	438
3	3	4	1	1	0	3,066
3	3	4	1	2	0	1,752
3	3	4	2	1	8	3,504
3	3	4	2	2	0	1,314
3	3	4	3	1	13	876
3	3	4	3	2	3	1,314
3	3	5	1	2	0	1,314
3	3	5	2	2	0	2,190
3	4	4	2	2	1	1,752
3	4	4	3	2	1	4,380
3	4	5	2	2	0	1,752
4	3	3	3	2	2	438
4	3	4	2	1	2	3,504
4	3	4	2	2	0	1,752
4	3	4	3	2	7	1,314
4	3	5	1	2	0	438
5	1	2	2	1	0	1,314
5	1	2	2	2	0	876
5	1	2	3	1	0	438
5	1	2	3	2	0	2,190
5	1	3	1	1	0	438
5	1	3	1	2	0	1,314
5	1	3	2	2	0	876
5	1	4	2	1	3	1,752
5	1	4	2	2	0	1,752
5	1	5	1	1	3	438
5	1	5	1	2	2	1,314
5	1	5	2	2	0	3,504

Table 1.2.11. (cont.)

System	Oper. Type	Valve Type	Size	Operating Mode	Failures	Time (100 Hrs)
5	1	6	1	1	0	438
5	1	6	2	2	0	876
5	2	3	2	2	0	4,818
5	2	4	1	1	0	438
5	3	2	2	1	0	438
5	3	2	2	2	0	876
5	3	2	3	1	2	1,752
5	3	2	3	2	0	876
5	3	4	2	1	2	2,190
5	3	4	2	2	1	6,132
5	3	5	2	2	0	876
5	4	3	1	1	1	2,190
5	4	3	1	2	0	876
5	4	3	2	1	0	1,314
5	4	4	1	2	0	438
5	4	4	2	1	0	438
5	4	5	2	2	0	438

1.3 Review of Discrete Distributions

This section summarizes the notation and properties of five families of discrete distributions.

Binomial Distribution

The notation $Y \sim B(n,p)$ means that Y follows a binomial distribution based on n independent trials with common success probability p. The probability mass function of $Y \sim B(n,p)$ is

$$P[Y=j] = \binom{n}{j} p^j (1-p)^{n-j}, \quad j = 0(1)n. \tag{1.3.1}$$

It is sometimes convenient to think in terms of the odds of success $p/(1-p)$ rather than the success probability p.

If $Y_1 \sim B(n_1,p_1)$ is independent of $Y_2 \sim B(n_2,p_2)$ then for $0 \le t \le n_1+n_2$

$$P[Y_1 = j \mid Y_1+Y_2=t] = \frac{\binom{n_1}{j}\binom{n_2}{t-j}\psi^j}{\sum_u \binom{n_1}{u}\binom{n_2}{t-u}\psi^u}, \tag{1.3.2}$$

for $\max\{0, t - n_2\} \le j \le \min\{n_1, t\}$ where the sum in the denominator is over u in the same range. The parameter $\psi = p_1(1-p_2)/\{(1-p_1)p_2\}$ is the ratio of the odds of success for Y_1 to the odds of success for Y_2. The conditional distribution (1.3.2) is called the *noncentral hypergeometric distribution.* An important special case of (1.3.2) occurs when $p_1 = p_2$ and

$$P[Y_1 = j \mid Y_1 + Y_2 = t] = \frac{\binom{n_1}{j}\binom{n_2}{t-j}}{\binom{n_1+n_2}{t}}$$

for $\max\{0, t - n_2\} \le j \le \min\{n_1, t\}$. This is the (central) hypergeometric distribution.

There are several "large n" approximations to the binomial distribution which are in common use.

1. For $Y \sim B(n, p)$, the central limit theorem gives

$$P[Y \le j] \simeq \Phi\left(\frac{j - np}{\sqrt{np(1-p)}}\right) \tag{1.3.3}$$

where $\Phi(\cdot)$ is the standard normal cumulative distribution function.

2. The so-called continuity corrected version of (1.3.3) is

$$P[Y \le j] \simeq \Phi\left(\frac{j + 1/2 - np}{\sqrt{np(1-p)}}\right). \tag{1.3.4}$$

3. Peizer and Pratt (1968) give an extremely accurate though more complicated normal approximation.

4. The Poisson approximation to the binomial distribution will be discussed later.

The continuity-corrected approximation (1.3.4) can be derived by comparing the approximating $N[np, np(1-p)]$ density to the histogram of the binomial mass function (1.3.1). Approximation (1.3.3) is best for central p and large n; i.e., $.2 \le p \le .8$ and $n \ge 30$. To illustrate, consider the following example. Suppose $Y \sim B(30, .1)$, then

$$\begin{aligned} P[Y \le 4] &= .8245 \quad \text{(exact)} \\ &\simeq .7286 \quad \text{from (1.3.3)} \\ &\simeq .8193 \quad \text{from (1.3.4).} \end{aligned}$$

Here (1.3.3) has an 11.6% relative error compared to the 0.6% relative error of (1.3.4). The Poisson approximation complements the "central" p approximations (1.3.3) and (1.3.4).

Negative Binomial Distribution

The notation $Y \sim NB(\alpha, \beta)$ is used to say that Y follows the negative binomial distribution with paramcters $\alpha > 0$ and $\beta > 0$. The probability mass function of Y is

$$P[Y = j] = \frac{\Gamma(j+\alpha)}{j!\Gamma(\alpha)} \left(\frac{1}{1+\beta}\right)^{\alpha} \left(\frac{\beta}{\beta+1}\right)^{j}, \quad j = 0, 1, \ldots$$

$$= \binom{\alpha+j-1}{j} \left(\frac{1}{1+\beta}\right)^{\alpha} \left(\frac{\beta}{\beta+1}\right)^{j} \tag{1.3.5}$$

where $\Gamma(x) := \int_0^\infty e^{-t}t^{x-1}dt$ is the gamma function and the second equality holds only if α is integral.

From equation (1.3.5) it can be immediately recognized that for integer α, $NB(\alpha, \beta)$ is the distribution of the number of failures until α successes in a sequence of independent trials with common probability $1/(1+\beta)$ of success. Thus "inverse sampling" from a $B(1, (1+\beta)^{-1})$ distribution is one way in which the negative binomial distribution arises. A second way in which the negative binomial distribution arises is as a mixture of Poisson distributions; this derivation will be given in the discussion of the Poisson distribution.

Multinomial Distribution

The multinomial distribution is a generalization of the binomial distribution which allows more than two outcomes. Write

$$\mathbf{Y}' = (Y_1, \ldots, Y_t) \sim M_t(n, \mathbf{p}' = (p_1, \ldots, p_t))$$

to indicate that $\mathbf{Y}$ has a t-cell multinomial distribution based on n trials having common cell probabilities $\mathbf{p}$ ($\sum_{i=1}^{t} p_i = 1$). Here and throughout the text the convention is adopted that all vectors are column vectors and prime denotes transpose. The probability mass function of the multinomial distribution is

$$P[\mathbf{Y} = \mathbf{y}] = \frac{n!}{y_1!y_2!\cdots y_t!} p_1^{y_1} p_2^{y_2} \cdots p_t^{y_t} \tag{1.3.6}$$

for $\mathbf{y}' = (y_1, \ldots, y_t)$ with nonnegative integral components satisfying $\sum_{i=1}^{t} y_i = n$. The first and second moments of $\mathbf{Y} \sim M_t(n, \mathbf{p})$ are $E[\mathbf{Y}] = n\mathbf{p}$, and

$$\operatorname{cov}(Y_i, Y_j) = \begin{cases} np_i(1-p_i) & \text{if } i = j \\ -np_ip_j & \text{if } i \neq j; \end{cases}$$

hence the variance–covariance matrix of $\mathbf{Y}$ is

$$n(\operatorname{Diag}(p_1, \ldots, p_t) - \mathbf{p}\mathbf{p}').$$

Poisson Distribution

That Y has a Poisson distribution with parameter $\lambda > 0$ will be denoted by $Y \sim P(\lambda)$. The probability mass function of Y is

$$f(y \mid \lambda) = \frac{e^{-\lambda}\lambda^j}{j!}, \quad j = 0, 1, \ldots . \tag{1.3.7}$$

Thus the range of possible values of a Poisson random variable is unbounded. The mean and variance of $Y \sim P(\lambda)$ are identical with $E[Y] = \text{Var}(Y) = \lambda$.

The Poisson distribution arises as the distribution of "random" events in time or space under mild assumptions (Karlin and Taylor, 1975, Chapter 4). Alternately, it occurs as an approximating distribution to the binomial distribution in the following way. If $W \sim P(\lambda)$, $\{Y_n\}_{n \geq 1}$ is a sequence of independent binomial random variables with $Y_n \sim B(n, p_n)$, and $np_n \to \lambda$ as $n \to \infty$, then

$$P[Y_n = j] \to P[W = j] = \frac{e^{-\lambda}\lambda^j}{j!} \quad \text{as} \quad n \to \infty.$$

This fact can be established by direct calculation. Recall that

$$P[Y_n = j] = \binom{n}{j} p_n^j (1 - p_n)^{n-j}, \quad j = 0(1)n.$$

Let $r_n := np_n - \lambda$; then $p_n = \lambda/n + r_n/n$ and $r_n \to 0$ as $n \to \infty$ by assumption. Thus

$$\begin{aligned} P[Y_n = j] &= \frac{n(n-1)\cdots(n-j+1)}{j!} p_n^j (1 - p_n)^{n-j} \\ &= np_n(np_n - p_n)\cdots(np_n - (j-1)p_n)\left(1 - \frac{\lambda}{n} - \frac{r_n}{n}\right)^n \\ &\quad \cdot \frac{(1 - p_n)^{-j}}{j!} \\ &\to \frac{\lambda^j e^{-\lambda}}{j!} \quad \text{as} \quad n \to \infty. \end{aligned}$$

As an example, recall the numerical illustration of the normal approximations to the binomial distribution $Y \sim B(30, .1)$. Let W be a Poisson random variable with mean $\lambda = 3 = 30 \times .1$, then

$$P[Y \leq 4] \simeq P[W \leq 4] = .8153$$

which gives a relative error of 1.1% for the Poisson approximation.

The Poisson distribution has the following relationship to the multinomial distribution. Let $Y_1, \ldots, Y_t$ be mutually independent random variables with $Y_i \sim P(\lambda_i)$ for $i = 1(1)t$; then $\sum_{i=1}^t Y_i \sim P(\sum_{i=1}^t \lambda_i)$. The conditional probability that $\mathbf{Y}' := (Y_1, \ldots, Y_t) = \mathbf{y}' := (y_1, \ldots, y_t)$ given that $\sum_{i=1}^t Y_i = s$ is

$$P\left[\mathbf{Y} = \mathbf{y} \mid \sum_{i=1}^t Y_i = s\right] = \frac{P[\mathbf{Y} = \mathbf{y}, \sum_{i=1}^t Y_i = s]}{P[\sum_{i=1}^t Y_i = s]}$$

$$= \begin{cases} \dfrac{P[\mathbf{Y} = \mathbf{y}]}{\exp(-\Sigma\lambda_j)(\Sigma\lambda_i)^s(1/s!)}, & \sum_{i=1}^t y_i = s \\ 0, & \text{otherwise} \end{cases}$$

$$= \begin{cases} \dfrac{\prod_{i=1}^t e^{-\lambda_i}\lambda_i^{y_i}(1/y_i!)}{\exp(-\Sigma\lambda_j)(\Sigma\lambda_i)^s(1/s!)}, & \sum_{i=1}^t y_i = s \\ 0, & \text{otherwise} \end{cases}$$

$$= \begin{cases} s! \displaystyle\prod_1^t \left(\frac{\lambda_i}{\Sigma\lambda_i}\right)^{y_i} \frac{1}{y_i!}, & \sum_{i=1}^t y_i = s \\ 0, & \text{otherwise.} \end{cases}$$

for $\mathbf{y}$ with nonnegative integral components. By comparison with (1.3.6), this proves that the conditional distribution of $\mathbf{Y}$ given $\sum_{i=1}^t Y_i = s$ is multinomial with s trials and

$$\mathbf{p}' = \left(\frac{\lambda_1}{\Sigma\lambda_i}, \ldots, \frac{\lambda_t}{\Sigma\lambda_i}\right).$$

Mixed Poisson Distribution

Fix a cumulative distribution function $F(\cdot)$ with $F(0) = 0$; i.e., which can only take positive values. The random variable Y has a mixed Poisson distribution with mixing distribution $F(\cdot)$ means

$$P[Y = j] = \int_0^\infty \frac{e^{-\lambda}\lambda^j}{j!} dF(\lambda) \tag{1.3.8}$$

for $j = 0, 1, \ldots$. Like the Poisson distribution, mixed Poisson distributions are supported on the non-negative integers.

The simplest case of the mixed Poisson distribution is the Poisson distribution which corresponds to a degenerate mixing distribution. If the mixing distribution is gamma then (1.3.8) can be integrated in closed form and gives the negative binomial distribution. Specifically, suppose the gamma is parametrized to have mean $\mu > 0$ and variance $\sigma^2 > 0$, then λ has density

$$f(\lambda) = \frac{\lambda^{(\mu^2/\sigma^2)-1}\exp\{-\mu\lambda/\sigma^2\}}{\Gamma(\mu^2/\sigma^2)(\sigma^2/\mu)^{\mu^2/\sigma^2}}, \quad 0 < \lambda < \infty.$$

It is straightforward to compute that (1.3.8) becomes

$$P[Y=j]=\frac{\Gamma(j+\mu^2/\sigma^2)}{j!\Gamma(\mu^2/\sigma^2)}\left(\frac{\mu}{\mu+\sigma^2}\right)^{\mu^2/\sigma^2}\left(\frac{\sigma^2}{\mu+\sigma^2}\right)^j \tag{1.3.9}$$

for $j=0,1,\ldots$ which is the negative binomial distribution (1.3.5) with parameters $\alpha=\mu^2/\sigma^2$ and $\beta=\sigma^2/\mu$.

Mixed Poisson distributions have at least as great variability as a Poisson distribution with the same mean, and thus are useful as models of "extra-Poisson variation" (Section 3.4). If Y is mixed Poisson with mixing distribution $F(\cdot)$, then

$$E[Y]=E[E[Y\mid\lambda]]=E_F[\lambda]$$

and

$$E[Y^2]=E[E[Y^2\mid\lambda]]=E_F[\lambda+\lambda^2].$$

Hence

$$\mathrm{Var}(Y)=E_F[\lambda]+E_F[\lambda^2]-(E_F[\lambda])^2=E_F[\lambda]+\mathrm{Var}_F(\lambda)\geq E[Y] \tag{1.3.10}$$

with equality if and only if $F(\cdot)$ is degenerate; i.e., Y is Poisson. In addition to being a model in its own right the negative binomial distribution has been used as an alternative distribution against which tests of the Poisson null model have been constructed (Section 2.3). The following example indicates the usefulness of the additional flexibility of the negative binomial family.

Example 1.3.1. Greenwood and Yule (1920) present the data in Table 1.3.1 on the number of accidents that occured in a 3-month period to each of 414 workers. In addition, the table lists estimates of the probabilities for each number of accidents based on (i) the maximum likelihood estimate of λ under the Poisson model (1.3.7), and (ii) a method of moments estimate of μ and σ^2 under the negative binomial distribution (1.3.9). The negative binomial distribution provides a substantially better fit to the data than the Poisson distribution. Intuitively one can motivate the mixing process in this example by assuming that the workers each have their own mean rate λ based on their personal susceptibility to accidents, and that the workers form a random sample from a population with three month accident rates λ distributed according to the gamma mixing distribution.

Table 1.3.1. Number of Accidents for Each of 414 Workers and Their Estimated Values Under the Poisson and Negative Binomial Models. (Reprinted with permission from Major Greenwood & G.V. Yule: "An Inquiry into the Nature of Frequency Distributions," *Journal of the Royal Statistical Society* 83, pp. 255–279, 1920.)

	0	1	2	3	4	5	Total
No. accidents	296	74	26	8	4	6	414
Poisson fit	256	122	30	5	1	0	414
Neg. Bin. fit	299	69	26	11	5	4	414

2

Univariate Discrete Responses

2.1 Binomial Responses

Perhaps the simplest discrete data problem involves single-sample binary responses. This section considers point and interval estimation for such data. The techniques are described in some detail since they are most easily understood in this simple setting and since analogs of these methods have been developed for many of the more complicated discrete data problems discussed in later sections.

Hypothesis testing is given relatively little emphasis because of the simplicity of the data and model in the binomial case. Formally, if p is the probability of success for each Bernoulli trial, then the point null hypothesis $H_{=}: p = p_0$ versus $H_{\neq}: p \neq p_0$ can be tested by determining if p_0 falls in a two-sided $100(1-\alpha)\%$ confidence interval for p. Similarly, one-sided null hypotheses $H_{\leq}: p \leq p_0$ versus $H_{>}: p > p_0$ can be tested by determining if p_0 falls above a $100(1-\alpha)\%$ confidence bound for p. In more complicated prolbems, the wider variety of possible alternatives suggests many different null and alternative hypotheses of potential interest, as Section 2.2 will illustrate for multinomial data.

A second reason for de-emphasizing testing in the binomial problem is that in many applications it is more informative to calculate a confidence interval (or upper or lower confidence bound) for p rather than to accept or reject a null hypothesis. For example, in clinical trials, even though the problem of determining treatment effectiveness may well be formulated as that of testing $H_{\leq}: p \leq p_0$ versus $H_{>}: p > p_0$ where p is the probability of a cure, a patient's willingness to use the treatment will be modified by its side effects. In this case knowing a confidence interval for p may be more informative than simply accepting or rejecting $H_{\leq}$. However for completeness, Problem 2.5 illustrates the construction of the (large-sample) Wald, score, and likelihood ratio tests of point null hypotheses for binomial data. Formulas for the power of these tests, which are useful in sample size calculations, are obtained by specializing the results in Section 2.2 on multinomial responses.

One important topic that is beyond the scope of this text is the description of multistage and sequential sampling plans. Ethical considerations in clinical trials, and cost considerations in sampling inspection make it critical to reduce the sample size required by tests of $H_{\leq}: p \leq p_0$ (given) versus $H_{>}: p > p_0$. Armitage (1975), Fleming (1982), and Whitehead (1983) discuss multistage tests in the clinical trials setting and MIL-STD 105D (1963) is a classical multistage testing plan developed for sampling inspection. The construction of confidence intervals following multistage tests is addressed in Jennison and Turnbull (1983), Atkinson and Brown (1985), and Duffy and Santner (1987b).

In the following, $Y \sim B(n,p)$ where n is known and p is unknown with $0 < p < 1$. Equivalent parametrizations in terms of the *odds of success* $\omega := p/(1-p)$ or the *log odds* $\lambda := \ln(\omega)$ are traditionally used in some disciplines. Figure 2.1.1 shows that both ω and λ are strictly increasing functions of p with λ mapping $(0,1)$ symmetrically about $p = 1/2$ onto $\mathbb{R}^1$. Chapter 5 discusses models for the odds and log-odds in more complicated covariate problems. The fact that p, ω, and λ are one-to-one functions of each other means that confidence intervals for any one of the three can be used to form confidence intervals for the remaining two.

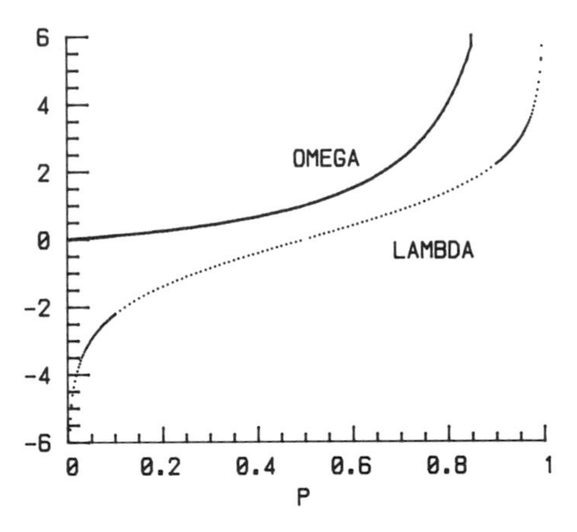

Figure 2.1.1. Plot of $\omega := p/(1-p)$ and $\lambda := \ln\{p/(1-p)\}$ versus p.

A. Point Estimation of p, ω, and λ

The naive estimator $\hat{p} := Y/n$ is the maximum likelihood estimator (MLE) of p (add 0 and 1 to the estimation space to make this statement technically correct for observed values of $Y = 0$ and n). The properties of $\hat{p}$ are discussed briefly below, and then several other estimators of p that have been proposed in the literature are introduced. Section 2.2 gives a more comprehensive treatment of loss functions and alternative point estimation methods for the single-sample case with arbitrary polychotomous (i.e., multinomial) responses.

The Estimator $\hat{p} = Y/n$

An application of the Cramer–Rao inequality shows that $\hat{p}$ is the uniformly minimum variance unbiased estimator (UMVUE) of p. The variance of $\hat{p}$ is

$p(1-p)/n$ which is also its mean squared error (MSE) since $\hat{p}$ is unbiased. The MSE is, up to a constant, the expected loss or risk of $\hat{p}$ under *squared error loss* (SEL) defined by

$$L_S(p, a) = n(p-a)^2.$$

The factor n in $L_S(\cdot, \cdot)$ makes the MSE of $\hat{p}$ at p, denoted by $R_S(p, \hat{p})$, independent of sample size; i.e.,

$$\begin{aligned} R_S(p, \hat{p}) &= E_p[n(\hat{p} - p)^2] \\ &= p(1-p). \end{aligned}$$

Geometrically, $R_S(p, \hat{p})$ is concave and symmetric about $p = 1/2$ with maximum value $(= 1/4)$ at $p = 1/2$ and minimum value $(= 0)$ when p is at either extreme (0 or 1). MSE is one of the most widely used general purpose risk functions for evaluating and comparing estimators. The reasons for this are its intuitiveness, mathematical tractability, and a historical momentum dating back to Gauss's and Legendres's use of least squares in the late 1700s. The disadvantage of MSE, or any other specific measured risk, is that it may not adequately reflect the costs of misspecification in a given application.

Recall that an estimator $\delta = \delta(Y)$ of p is *inadmissible* with respect to SEL if there exists a $\delta^* = \delta^*(Y)$ such that $R_S(p, \delta^*) \leq R_S(p, \delta)$ for all p with strict inequality for some p. The estimator δ is *admissible* if it is not admissible. While there is agreement that inadmissible estimators should not be used, admissibility is a weak optimality property that does not, by itself, either identify a single estimator as best or justify using an estimator. As an extreme example, estimators which disregard the data completely and guess a constant value, say p_0, are admissible since they have (near) zero risk properties when the parameter is (near) p_0.

It is a textbook exercise to establish that $\hat{p}$ is an admissible estimator under SEL (Berger, 1985, p. 165). The MLE owes its admissibility to its near-zero risk for p near 0 and unity. Thus it is of interest to consider alternative estimators with lower MSE for central p. Since $\hat{p}$ is the UMVUE, any competing estimator with lower MSE than $\hat{p}$ will necessarily be biased.

Before deriving alternative estimators, the performance of $\hat{p}$ will be considered under a second natural loss function for the binomial problem, namely, *relative squared error loss* (RSEL) defined by

$$L_R(p, a) = n(p-a)^2/p(1-p)$$

with $0/0 := 0$ and $+/0 := +\infty$. The loss $L_R(\cdot, \cdot)$ weights the squared error by the inverse of the variance; it places a premium on correctly identifying extreme p since $L_R(p, a) = \infty$ for $p = 0$ or 1 when $a \neq p$. The risk of $\hat{p}$ with respect to L_R is

$$R_R(p, \hat{p}) = \frac{1}{p(1-p)} R_S(p, \hat{p}) = 1.$$

Constant risk estimators are often minimax; recall that $\hat{p}$ is minimax with respect to $L_R(\cdot,\cdot)$ if it satisfies

$$\sup_p R_R(p,\hat{p}) = \inf_\delta \sup_p R_R(p,\delta)$$

where the infimum is over all estimators $\delta = \delta(Y)$ of p. Olkin and Sobel (1979) apply the divergence theorem (Kaplan, 1952, Sec. 5.11) to prove that $\hat{p}$ is unique minimax under $L_R(\cdot,\cdot)$ and hence admissible. An alternate proof of this fact and further discussion of RSEL for the general multinomial case is given in Section 2.2.

The primary motivation for the development of alternatives to $\hat{p}$ is the availability of external information. Two classes of methods which incorporate additional information are: (1) Bayes (and related) methods, and (2) smoothing techniques. Both incorporate prior knowledge about the unknown p. The paragraphs below describe Bayes estimators first with respect to a completely specified prior and loss, and then generalizations including gamma minimax, hierarchical Bayes, empirical Bayes, and pseudo Bayes estimators. The setting appropriate for estimators developed by each of these methods is discussed for the problem of estimating a binomial p. Smoothing techniques are more naturally introduced in the case of a multinomial response where one might have reason to believe that the probabilities of adjacent (or other groups of) cells should vary smoothly. Hence the discussion of smoothing methods will be deferred until Section 2.2.

Bayes Estimators of p

The mathematically simplest Bayes estimators are those computed with respect to conjugate prior distributions. A prior chosen from a family $\mathcal{F}$ is a *conjugate prior* if the posterior distribution (the conditional distribution of the parameter given the data) also belongs to $\mathcal{F}$. The beta family is conjugate for the binomial problem. The beta density with parameters $\alpha,\beta > 0$, denoted by $\text{Be}(\alpha,\beta)$, is defined by

$$\frac{\Gamma(\alpha+\beta)}{\Gamma(\alpha)\Gamma(\beta)} p^{\alpha-1}(1-p)^{\beta-1}, \quad 0 < p < 1.$$

The mean and variance of the $\text{Be}(\alpha,\beta)$ distribution are $\mu = \alpha/(\alpha+\beta)$ and $(\mu(1-\mu))/(\alpha+\beta+1)$, respectively. If $\min\{\alpha,\beta\} > 1$, the beta distribution is unimodal with mode equal to $(\alpha-1)/(\alpha+\beta-2)$; in particular if $\alpha = \beta > 1$, the beta distribution is also symmetric about $1/2$ and both the mean and mode are equal to $1/2$. Lastly, if p is distributed as $\text{Be}(\alpha,\beta)$, algebra shows that the conditional distribution of p given $Y = y$ is $\text{Be}(\alpha+y,\ \beta+n-y)$ which proves the beta family is conjugate.

It is an elementary result of decision theory that the Bayes estimator of p with respect to SEL is the mean of the posterior distribution. Applying the formula for the mean of the beta distribution to the $\text{Be}(\alpha+y, \beta+n-y)$ posterior distribution gives

$$\hat{p}^B = \frac{Y+\alpha}{n+\alpha+\beta} = \frac{n}{a+\beta+n}\hat{p} + \frac{\alpha+\beta}{\alpha+\beta+n}\mu. \tag{2.1.1}$$

Equation (2.1.1) expresses $\hat{p}^B$ as a convex combination of the MLE $\hat{p}$ and the prior mean μ. The weights depend on the sample size n and $K := \alpha+\beta$ in such a way that $\hat{p}^B \to \hat{p}$ as $n \to \infty$ for fixed K, and $\hat{p}^B \to \mu$ as $K \to \infty$ for fixed n. The form of (2.1.1), as well as the expression for the variance of the $\text{Be}(\alpha, \beta)$ distribution, suggest that K can be interpreted as the "prior sample size." A second interpretation of (2.1.1) is the "fake data viewpoint": $\hat{p}^B$ is the MLE for data obtained by supplementing the real data (y successes out of n trials) by "fictitious data" consisting of α success in K trials. The quantities α and K need not be integers. Again K plays the role of the prior sample size.

The estimator $\hat{p}^B$ is biased since

$$E_p[\hat{p}^B] = \frac{n}{\alpha+\beta+n}p + \frac{\alpha+\beta}{\alpha+\beta+n}\mu.$$

Its MSE is

$$R_S(p, \hat{p}^B) = n\left(\frac{K}{K+n}\right)^2 (\mu-p)^2 + \left(\frac{n}{K+n}\right)^2 p(1-p). \tag{2.1.2}$$

For p sufficiently near μ, $\hat{p}^B$ has smaller mean squared error than $\hat{p}$; i.e.,

$$R_S(p, \hat{p}^B) < p(1-p) = R_S(p, \hat{p})$$

(and the opposite inequality holds otherwise). Lastly, the estimator $\hat{p}^B$ is unique positive Bayes and thus admissible (Berger, 1985, p. 253).

One important case of the Bayes estimator (2.1.1) is $\alpha = \beta = \sqrt{n}/2$ for which

$$\hat{p}^B = \frac{Y+\sqrt{n}/2}{n+\sqrt{n}}.$$

Calculation of (2.1.2) for this case shows that $\hat{p}^B$ has constant risk and must therefore be minimax since it is Bayes (Problem 2.1).

A second special case is $\alpha = \beta = 1/2$ which yields the p-estimator implicitly used in Anscombe's (1956) $o(n^{-1})$ biased estimators of ω and λ. (See the discussion below on estimation for ω and λ.) The $\text{Be}(1/2, 1/2)$ distribution is also Jeffreys's (1961) choice of a prior for expressing little or no information regarding p. However there is not universal agreement about this selection, and Geisser (1984) contains a critique of the various noninformative priors that have been used in the literature. Geisser advocates the $\text{Be}(1,1)$ distribution (uniform over the interval $(0,1)$) as a noninformative prior.

A typical Bayes estimator is illustrated in Example 2.1.1.

Example 2.1.1. Segaloff (1961) analyzes the results of many studies of the effectiveness of different estrogen and androgen chemotherapies for treating advanced breast cancer. The overall goal of the studies was to reject treatments less effective than testosterone propionate (TP), the standard chemotherapy at the time. In one of the studies, 3 out of 10 women who had metastases to the bone tissue and were treated with TP improved. Let $p := P[\text{Improvement}]$; then the MLE for p is $\hat{p} = .30$. If p has the Be($\alpha = 2$, $\beta = 2$) prior, then the Bayes estimate is $\hat{p}^B = .36$. This prior is symmetric about $\mu = 1/2$, and can be interpreted as a prior sample of 2 improvements in a group of 4 women.

More generally for $n = 10$, the MSE curves of the estimators $\hat{p}$ and $\hat{p}^B$ (for the Be(2, 2) prior), can be compared as a function of p. Figure 2.1.2 shows the extent to which $\hat{p}^B(\hat{p})$ has lower MSE for central (extreme) p. In particular, $\hat{p}^B$ has slightly lower risk than $\hat{p}$ at $p = .3$. The figure also shows that the risk curve of $\hat{p}^B$ does not converge to zero as p converges to its extremes. For an arbitrary estimator $\hat{p}^*(\cdot)$, $R_S(p, \hat{p}^*) \to (\hat{p}^*(0) - 0)^2$ as $p \to 0$ and $R_S(p, \hat{p}^*) \to (\hat{p}^*(n) - 1)^2$ as $p \to 1$. Thus only estimators which guess zero at $Y = 0$ and unity at $Y = n$ will have zero risk as p converges to its limits. However as n increases (with α and β fixed), the second term of $R_S(p, \hat{p}^B)$ in (2.1.2) dominates and the two risk curves $R_S(p, \hat{p})$ and $R_S(p, \hat{p}^B)$ become indistinguishable.

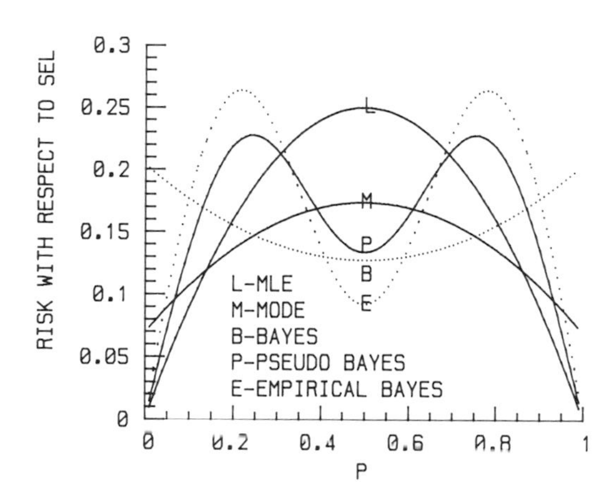

Figure 2.1.2. Risk functions (SEL) of maximum likelihood, Bayes, mode, pseudo Bayes, and empirical Bayes estimators for $B(10, p)$ data.

Bayes estimators have been constructed in the literature for nonconjugate priors. One class of nonconjugate priors arises by modeling the distribution of the log odds λ (on $\mathbb{R}^1$); this specification induces a prior distribution on p (Leonard, 1972). For example, if $\lambda \sim N(0, \sigma^2)$ then the prior density for

p is

$$\frac{\phi(\ln\{p/(1-p)\}/\sigma)}{\sigma p(1-p)}, \quad 0 < p < 1$$

where $\phi(x) = (2\pi)^{-1/2}\exp\{-x^2/2\}$ is the standard normal density. The posterior mean for the above density can only be determined by numerical techniques although this is not difficult. Bayesian parameter estimation based on postulating priors on the log odds scale will be considered further in Section 5.4 for logistic regression models.

Changing the loss from $L_S(\cdot,\cdot)$ changes the form of the Bayes estimator. If the loss is not well-formulated, some functional of the posterior distribution such as its mode can be used as a non-loss-specific Bayesian point estimator. The mode of the posterior can be regarded as a Bayesian MLE; it also has the practical advantage of being easier to compute than the posterior mean in many problems.

To illustrate the computation and interpretation of the posterior mode, consider again the binomial problem with beta prior. The log posterior likelihood with respect to the Be(α, β) prior is, apart from constants,

$$(\alpha - 1 + y)\ln(p) + (\beta - 1 + n - y)\ln(1-p).$$

If $\min\{\alpha + y, \beta + n - y\} > 1$, then the posterior mode is

$$\hat{p}^M = \frac{\alpha - 1 + Y}{K - 2 + n} = \frac{n}{K-2+n}\hat{p} + \frac{K-2}{K-2+n}\left(\frac{\alpha-1}{K-2}\right). \tag{2.1.3}$$

Equation (2.1.3) shows that $\hat{p}^M$ is a convex combination of $\hat{p}$ and an expression involving α and β. If $\min\{\alpha, \beta\} > 1$, $\hat{p}^M$ is formally the Bayes rule corresponding to the prior Be$(\alpha - 1, \beta - 1)$. The character of $\hat{p}^M$ with respect to this modified prior has two implications: (i) $\hat{p}^M$ can be regarded as a Bayes rule with respect to $L_S(\cdot,\cdot)$ based on *less certain* prior information than the Bayes rule $\hat{p}^B$ with respect to Be(α, β) since $\hat{p}^M$ has prior sample size $\alpha + \beta - 2$ rather than $\alpha + \beta$, and (ii) the region where $\hat{p}^M$ is superior to $\hat{p}$ is less "central" than that of $\hat{p}^B$ since

$$\left|\frac{\alpha - 1}{\alpha + \beta - 2} - \frac{1}{2}\right| \geq \left|\frac{\alpha}{\alpha + \beta} - \frac{1}{2}\right|. \tag{2.1.4}$$

Equality holds in (2.1.4) if and only if $\alpha = \beta$ in which case the prior mean for both Be(α, β) and Be$(\alpha - 1, \beta - 1)$ is $\mu = 1/2$. Furthermore, $\hat{p}^M$ is admissible since it is positive Bayes.

Example 2.1.1 (continued). The posterior mode of p for the Segaloff data and the Be(2,2) prior is $\hat{p}^M = .33$. This value is intermediate between $\hat{p} = .30$ and $\hat{p}^B = .36$ because $\hat{p}^B$ is associated with a larger prior sample size K than $\hat{p}^M$.

In terms of operating characteristics, Figure 2.1.2 plots the MSE of $\hat{p}^M$ ($R_S(p, \hat{p}^M)$) for $n = 10$ and the Be(2, 2) prior. Note that for extreme (central) p

$$R_S(p, \hat{p}) \leq (\geq) R_S(p, \hat{p}^M) \leq (\geq) R_S(p, \hat{p}^B).$$

Both $\hat{p}^B$ and $\hat{p}^M$ dominate $\hat{p}$ for p in symmetric intervals about $p = 1/2$ since the mean of the prior Be(2, 2) distribution is 1/2. However, $\hat{p}^B$ dominates $\hat{p}$ by a *larger* amount than $\hat{p}^M$ for p near 1/2 reflecting its larger prior sample size. Conversely, $\hat{p}^M$ dominates $\hat{p}$ over a *wider* interval than $\hat{p}^B$ ((.16, .84) versus (.19, .81)) since $\hat{p}^B$ pulls $\hat{p}$ toward 1/2 more strongly than $\hat{p}^M$. Finally, $R_S(p, \hat{p}^M)$ does not converge to zero as p approaches its extreme values since $\hat{p}^M(0) > 0$ and $\hat{p}^M(10) < 1$.

Gamma Minimax and Hierarchical Bayes Estimators of p

Both $\hat{p}^B$ and $\hat{p}^M$ require the complete specification of a prior distribution. If the elucidation of prior information is difficult or impossible then several approaches have been proposed for estimating p. The gamma minimax principle assumes that it is possible to specify a class $\mathcal{G}$ of priors for p and uses the estimator $\hat{p}^{\mathcal{G}}(\cdot)$ which minimizes the maximum Bayes risk over $\mathcal{G}$:

$$\max_{G \in \mathcal{G}} r(G, \hat{p}^{\mathcal{G}}) = \min_{\hat{p}^*} \max_{G \in \mathcal{G}} r(G, \hat{p}^*)$$

where $r(G, \hat{p}^*) = \int R(p, \hat{p}^*) dG(p)$ is the Bayes risk with respect to the prior $G(\cdot)$. For example, Copas (1972) proves that when $\mathcal{G}$ consists of all priors with a *given* mean μ, then the gamma minimax estimator is

$$\hat{p}^{\mathcal{G}}(y) = \frac{\hat{p} + \mu\sqrt{n}}{1 + \sqrt{n}}.$$

Hierarchical Bayes estimation assumes the prior is parametric and known up to the specification of parameters, and the data analyst is able to *model* the prior parameters by a prior distribution. This formulation is called hierarchical because there is a hierarchy of models—one for the parameters in the data distribution and one for the parameters in the prior distribution. For example, Good (1967, 1983), Lee and Sabavala (1987), and other authors have adopted this approach when estimating p based on the Be(α, β) prior with *unknown* (α, β). Formally, their hierarchical Bayes estimators are Bayes rules with respect to a mixture of beta distributions. Good (1983) emphasizes (i) the robustness of such estimators against misspecification of the prior and (ii) that even improper priors (i.e., nonnegative functions which are not integrable) can be used at the second stage.

To continue the binomial example, Lee and Sabavala (1987) reparametrize the Be(α, β) prior to $\mu = (\alpha/\alpha+\beta)$ and $\rho := 1/(1+\alpha+\beta) = (1+K)^{-1}$ where μ is the mean of each Bernoulli trial and ρ is the correlation of any two Bernoulli trials based on the distribution of Y given (α, β). They calculate hierarchical Bayes estimators of p based on independent Beta marginals for

μ and ρ. Good (1967, 1983) considers choosing μ and ρ from an arbitrary joint distribution $G(\mu, \rho)$ (and p according to the Beta(α, β) distribution where $\alpha := \mu(\rho^{-1} - 1)$ and $\beta := (1 - \mu)(\rho^{-1} - 1)$). See also Chapter 12 of Bishop, Fienberg, and Holland (1975). Leonard (1972) constructs hierarchical Bayes estimators for a normal first stage prior on λ with unknown mean and variance.

Empirical Bayes and Pseudo Bayes Estimators of p

Both the empirical Bayes (EB) and pseudo Bayes (PB) principles also assume the prior information is only partially known; they use the data to estimate the prior. For this discussion, it will be assumed that the prior belongs to a known *parametric* family. The parametric empirical Bayes method will be explained in general terms and then in the specific case of the binomial example. This method will also be used in Chapter 5 to derive alternatives to the maximum likelihood estimator for logistic regression parameters. Nonparametric versions of the EB problem described below have been considered in the literature (see Griffin and Krutchkoff, 1971; Copas, 1972; Maritz and Lian, 1974; Walter and Hamdani, 1987). PB methodology has received extensive attention in the multinomial literature; it is described in detail in Chapter 12 of Bishop et al. (1975). One PB estimator for the binomial problem is described below. Another application of PB estimation is found in Section 2.2.

Empirical Bayes estimation assumes a sequence of similar experiments is conducted each having its own parameter. In sampling inspection, suppose Y_i is the number of defective items in a sample of n_i items from the ith production lot. EB methods are used to estimate the proportion of defectives in future lots or to assess some other property of the distribution of defective items. Another application is in long-term rodent carcinogenesis experiments. Consider a series of bioassay experiments each performed with its own control group and let Y_i be the number of rodents that develop cancer from among the n_i control animals in the ith experiment. EB methods have been applied to improve estimation of the carcinogenesis rate in the control population (Tarone, 1982; Hoel and Yanagawa, 1986).

Formally, suppose $\pi(\cdot \mid \boldsymbol{\tau})$ is the prior where the parameter $\boldsymbol{\tau}$ is unknown. Let θ_1 be the parameter for the first problem (chosen from $\pi(\cdot \mid \boldsymbol{\tau})$), Y_1 be the corresponding data (with distribution $F(\cdot \mid \theta_1)$), θ_2 be the parameter for the second problem (from $\pi(\cdot \mid \boldsymbol{\tau})$), Y_2 be the data for the second problem (from $F(\cdot \mid \theta_2)$), and so forth. After the mth experiment the data $\mathbf{Y}' = (Y_1, \ldots, Y_m)$ are available which can be viewed as arising from the *marginal* distribution $\mathbf{Y} \mid \boldsymbol{\tau}$

$$\prod_{i=1}^{m} \int F(\cdot \mid \theta_i) \pi(d\theta_i \mid \boldsymbol{\tau}). \tag{2.1.5}$$

Let $\boldsymbol{\theta}' := (\theta_1, \ldots, \theta_m)$ be the vector of unobserved parameters. The problem is to make inferences about (some functional of) $\pi(\cdot \mid \boldsymbol{\tau})$ such as the mean of the posterior distribution given a new observation $Y_{m+1} = y_{m+1}$. EB procedures use $\mathbf{Y}$ to estimate $\pi(\cdot \mid \boldsymbol{\tau})$ by $\hat{\pi} = \pi(\cdot \mid \hat{\boldsymbol{\tau}})$ where $\hat{\boldsymbol{\tau}}$ is an estimator of $\boldsymbol{\tau}$ based on the marginal distribution (2.1.5) of $\mathbf{Y}$. The estimated prior $\hat{\pi}$ is used in standard Bayesian calculations. Berger (1985, Sec. 4.5) gives a more detailed introduction to this subject.

To illustrate the calculations in a simple case, consider estimating a binomial probability p based on $Y \sim B(n,p)$ with a beta prior distribution where the prior mean μ is *known* but the prior sample size K is *unknown;* i.e., $\alpha = \mu K$ and $\beta = K(1-\mu)$. In terms of the general framework of the preceding paragraph, $m = 1$ and $\tau = K$ is one dimensional. In the more realistic case there are independent observations $Y_i \sim B(n_i, p_i)$, $1 \le i \le m$, with the $\{n_i\}$ known and $p_1, \ldots, p_m$ iid.

Recall that, in terms of μ and K, the Bayesian estimator of p with respect to SEL is

$$\hat{p}^B = \hat{p}\frac{n}{n+K} + \mu\frac{K}{n+K}.$$

The marginal distribution of $Y \mid K$ is

$$\begin{aligned} m(y \mid K) &= \int_0^1 \binom{n}{y} p^y (1-p)^{n-y} \frac{\Gamma(K)}{\Gamma(\mu K)\Gamma(\mu\{1-K\})} \\ &\quad \cdot p^{\mu K - 1}(1-p)^{K(1-\mu)-1} dp \\ &= \binom{n}{y} \frac{\Gamma(K)\Gamma(\mu K + y)\Gamma(K\{1-\mu\}+n-y)}{\Gamma(\mu K)\Gamma(\mu\{1-K\})\Gamma(K+n)}. \end{aligned} \tag{2.1.6}$$

Consider maximum likelihood estimation of K. Differentiating the logarithm of (2.1.6) with respect to K yields the likelihood equation

$$\begin{aligned} \mu \sum_{j=1}^{y} (\mu K + y - j)^{-1} + (1-\mu) \sum_{j=1}^{n-y} ((1-\mu)K + n - y - j)^{-1} \\ = \sum_{j=1}^{n} (K + n - j)^{-1}. \end{aligned} \tag{2.1.7}$$

The solution $\hat{K}^E$ of (2.1.7) does not have a closed form but is easy to compute. The resulting EB estimator is

$$\hat{p}^E = \hat{p}\frac{n}{n+\hat{K}^E} + \mu\frac{\hat{K}^E}{n+\hat{K}^E}. \tag{2.1.8}$$

Other EB estimators can be derived by using alternative principles, such as the method of moments based on (2.1.6), to estimate K.

In the general case of mutually independent $Y_i \sim B(n_i, p_i)$, $1 \le i \le m$, and iid beta distributed $\{p_i\}$, Griffiths (1973) gives the likelihood equations for finding the joint MLE of both μ and K based on the marginal distribution of $\mathbf{Y}' = (Y_1, \ldots, Y_m)$. (See Problem 2.2 for the special case $n_i = 1$.) Tamura and Young (1987) and the references therein consider moment estimators of μ and K. Sabavala and Lee (1981) and Lee and Sabavala (1982) consider similar estimation methods based on a different motivation. Lastly, as will be discussed in Chapter 4, the EB method has also been applied to analyze cross-classified data (Leonard, 1975; Laird, 1978a; and Nazaret, 1987).

To describe the PB method, again consider estimation of the binomial p under SEL with a Be(α, β) prior where $\mu = \alpha/(\alpha + \beta)$ is *known* and $K = \alpha + \beta$ is *unknown*. The goal of the PB method is to use the Bayes rule $\hat{p}^B(\mu, K)$ with K chosen to minimize $R_S(p, \hat{p}^B(\mu, K))$. Fixing p and differentiating equation (2.1.2), it is easy to calculate that

$$K = K(p, \mu) = \frac{p(1-p)}{(p-\mu)^2} \tag{2.1.9}$$

is the desired optimum. Formula (2.1.9) can be intuitively verified in several cases. If $p = 0$ or 1 and $\mu \neq p$, then (2.1.9) yields $K = 0$ and $\hat{p}^B = \hat{p} = Y/n$ which correctly guesses the true p with probability 1 in both cases; hence $R_S(p, \hat{p}^B) = 0$ which is obviously the global minimum. If $0 < p = \mu < 1$, then (2.1.9) gives $K = \infty$ and $\hat{p}^B = \mu$ for which $R_S(p, \hat{p}^B)$ is again at its minimum of zero.

The PB method estimates the unknown K (it depends on the unknown p) via the *conditional* distribution $Y \mid p$; i.e., by the $B(n, p)$ distribution. Many estimators of K are possible. For example, if one estimates p by $\hat{p}$, the corresponding estimate of K is

$$\hat{K}^O := \frac{\hat{p}(1-\hat{p})}{(\hat{p}-\mu)^2}$$

which gives rise to the PB estimate of p,

$$\hat{p}^O := \hat{p}^B(\hat{K}, \mu) = \frac{n}{n + \hat{K}^O}\hat{p} + \frac{\hat{K}^O}{n + \hat{K}^O}\mu. \tag{2.1.10}$$

Other pseudo Bayes estimates of p result from different K-estimators. (Section 2.2 gives additional examples and some comparisons.)

Example 2.1.1 (continued). When the beta prior has mean $\mu = 1/2$ and unknown K, the EB and PB estimators for the breat cancer data are $\hat{p}^E = .41$ ($\hat{K}^E = 13.01$) and $\hat{p}^O = .37$ ($\hat{K}^O = 5.25$). The empirical Bayes estimator is somewhat suspect because the value $n = 10$ is small relative to the estimate of K. However, the MSE curves for (2.1.8) and (2.1.10) for $n = 10$, displayed in Figure 2.1.2, show both estimators pull $\hat{p}$ toward 1/2 except at the outcomes $Y = 0$ and 10 for which $\hat{p}^E(0) = \hat{p}^O(0) = 0$ and $\hat{p}^E(10) = \hat{p}^O(10) = 1$. From the previous analysis of this example, the latter insures that the risk curves of both estimators tend to zero as p tends to either extreme. For the outcomes $Y \in \{1, \ldots, 9\}$, the EB estimator pulls $\hat{p}$ toward 1/2 more strongly than the PB estimator because the estimate of K is larger for the EB estimator. As a result, $\hat{p}^E$ has lower risk than $\hat{p}^P$ for approximately $p \in (.3, .7)$. Note that at $p = .3$ (the MLE for the breast cancer data) both $\hat{p}^E$ and $\hat{p}^O$ have larger MSE than $\hat{p}$.

Estimating ω and λ

The estimators $\hat{\omega} := \hat{p}/(1-\hat{p})$ and $\hat{\lambda} := \ln(\hat{\omega})$ are the MLEs of ω and λ for outcomes $y \in \{1, \ldots, n-1\}$; the MLE's of ω and λ do not exist for the outcomes $y = 0$ and n. Furthermore there is no way of defining $\hat{\omega}$ and $\hat{\lambda}$ at these two values to produce unbiased estimators of ω or λ (Problem 2.4).

Classical work on the estimation of ω and λ focused on approximate unbiased estimators. For example, Haldane (1955) and Anscombe (1956) considered the class of estimators called *empiric log odds* given by $\hat{\lambda}(c) := \ln\{\frac{Y+c}{n-Y+c}\}$. The empiric log odds comes from using the MLE $\hat{p}$ of p in the expression for λ based on augmented data obtained by adding c "fake successes" and c "fake failures" to the real data Y; c need not be an integer. Anscombe showed that

$$E_p[\hat{\lambda}(c)] = \lambda + \frac{(1-2p)(c-(1/2))}{p(1-p)n} + o(1/n).$$

The estimator $\hat{\lambda}(1/2)$ is an $o(1/n)$-unbiased estimator *for all p*.

Gart, Pettigrew, and Thomas (1985) report the bias and first four cumulants of $\hat{\lambda}(c)$ for a range of c from $-1/2$ to $1/2$ as part of a study of more complicated regression problems involving log odds. They find that no single correction such as $c = 1/2$ is adequate in all such cases.

Gart, Pettigrew, and Thomas (1985) and Gart and Zweifel (1967) study estimation of the variance of $\hat{\lambda}(1/2)$. They suggest using

$$\hat{V} := \frac{(n+1)(n+2)}{n(Y+1)(n-Y+1)}. \tag{2.1.11}$$

In a similar spirit Haldane (1957) considers "almost" unbiased estimation where "almost" unbiased is defined below.

Definition. $\hat{g} = \hat{g}(Y)$ is an *almost unbiased estimator* of a real function $g(\theta)$ of the unknown parameter θ means $E_\theta[\hat{g}(Y)] = g(\theta) + o(n^{-k})$ for all $k \in \{1, 2, \ldots\}$. Haldane (1957) proved that $\hat{g}(Y) := D(Y) - D(n-Y)$ is almost unbiased for λ where $D(y) := (d/dy)\ln(\Gamma(y+1))$ is the digamma function.

B. Interval Estimation

This subsection considers confidence interval construction. The discussion is limited in several ways. Large and small sample confidence intervals are determined only for p. As noted earlier, confidence intervals for either ω or λ follow directly from those for p. Second, only two-sided intervals are considered. One-sided small sample confidence bounds can be obtained from the Clopper/Pearson tail intervals described below; one-sided large sample confidence bounds are discussed in Fujino (1980) and Blyth (1986).

Small Sample Intervals

By way of review, recall that a level $100(1-\alpha)\%$ confidence set for p can be constructed by inverting a family of acceptance regions of size less than or equal to α for the hypotheses $H_0\colon p = p_0$ versus $H_A\colon p \neq p_0$ (as p_0 ranges over $(0,1)$). Formally, if $A(p_0)$ is the acceptance region of H_0, then $P_{p_0}[Y \in A(p_0)] \geq 1-\alpha$. If $I(j) := \{p_0 : j \in A(p_0)\}$ is defined for $j = 0(1)n$, then it is a tautology that

$$P_{p_0}[p_0 \in I(Y)] = P_{p_0}[Y \in A(p_0)] \geq 1-\alpha$$

so that $I(Y)$ is a level $100(1-\alpha)\%$ confidence set. Confidence sets corresponding to three families of acceptance regions considered in the literature are discussed.

(i) *Uniformly Most Accurate Unbiased Intervals:* Blyth and Hutchinson (1960) table the intervals corresponding to the acceptance regions $A(p_0)$ of the uniformly most powerful unbiased tests of H_0 versus H_A. These $A(p_0)$ are of the form $\{L(p_0), \ldots, U(p_0)\}$ with the endpoints randomized so that each "tail" of the region has probability $\alpha/2$. Thus $A(p_0)$ has exactly $1-\alpha$ coverage. The UMAU intervals are not used in practice because of their randomization. Other methods reviewed produce non-randomized intervals.

(ii) *Tail Intervals:* Clopper and Pearson (1934) proposed a set of nonrandomized intervals which attempt, in a conservative fashion, to mimic the equi-tailed property of the UMAU intervals. Fix $0 < \alpha < 1$ and define the rejection region $R(p_0) := \underline{R}(p_0) \cup \overline{R}(p_0)$ where

$$\underline{R}(p_0) := \{\{0, \ldots, L\} : P_{p_0}[Y \leq L] \leq \alpha/2 < P_{p_0}[Y \leq L+1]\}$$

and

$$\overline{R}(p_0) := \{\{U, \ldots, n\} : P_{p_0}[Y \geq U] \leq \alpha/2 < P_{p_0}[Y \leq U-1]\}.$$

In words, the lower and upper tails $\underline{R}(p_0)$ and $\overline{R}(p_0)$ of the rejection region have p_0-probability as close to $\alpha/2$ as possible but less than or equal to it. However $\underline{R}(p_0)$ ($\overline{R}(p_0)$) can be quite small—even empty for sufficiently small (large) p_0 since $P_{p_0}[Y = 0]$ ($P_{p_0}[Y = n]$) can be greater than $\alpha/2$. This leads to conservative tests (and long intervals).

Define the acceptance regions to be $A(p_0) := \{0, \ldots, n\} \setminus R(p_0)$; $A(p_0)$ is of the form $\{L(p_0), \ldots, U(p_0)\}$ and has coverage probability at least $1 - \alpha$. To see this, note that $\overline{R}(p_0) \cap \underline{R}(p_0) = \emptyset$ since otherwise $R(p_0) = \{0, \ldots, n\}$ which gives the contradiction

$$1 = P_{p_0}[\overline{R}(p_0) \cup \underline{R}(p_0)] \leq P_{p_0}[\overline{R}(p_0)] + P_{p_0}[\underline{R}(p_0)] \leq \alpha/2 + \alpha/2 < 1.$$

Hence

$$\begin{aligned} P_{p_0}[Y \in A(p_0)] &= 1 - P_{p_0}[Y \in R(p_0)] \\ &= 1 - P_{p_0}[Y \in \underline{R}(p_0)] - P_{p_0}[Y \in \overline{R}(p_0)] \\ &\geq 1 - \alpha/2 - \alpha/2 \\ &= 1 - \alpha. \end{aligned}$$

The confidence intervals resulting from inversion of $\{A(p_0) : 0 < p_0 < 1\}$ are characterized as follows.

Proposition 2.1.1. *The system of* $100(1 - \alpha)\%$ *tail intervals for* p *are given by*

$$I(j) := \{p_0 : j \in A(p_0)\} = (\underline{p}(j), \overline{p}(j))$$

where $\underline{p}(0) = 0$, $P_{\underline{p}(j)}[Y \geq j] = \alpha/2$ *for* $1 \leq j \leq n$, *and* $\overline{p}(n) = 1$, $P_{\overline{p}(j)}[Y \leq j] = \alpha/2$ *for* $0 \leq j < n$.

Proof. The claim is proved below for $1 \leq j \leq n - 1$; the cases $j = 0$ and $j = n$ follow by similar arguments. By the definition of $\underline{p}(j)$ and $\overline{p}(j)$ and the monotonicity of $P_p[Y \leq j]$ as a function of p, the point p_0 satisfies $\underline{p}(j) < p_0 < \overline{p}(j)$ if and only if (i) $P_{p_0}[Y \geq j] > P_{\underline{p}(j)}[Y \geq j] = \alpha/2$ and (ii) $P_{p_0}[Y \leq j] > P_{\overline{p}(j)}[Y \leq j] = \alpha/2$. These two statements imply $j \notin \underline{R}(p_0)$ and $j \notin \overline{R}(p_0)$, respectively, so that $j \in A(p_0)$ which proves the result. □

Tail intervals have many advantages including the following closed form expressions for $\underline{p}(j)$ and $\overline{p}(j)$. Expressing the binomial tail probabilities in terms of the incomplete beta function gives

$$\begin{aligned} &\int_{\overline{p}(j)}^{1} \frac{n!}{j!(n-j-1)!} t^j (1-t)^{n-j-1} dt = \alpha/2, \qquad 0 \leq j \leq n-1 \\ &\int_{0}^{\underline{p}(j)} \frac{n!}{(j-1)!(n-j)!} t^{j-1} (1-t)^{n-j} dt = \alpha/2, \quad 1 \leq j \leq n. \end{aligned} \tag{2.1.12}$$

Alternatively one can use the relationship between the beta and F distributions, $\text{Be}(\nu_1, \nu_2) \sim [1 + \nu_2/\nu_1 F_{2\nu_1, 2\nu_2}]^{-1}$, to obtain the formulae

$$\underline{p}(j) = \frac{1}{1 + \frac{(n-j+1)}{j} F_{\alpha/2, 2(n-j+1), 2j}}, \quad 1 \leq j \leq n$$
$$\overline{p}(j) = \frac{\frac{j+1}{n-j} F_{\alpha/2; 2(j+1), 2(n-j)}}{1 + \frac{j+1}{n-j} F_{\alpha/2, 2(j+1), 2(n-j)}}, \quad 0 \leq j \leq n-1. \tag{2.1.13}$$

The symbol F_{α, ν_1, ν_2} is the upper α percentile of the F distribution with ν_1 and ν_2 degrees of freedom; i.e., if $Y \sim F_{\nu_1, \nu_2}$; then $P[Y > F_{\alpha, \nu_1, \nu_2}] = \alpha$.

Tail intervals have many attractive symmetry properties including:

(i) Invariance under relabeling of successes and failures: if $Y \to n - Y$, then $(\underline{p}, \overline{p}) \to (1 - \overline{p}, 1 - \underline{p})$ (Problem 2.6).

(ii) Symmetry about 1/2 when n is even and $j = n/2$ ($\hat{p} = 1/2$): $\underline{p}(n/2) = 1 - \overline{p}(n/2)$.

(iii) Monotonicity in Y: for fixed n, $\underline{p}(j) \leq \underline{p}(j+1)$ and $\overline{p}(j) \leq \overline{p}(j+1)$ for $0 \leq j \leq n-1$.

(iv) Monotonicity in n: for fixed $0 \leq j \leq n$, $\overline{p}(j)$ is nonincreasing in n.

(v) Monotonicity in α: for fixed n and $0 \leq j \leq n$, $\underline{p}(j)$ is nondecreasing in α and $\overline{p}(j)$ is nonincreasing in α.

These properties all have intuitive appeal. For example, (iii) says that when a success is observed in place of a failure the interval estimate is not revised downward. Property (iv) says that when a failure is observed for an additional observation, then the interval is not revised upward.

Tail intervals have been adapted to many other problems because of their intuitive appeal and computational ease. For example, they have been calculated for the mean of a Poisson distribution based on $Y \sim P(\lambda)$ (Garwood, 1936), for the odds ration $\psi := \frac{p_1(1-p_2)}{p_2(1-p_1)}$ and difference $\Delta := p_1 - p_2$ of the success probabilities of two independent binomial distributions (Cornfield, 1956; Fisher, 1962; Santner and Snell, 1980), and for a binomial success probability based on a multistage stopping boundary (Jennison and Turnbull, 1983).

Unfortunately, tail intervals have one overriding deficiency—they are extremely conservative. This is because $\underline{R}(p_0)$ and $\overline{R}(p_0)$ are often too "small"; in fact, the coverage probability is at least $1 - \alpha/2$ (instead of $1 - \alpha$) for p values near 0 or 1. Figure 2.1.3 illustrates a typical example; see also Problem 2.7. Angus and Schafer (1984) give general results on the minimum achieved confidence coefficient of the Clopper–Pearson intervals.

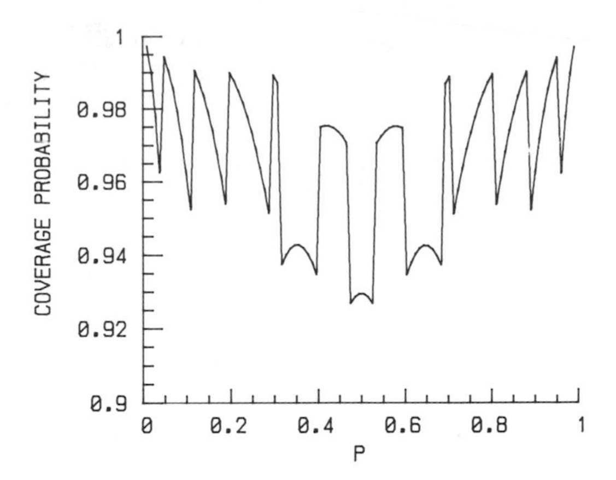

Figure 2.1.3. Achieved coverage probability of nominal 90% Clopper–Pearson tail intervals for $n = 8$.

Several suggestions have been made in the literature for constructing less conservative tail intervals. Vos (1978, 1979) proposed a system he called average confidence intervals for which the average achieved coverage probability over p in $(0, 1)$ is equal to the nominal level $1 - \alpha$. Fujino and Okuno (1984) proposed a modification of the Vos criteria by determining the system of intervals which maximize the minimum achieved coverage probability among those for which the average achieved coverage probability over p in $(0, 1)$ is equal to the nominal level $1 - \alpha$. Obviously, both of these proposals can be given Bayesian interpretations.

Other small sample methods work directly with the acceptance regions to force $P_{p_0}[A(p_0)]$ nearer to its nominal value $1 - \alpha$. The approach described next is one such construction.

(iii) *Sterne/Crow/Blyth/Still Intervals:* The intuition behind this method is that short intervals should result from small $A(p_0)$. Sterne (1954) proposed putting the "most likely outcomes" in $A(p_0)$. Formally, fix $\alpha \in (0, 1)$ and $p_0 \in (0, 1)$. Let $\ell_1, \ldots, \ell_{n+1}$ denote the $n+1$ outcomes $0, 1, \ldots, n$ ranked according to the criteria

$$P_{p_0}[Y = \ell_1] \geq P_{p_0}[Y = \ell_2] \geq \ldots \geq P_{p_0}[Y = \ell_{n+1}].$$

Then define $A(p_0) := \{\ell_1, \ell_2, \ldots, \ell_J\}$ where J satisfies

$$\sum_{q=1}^{J} P_{p_0}[Y = \ell_q] \geq 1 - \alpha > \sum_{q=1}^{J-1} P_{p_0}[Y = \ell_q]. \tag{2.1.14}$$

In words, $A(p_0)$ contains as many of the most likely outcomes needed to achieve the desired coverage; thus $A(p_0)$ is of minimal possible cardinality.

The goal of minimizing $A(p_0)$ tacitly assumes that small $A(p_0)$ will generate small confidence sets. However the only property that Sterne's confidence sets can analytically be shown to possess is that they minimize

$$\sum_{j=0}^{n} \{\text{Lebesgue measure of } I(j)\}. \tag{2.1.15}$$

(Here Lebesgue measure must be used rather than length since $I(j)$ need not be an interval as will be shown by the example below.)

There are two problems associated with Sterne's acceptance regions. The first is minor; it concerns an ambiguity in the definition of $A(p_0)$. Sets of (n, α, p_0) for which $P_{p_0}[Y = \ell_j] = P_{p_0}[Y = \ell_{j+1}]$ and one but *not* both outcomes are required to satisfy (2.1.14) lead to non-uniquely defined $A(p_0)$. Fortunately this is not a practical problem as the set of p_0 for which this phenomenon holds has Lebesgue measure zero for given n and α.

Crow (1956) pointed out the more serious problem alluded to above; namely, that the Sterne confidence sets $I(j) = \{p_0 : j \in A(p_0)\}$ need not be intervals. To see why this happens, first note that the unimodality of the binomial probability mass function implies that $A(p_0)$ must always be an interval of integers of the form $\{L(p_0), \ldots, U(p_0)\}$. Table 2.1.1 lists the 95% acceptance regions $A(p_i)$ for $n = 20$ and a selected set of p_i values. They show the inverted confidence sets are not intervals. Specifically for $y = 0, 1$ the Sterne confidence sets are $I(0) = (0, .127) \cup (.141, .147)$ and $I(1) = (.006, .203) \cup (.221, .223)$.

Table 2.1.1. Sterne Acceptance Regions for $(\alpha, n) = (.05, 20)$ and Selected p_i

p_i	$A(p_i)$
.001	$\{0\}$
.006	$\{0, 1\}$
.127	$\{0, \ldots, 5\}$
.128	$\{1, \ldots, 5\}$
.140	$\{1, \ldots, 5\}$
.141	$\{0, \ldots, 5\}$

It is easy to show that the sets $\{I(j) : j = 0(1)n\}$ are intervals if and only if the endpoints $L(p_0)$ and $U(p_0)$ of $A(p_0)$ are nondecreasing in p_0. Thus Crow (1956) proposed an algorithmic construction of acceptance sets $A(p_i) = \{L(p_i), \ldots, U(p_i)\}$ for a partition $0 < p_1 < \ldots < p_M < 1$ of $[0, 1]$ in such a way to force $L(p_i)$ and $U(p_i)$ to be nondecreasing. He simultaneously attempted to force $A(p_i)$ to contain the same number of points as Sterne's acceptance regions. Hence, the Crow intervals also minimize (2.1.15) in most cases.

Blyth and Still (1983) note that the Crow intervals violate several of the symmetry properties (i)–(v) listed earlier. For example, when $(\alpha, y) =$

(.05, 2) Table 2.1.2 shows the upper Sterne/Crow limit $\overline{p}(2)$ increases as n goes from 9 to 10 violating Property (iv) above. In addition there are cases where the length $(\overline{p}(j) - \underline{p}(j))$ *increases* as n increases $((\alpha, j)$ fixed). This contradicts the intuitive notion that the precision of the interval should improve with increasing data.

Table 2.1.2. Upper Tail Limit for $(\alpha, j) = (.05, 2)$, and $n = 8(1)11$

n	$\overline{p}(2)$
8	.685
9	.558
10	.603
11	.500

Blyth and Still construct a set of modified Crow intervals which satisfy symmetry properties (i) to (v) with as regular size monotonicity steps in (iii) and (iv) as possible, and which contain the same number of points as Sterne's $A(p_0)$ for most (j, n, α). Table 2.1.3 lists these intervals for $n \leq 30$ and $\alpha = .01, .05$.

Table 2.1.3 (a). Blyth and Still 95% Confidence Intervals for $n \leq 30$ (Reprinted with permission from C.R. Blyth and H.A. Still: "Binomial Confidence Intervals," *Journal of the American Statistical Association,* vol. 78, no. 381, 1983.)

X	$n = 1$		$n = 2$		$n = 3$		$n = 4$		$n = 5$		$n = 6$		X
0	0	.95	0	.78	0	.63	0	.53	0	.50	0	.41	0
1	.05	1	.03	.97	.02	.86	.01	.75	.01	.66	.01	.59	1
2			.22	1	.14	.98	.10	.90	.08	.81	.06	.73	2
3					.37	1	.25	.99	.19	.92	.15	.85	3

X	$n = 7$		$n = 8$		$n = 9$		$n = 10$		$n = 11$		$n = 12$		X
0	0	.38	0	.36	0	.32	0	.29	0	.26	0	.24	0
1	.01	.55	.01	.50	.01	.44	.01	.44	.00	.40	.00	.37	1
2	.05	.66	.05	.64	.04	.56	.04	.56	.03	.50	.03	.46	2
3	.13	.77	.11	.71	.10	.68	.09	.62	.08	.60	.07	.54	3
4	.23	.87	.19	.81	.17	.75	.15	.70	.14	.67	.12	.63	4
5	.34	.95	.29	.89	.25	.83	.22	.78	.20	.74	.18	.71	5
6	.45	.99	.36	.95	.32	.90	.29	.85	.26	.80	.24	.76	6

Table 2.1.3(a). (cont.)

X	$n = 13$		$n = 14$		$n = 15$		$n = 16$		$n = 17$		$n = 18$		X
0	0	.23	0	.23	0	.22	0	.20	0	.19	0	.18	0
1	.00	.34	.00	.32	.00	.30	.00	.30	.00	.28	.00	.27	1
2	.03	.43	.03	.42	.02	.39	.02	.37	.02	.35	.02	.33	2
3	.07	.52	.06	.50	.06	.47	.05	.44	.05	.42	.05	.41	3
4	.11	.59	.10	.58	.10	.53	.09	.50	.08	.49	.08	.47	4
5	.17	.66	.15	.63	.14	.61	.13	.56	.12	.54	.12	.53	5
6	.22	.74	.21	.68	.19	.67	.18	.63	.17	.59	.16	.59	6
7	.26	.78	.24	.76	.22	.71	.20	.70	.19	.65	.18	.63	7
8	.34	.83	.32	.79	.29	.78	.27	.73	.25	.72	.24	.67	8
9	.41	.89	.37	.85	.33	.81	.30	.80	.28	.75	.27	.73	9

X	$n = 19$		$n = 20$		$n = 21$		$n = 22$		$n = 23$		$n = 24$		X
0	0	.17	0	.16	0	.15	0	.15	0	.14	0	.13	0
1	.00	.25	.00	.24	.00	.23	.00	.22	.00	.21	.00	.20	1
2	.02	.32	.02	.32	.02	.30	.02	.29	.02	.27	.02	.26	2
3	.04	.39	.04	.37	.04	.35	.04	.34	.04	.32	.03	.31	3
4	.08	.45	.07	.42	.07	.40	.06	.39	.06	.39	.06	.37	4
5	.11	.50	.10	.47	.10	.46	.09	.45	.09	.43	.09	.41	5
6	.15	.55	.14	.53	.13	.51	.13	.50	.12	.48	.11	.46	6
7	.17	.61	.16	.58	.15	.55	.15	.55	.14	.52	.13	.50	7
8	.22	.66	.21	.63	.20	.60	.19	.58	.18	.57	.17	.54	8
9	.25	.69	.24	.68	.23	.65	.22	.62	.21	.61	.20	.59	9
10	.31	.75	.29	.71	.28	.70	.26	.66	.25	.64	.23	.63	10
11	.34	.78	.32	.76	.30	.72	.29	.71	.27	.68	.26	.66	11
12	.39	.83	.37	.79	.35	.77	.34	.74	.32	.73	.31	.69	12

X	$n = 25$		$n = 26$		$n = 27$		$n = 28$		$n = 29$		$n = 30$		X
0	0	.13	0	.12	0	.12	0	.12	0	.11	0	.11	0
1	.00	.19	.00	.19	.00	.18	.00	.17	.00	.17	.00	.16	1
2	.01	.25	.01	.24	.01	.23	.01	.23	.01	.22	.01	.21	2
3	.03	.30	.03	.30	.03	.29	.03	.28	.03	.27	.03	.26	3
4	.06	.36	.05	.34	.05	.33	.05	.32	.05	.31	.05	.30	4
5	.08	.40	.08	.38	.08	.37	.07	.36	.07	.36	.07	.35	5
6	.11	.44	.11	.42	.10	.41	.10	.41	.09	.39	.09	.38	6
7	.13	.48	.12	.47	.12	.46	.12	.44	.11	.43	.11	.41	7
8	.16	.52	.15	.51	.15	.50	.14	.48	.14	.46	.13	.45	8
9	.19	.56	.19	.54	.18	.54	.17	.52	.17	.50	.16	.48	9
10	.22	.60	.21	.58	.20	.57	.19	.56	.18	.54	.18	.52	10
11	.25	.64	.24	.62	.23	.60	.23	.59	.22	.57	.21	.55	11
12	.30	.68	.28	.66	.27	.63	.26	.62	.25	.61	.24	.59	12
13	.32	.70	.30	.70	.29	.67	.28	.65	.27	.64	.26	.62	13
14	.36	.75	.34	.72	.33	.71	.32	.68	.31	.66	.30	.65	14
15	.40	.78	.38	.76	.37	.73	.35	.72	.34	.69	.32	.68	15

Table 2.1.3 (b). Blyth and Still 99% Confidence Intervals for $n \leq 30$ (Reprinted with permission from C.R. Blyth and H.A. Still: "Binomial Confidence Intervals," *Journal of the American Statistical Association*, vol. 78, no. 381, 1983.)

X	$n=1$		$n=2$		$n=3$		$n=4$		$n=5$		$n=6$		X
0	0	.99	0	.90	0	.78	0	.68	0	.60	0	.54	0
1	.01	1	.01	.99	.00	.94	.00	.86	.00	.78	.00	.71	1
2			.10	1	.06	1	.04	.96	.03	.89	.03	.83	2
3					.22	1	.14	1	.11	.97	.08	.92	3

X	$n=7$		$n=8$		$n=9$		$n=10$		$n=11$		$n=12$		X
0	0	.50	0	.45	0	.43	0	.38	0	.36	0	.35	0
1	.00	.64	.00	.59	.00	.57	.00	.51	.00	.50	.00	.45	1
2	.02	.76	.02	.71	.02	.66	.02	.62	.01	.59	.01	.55	2
3	.07	.86	.06	.80	.05	.75	.05	.70	.04	.66	.04	.65	3
4	.14	.93	.12	.88	.11	.83	.09	.78	.08	.74	.08	.70	4
5	.24	.98	.20	.94	.17	.89	.15	.85	.13	.81	.12	.77	5
6	.36	1	.29	.98	.25	.95	.22	.91	.19	.87	.17	.83	6

X	$n=13$		$n=14$		$n=15$		$n=16$		$n=17$		$n=18$		X
0	0	.32	0	.30	0	.28	0	.26	0	.26	0	.25	0
1	.00	.43	.00	.42	.00	.39	.00	.36	.00	.35	.00	.34	1
2	.01	.52	.01	.50	.01	.46	.01	.45	.01	.43	.01	.41	2
3	.04	.59	.03	.58	.03	.54	.03	.52	.03	.50	.03	.47	3
4	.07	.68	.06	.64	.06	.61	.06	.58	.05	.57	.05	.53	4
5	.11	.73	.10	.70	.09	.67	.09	.64	.08	.62	.08	.59	5
6	.16	.79	.15	.75	.13	.72	.13	.70	.12	.66	.11	.66	6
7	.21	.84	.19	.81	.18	.77	.17	.74	.16	.73	.15	.69	7
8	.27	.89	.25	.85	.23	.82	.21	.79	.20	.76	.18	.75	8
9	.32	.93	.30	.90	.28	.87	.26	.83	.24	.80	.23	.77	9

X	$n=19$		$n=20$		$n=21$		$n=22$		$n=23$		$n=24$		X
0	0	.24	0	.22	0	.21	0	.20	0	.19	0	.19	0
1	.00	.32	.00	.31	.00	.29	.00	.28	.00	.27	.00	.26	1
2	.01	.39	.01	.37	.01	.37	.01	.35	.01	.33	.01	.32	2
3	.02	.46	.02	.44	.02	.42	.02	.40	.02	.39	.02	.39	3
4	.05	.52	.04	.50	.04	.47	.04	.45	.04	.45	.04	.43	4
5	.07	.56	.07	.56	.07	.53	.06	.50	.06	.50	.06	.48	5
6	.10	.61	.10	.60	.09	.58	.09	.55	.08	.55	.08	.52	6
7	.14	.68	.13	.64	.12	.63	.12	.60	.11	.58	.11	.57	7
8	.17	.71	.16	.69	.15	.66	.15	.65	.14	.62	.13	.61	8
9	.21	.76	.20	.73	.19	.71	.18	.68	.17	.67	.16	.64	9
10	.24	.79	.22	.78	.21	.74	.20	.72	.19	.70	.19	.68	10
11	.29	.83	.27	.80	.26	.79	.24	.76	.23	.73	.22	.72	11
12	.32	.86	.31	.84	.29	.81	.28	.80	.27	.77	.26	.74	12

Table 2.1.3(b). (cont.)

X	$n=25$		$n=26$		$n=27$		$n=28$		$n=29$		$n=30$		X
0	0	.18	0	.17	0	.17	0	.16	0	.16	0	.16	0
1	.00	.26	.00	.25	.00	.24	.00	.23	.00	.22	.00	.22	1
2	.01	.31	.01	.30	.01	.30	.01	.29	.01	.28	.01	.27	2
3	.02	.37	.02	.36	.02	.34	.02	.33	.02	.32	.01	.31	3
4	.03	.41	.03	.40	.03	.38	.03	.38	.03	.37	.03	.36	4
5	.05	.46	.05	.44	.05	.44	.05	.42	.05	.41	.04	.39	5
6	.08	.50	.07	.49	.07	.48	.07	.46	.07	.44	.06	.43	6
7	.10	.54	.10	.53	.09	.52	.09	.50	.09	.48	.08	.47	7
8	.13	.59	.12	.56	.12	.56	.11	.54	.11	.52	.10	.51	8
9	.16	.63	.15	.60	.14	.59	.14	.58	.13	.56	.13	.54	9
10	.18	.66	.17	.64	.17	.62	.16	.62	.16	.59	.15	.57	10
11	.21	.69	.19	.68	.18	.66	.18	.64	.17	.63	.16	.61	11
12	.25	.74	.23	.70	.22	.70	.21	.67	.21	.65	.20	.64	12
13	.26	.75	.25	.75	.24	.72	.23	.71	.22	.68	.22	.67	13
14	.31	.79	.30	.77	.28	.76	.27	.73	.26	.72	.25	.69	14
15	.34	.82	.32	.81	.30	.78	.29	.77	.28	.74	.27	.73	15

Casella (1986a) considers decision theoretic analysis of the confidence interval problem in which the loss is interval length. He determines a complete class of invariant $100(1-\alpha)\%$ confidence intervals for p for $\alpha = .01$, .05 and $n = 1(1)30$. The Blyth–Still intervals are members of this class.

Large Sample Intervals

The two commonly used large sample intervals are $I_N = I_N(Y, n, \alpha)$ defined as

$$\left(\hat{p} - c\{\hat{p}(1-\hat{p})\}^{1/2}, \hat{p} + c\{\hat{p}(1-\hat{p})\}^{1/2}\right),$$

and a more sophisticated version $I_S = I_S(Y, n, \alpha)$ defined as

$$\left(\frac{2\hat{p} + c^2 - c\{c^2 + 4\hat{p}\hat{q}\}^{1/2}}{2(1+c^2)}, \frac{2\hat{p} + c^2 + c\{c^2 + 4\hat{p}\hat{q}\}^{1/2}}{2(1+c^2)}\right),$$

where $c := z_{\alpha/2}/\sqrt{n}$, $\hat{q} := 1 - \hat{p}$, and $z_{\alpha/2}$ is the upper $\alpha/2$ percentile of the standard normal distribution.

The justification for the I_N interval is that

$$\frac{\sqrt{n}(\hat{p} - p)}{(\hat{p}(1-\hat{p}))^{1/2}} \overset{\text{app}}{\sim} N(0,1)$$

for large n which yields $\overline{p}$ and $\underline{p}$ as the solutions of the quadratic equation (in p)

Table 2.1.4. Achieved Coverage Probabilities of I_N and I_S (Reprinted with permission from B.K. Ghosh: "A Comparison of Some Approximate Confidence Intervals," *Journal of the American Statistical Association,* vol. 74, no. 368, 1979.)

	$n = 15$				$n = 20$			
	$\alpha = .05$		$\alpha = .01$		$\alpha = .05$		$\alpha = .01$	
p	C_S	C_N	C_S	C_N	C_S	C_N	C_S	C_N
.01, .99	.860	.140	.990	.140	.983	.182	.983	.182
.05, .95	.964	.536	.964	.537	.925	.639	.984	.641
.10, .90	.944	.792	.987	.794	.957	.876	.989	.878
.20, .80	.982	.815	.982	.961	.956	.921	.990	.928
.30, .70	.915	.949	.996	.961	.975	.947	.994	.959
.40, .60	.939	.939	.985	.964	.963	.928	.990	.978
.50	.964	.882	.993	.965	.959	.959	.988	.959

	$n = 30$				$n = 50$			
	$\alpha = .05$		$\alpha = .01$		$\alpha = .05$		$\alpha = .01$	
p	C_S	C_N	C_S	C_N	C_S	C_N	C_S	C_N
.01, .99	.964	.260	.964	.260	.911	.395	.966	.395
.05, .95	.939	.782	.984	.785	.962	.920	.988	.923
.10, .90	.974	.809	.992	.957	.970	.879	.991	.965
.20, .80	.964	.946	.989	.953	.951	.938	.992	.979
.30, .70	.930	.953	.992	.968	.957	.935	.992	.979
.40, .60	.962	.935	.986	.975	.941	.941	.987	.979
.50	.957	.957	.995	.984	.935	.935	.993	.985

	$n = 100$				$n = 200$			
	$\alpha = .05$		$\alpha = .01$		$\alpha = .05$		$\alpha = .01$	
p	C_S	C_N	C_S	C_N	C_S	C_N	C_S	C_N
.01, .99	.921	.633	.982	.634	.948	.862	.984	.866
.05, .95	.966	.877	.989	.962	.967	.926	.986	.973
.10, .90	.936	.932	.988	.975	.956	.927	.967	.982
.20, .80	.941	.933	.992	.986	.958	.941	.990	.987
.30, .70	.937	.950	.988	.987	.947	.944	.989	.989
.40, .60	.948	.948	.990	.986	.949	.949	.989	.988
.50	.943	.943	.988	.988	.944	.944	.991	.987

$$n(\hat{p} - p)^2 = \hat{p}(1 - \hat{p})z_{\alpha/2}^2.$$

The I_S interval is justified by the fact

$$\frac{\sqrt{n}(\hat{p} - p)}{(p(1 - p))^{1/2}} \overset{\text{app}}{\sim} N(0, 1)$$

for large n which yields the I_S limits as the solutions of the quadratic equation (in p)

$$n(\hat{p}-p)^2 = p(1-p)z^2_{\alpha/2}.$$

The difference between the two intervals is that I_N estimates the standard deviation of $\hat{p}$ whereas I_S does not.

Ghosh (1979) compares I_S and I_N via three criteria: (i) coverage probability, (ii) interval length, and (iii) bias. Table 2.1.4, reproduced from Ghosh (1979), lists the achieved coverages $C_S(p,n,\alpha) = P_p[p \in I_S(Y,n,\alpha)]$ and $C_N(p,n,\alpha) = P_p[p \in I_N(Y,n,\alpha)]$ for $\alpha = .01$, $.05$; $n = 15, 20, 30, 50, 100, 200$; and a grid of p values. In almost every instance $C_S > C_N$ with C_S being closer to the nominal $1-\alpha$ coverage probability. The lower C_N values are disturbingly inadequate for p near 0 or 1. Similar exact calculations show that I_S tends to be shorter and less biased than I_N. Thus by all three measures I_S has better operating characteristics than I_N indicating that the more naive application of the Central Limit Theorem giving rise to I_N is inadequate.

Several modifications of I_S which further enhance its operating characteristics have been proposed in the literature. Blyth and Still (1983) propose the continuity corrected interval I_{BS} defined by $\underline{p}(0) = 0$, $\overline{p}(n) = 1$ and otherwise as

$$\left(\frac{n\hat{p} + .5 + z^2_{\alpha/2}/2 + z_{\alpha/2}\{n\hat{p} + .5 - (n\hat{p}+.5)^2/n + z^2_{\alpha/2}/4\}^{1/2}}{n + z^2_{\alpha/2}},\right.$$

$$\left.\frac{n\hat{p} - .5 + z^2_{\alpha/2}/2 - z_{\alpha/2}\{n\hat{p} - .5 - (n\hat{p}-.5)^2/n + z^2_{\alpha/2}/4\}^{1/2}}{n + z^2_{\alpha/2}}\right). \quad (2.1.16)$$

Their numerical work shows that the I_{BS} intervals have achieved confidence coefficients which are nearly $(1-\alpha)$ for moderate and large values of n ($n \geq 30$).

2.2 Multinomial Responses

This section considers statistical problems for single-sample multinomial data. Throughout, the t-category multinomial distribution for $\mathbf{Y} = (Y_1, \ldots, Y_t)'$ based on n independent trials with vector of cell probabilities $\mathbf{p} = (p_1, \ldots, p_t)'$ is denoted by $M_t(n, \mathbf{p})$. The vector $\mathbf{p}$ is assumed to be an element of the $(t-1)$-dimensional simplex

$$\mathcal{S} := \left\{ \mathbf{w} \in \mathbb{R}^t : w_i \geq 0 \text{ for } i = 1(1)t \text{ and } \sum_{i=1}^{t} w_i = 1 \right\},$$

and thus $\mathbf{Y}$ has support

$$\mathcal{Y} := \left\{ \mathbf{y} \in \mathbb{R}^t : y_i \geq 0 \text{ and integral for } i = 1(1)t, \sum_{i=1}^{t} y_i = n \right\}.$$

The problems discussed are first point estimation of $\mathbf{p}$, followed by hypothesis tests about $\mathbf{p}$, and then simultaneous confidence intervals for linear combinations of $p_1, \ldots, p_t$. The section concludes with an introduction to selection problems.

A. Point Estimation of p

Our first objective is to prove the well-known result that the vector of sample cell proportions is the maximum likelihood estimator (MLE) of $\mathbf{p}$. Then the MLE's classical statistical properties are stated and its performance with respect to several intuitive loss functions is analyzed. Lastly alternatives to the MLE are considered.

Maximum Likelihood Estimation

Proposition 2.2.1. *The maximum likelihood estimator of* $\mathbf{p}$ *is* $\hat{\mathbf{p}} := \mathbf{Y}/n$.

Proof. Given $\mathbf{y} \in \mathcal{Y}$, the likelihood at $\mathbf{p} \in \mathcal{S}$ is

$$L(\mathbf{p}) = n! \prod_{i=1}^{t} \left(\frac{p_i^{y_i}}{y_i!} \right)$$

where $0^0 = 1$ by continuity. It suffices to show that

$$\prod_{i=1}^{t} p_i^{y_i} \leq \prod_{i=1}^{t} \left(\frac{y_i}{n} \right)^{y_i} \quad \text{for} \quad \begin{cases} \mathbf{y} \in \mathcal{Y} \text{ with } y_i > 0, & i = 1(1)t \\ \text{and} & \\ \mathbf{p} \text{ with } p_i > 0, & i = 1(1)t, \sum_{i=1}^{t} p_i \leq 1. \end{cases} \tag{2.2.1}$$

The reason is that for all $\mathbf{p} \in \mathcal{S}$ and $\mathbf{y} \in \mathcal{Y}$, exactly one of the following holds:

(i) there exists a j such that $p_j = 0$ and $y_j \geq 1$, or

(ii) there exists a j such that $y_j = 0$ (with corresponding $p_j \geq 0$), or

(iii) $p_i > 0$ and $y_i > 0$ for $i = 1(1)t$.

If $\mathbf{p}$ and $\mathbf{y}$ satisfy (i), then $p_j^{y_j} = 0$ implying that $0 = L(\mathbf{p}) \leq L(\hat{\mathbf{p}})$. If $\mathbf{p}$ and $\mathbf{y}$ satisfy (ii), then $p_j^{y_j} = 1 = (y_j/n)^{y_j}$ for all j satisfying $y_j = 0$ so that this factor does not contribute to either $L(\mathbf{p})$ or $L(\hat{\mathbf{p}})$ reducing the problem to (2.2.1). Lastly, (2.2.1) obviously implies $L(\mathbf{p}) \leq L(\hat{\mathbf{p}})$ for any $\mathbf{p}$

and $\mathbf{y}$ satisfying (iii). To prove (2.2.1) recall that the geometric mean and the arithmetic mean satisfy

$$\left(\prod_{i=1}^{m} a_i\right)^{1/m} \leq \frac{1}{m}\sum_{i=1}^{m} a_i \quad \text{for} \quad a_i > 0, i = 1(1)m \tag{2.2.2}$$

(by the convexity of $\ln(\cdot)$ on $(0,\infty)$). Applying (2.2.2) with $m = n = \sum_{i=1}^{t} y_i$ and

$$\mathbf{a}' = \left(\frac{p_1}{y_1}, \ldots, \frac{p_1}{y_1}, \frac{p_2}{y_2}, \ldots, \frac{p_2}{y_2}, \ldots, \frac{p_t}{y_t}, \ldots, \frac{p_t}{y_t}\right),$$

where there are y_i copies of the factor p_i/y_i for $i = 1(1)t$, yields

$$\left[\prod_{i=1}^{t}\left(\frac{p_i}{y_i}\right)^{y_i}\right]^{1/n} \leq \frac{1}{n}\sum_{i=1}^{t}\frac{y_i p_i}{y_i} \leq \frac{1}{n}.$$

This implies

$$\prod_{i=1}^{t}\left(\frac{p_i}{y_i}\right)^{y_i} \leq \frac{1}{n^n}$$

which gives

$$\prod_{i=1}^{t} p_i^{y_i} \leq \frac{1}{n^n}\prod_{i=1}^{t} y_i^{y_i} = \prod_{i=1}^{t}\left(\frac{y_i}{n}\right)^{y_i}$$

and completes the proof. □

For fixed t the estimator $\hat{\mathbf{p}}$ has the following classical optimality properties:

(i) Coordinatewise $\hat{\mathbf{p}}$ is the UMVUE of $\mathbf{p}$.

(ii) $\hat{\mathbf{p}}$ is efficient; i.e., as $n \to \infty$, the variance–covariance matrix of $n^{1/2}(\hat{\mathbf{p}} - \mathbf{p})$ approaches the inverse of the Fisher information.

Both (i) and (ii) follow directly from consideration of the multinomial distribution as a regular exponential family. Additional insight into the small sample performance of $\hat{\mathbf{p}}$ is obtained by examining its risk function under some meaningful loss function. Three loss functions will be considered here; in practice the application will determine the most appropriate loss and none of the three described below may be useful.

Loss Functions

The *squared error loss* (SEL) of estimating $\mathbf{p}$ by $\mathbf{a}' = (a_1, \ldots, a_t)$ is defined by

$$L_S(\mathbf{p}, \mathbf{a}) = n\sum_{i=1}^{t}(a_i - p_i)^2. \tag{2.2.3}$$

There is a slight inconsistency between (2.2.3) and our definition of SEL for the binomial problem in Section 2.1. Applying (2.2.3) to the case $t = 2$ for which $\mathbf{p}' = (p_1, 1 - p_1)$ and $\mathbf{a}' = (a_1, 1 - a_1)$ yields

$$L_S(\mathbf{p}, \mathbf{a}) = n(a_1 - p_1)^2 + n(1 - a_1 - [1 - p_1])^2 = 2n(a_1 - p_1)^2$$

which is a factor of 2 larger than the analogous expression in Section 2.1. A similar relationship also holds for relative squared error loss as defined in (2.2.4) below and in Section 2.1. Since constant factors do not affect an estimator's decision theoretic properties, the choice is simply a matter of convenience.

Some properties of SEL are:

(1) $0 \leq L_S(\mathbf{p}, \mathbf{a}) < \infty$ for all $\mathbf{p}, \mathbf{a} \in \mathcal{S}$; $L_S(\mathbf{p}, \mathbf{a}) = 0$ if and only if $\mathbf{p} = \mathbf{a}$.

(2) $L_S(\cdot, \cdot)$ is symmetric in the sense that it is invariant under a common permutation of the coordinates of $\mathbf{p}$ and $\mathbf{a}$. Intuitively, it treats all components p_i equally.

(3) For all $\mathbf{p}$, $L_S(\mathbf{p}, \mathbf{a})$ is convex in $\mathbf{a}$. Thus there is a greater loss the further the guess $\mathbf{a}$ is from $\mathbf{p}$. In particular, the iso-loss curves of $L_S(\mathbf{p}, \cdot)$ are spheres; i.e., $\{\mathbf{a} : L_S(\mathbf{p}, \mathbf{a}) \text{ is constant}\}$ is a sphere.

Perhaps the second most frequently adopted loss function for the multinomial problem is *relative squared error loss* (RSEL) defined as:

$$L_R(\mathbf{p}, \mathbf{a}) = n \sum_{i=1}^{t} \frac{[a_i - p_i]^2}{p_i} \tag{2.2.4}$$

where $0/0 := 0$ and $+/0 := +\infty$. The motivation for using $L_R(\cdot, \cdot)$ is that it puts a premium on estimating small p_i. In the extreme, if $p_i = 0$ then $L_R(\mathbf{p}, \mathbf{a}) = +\infty$ unless $a_i = 0$.

Some properties of $L_R(\cdot, \cdot)$ are:

(1) $0 \leq L_R(\mathbf{p}, \mathbf{a}) \leq \infty$ for all $\mathbf{p}, \mathbf{a} \in \mathcal{S}$; $L_R(\mathbf{p}, \mathbf{a}) = \infty$ if and only if there exists an i with $a_i > 0$ and $p_i = 0$; $L_R(\mathbf{p}, \mathbf{a}) = 0$ if and only if $\mathbf{p} = \mathbf{a}$.

(2) $L_R(\cdot, \cdot)$ is invariant under a common permutation of the coordinates of $\mathbf{p}$ and $\mathbf{a}$.

(3) For all $\mathbf{p}$, $L_R(\mathbf{p}, \mathbf{a})$ is convex in $\mathbf{a}$. The iso-loss curves for $L_R(\cdot, \cdot)$ are ellipses.

One potential problem with using $L_S(\cdot, \cdot)$ or $L_R(\cdot, \cdot)$ is that they do *not* distinguish between positive and zero estimates a_i of p_i. In particular if $p_i > 0$, then provided $2p_i \leq 1$, both $L_S(\cdot, \cdot)$ and $L_R(\cdot, \cdot)$ say that it is equivalent to estimate p_i by $a_i = 0$ or by $a_i = 2p_i > 0$. This equivalence would make both losses inappropriate for problems such as low dose extrapolation where it is desired to determine the carcinogenic effect of exposure to low doses of a chemical. A zero estimate of a positive carcinogenic effect presumably is a more serious error (loss) than a positive estimate.

There are loss functions which distinguish between positive and zero guesses of $p_i > 0$, one of which is *entropy loss* (EL) defined as

$$L_E(\mathbf{p}, \mathbf{a}) = n \sum_{i=1}^{t} p_i[\ln(p_i) - \ln(a_i)] \tag{2.2.5}$$

where

$$b \ln(0) := \begin{cases} 0 & \text{if } b = 0 \\ +\infty & \text{if } b < 0. \end{cases}$$

The entropy loss is *infinite* for any guess $\mathbf{a}$ of $\mathbf{p}$ that does not have $a_i > 0$ whenever $p_i > 0$. Entropy loss can be derived as the Kullback–Liebler measure of distance between the $M_t(n, \mathbf{p})$ and $M_t(n, \mathbf{a})$ distributions. To define this distance let $f(\cdot \mid \mathbf{p})$ and $E_{\mathbf{p}}[\cdot]$ denote the probability mass function and the expectation for the $M_t(n, \mathbf{p})$ distribution, respectively. The Kullback–Liebler distance between the $M_t(n, \mathbf{p})$ and $M_t(n, \mathbf{a})$ distributions is

$$E_{\mathbf{p}} \left[\ln \left(\frac{f(\mathbf{Y} \mid \mathbf{p})}{f(\mathbf{Y} \mid \mathbf{a})} \right) \right]$$

which is equal to (2.2.5).

Entropy loss has the following properties:

(1) $0 \leq L_E(\mathbf{p}, \mathbf{a}) \leq \infty$ for all $\mathbf{p}, \mathbf{a} \in \mathcal{S}$; $L_E(\mathbf{p}, \mathbf{a}) = \infty$ if and only if there is an i with $a_i = 0$ and $p_i > 0$; $L_E(\mathbf{p}, \mathbf{a}) = 0$ if and only if $\mathbf{p} = \mathbf{a}$.

(2) $L_E(\cdot, \cdot)$ is invariant under a common permutation of the coordinates of $\mathbf{p}$ and $\mathbf{a}$.

(3) For all $\mathbf{p} \in \mathcal{S}$, $L_E(\mathbf{p}, \mathbf{a})$ is convex in $\mathbf{a}$. The actions $\mathbf{a} \in \mathcal{S}$ which have equal $L_E(\mathbf{p}, \mathbf{a})$ values are those with equivalent likelihood of producing "data" $\mathbf{y} := n\mathbf{p}$.

To facilitate comparison among these loss functions, Figure 2.2.1 (a)–(c) displays the iso-loss curves of SEL, RSEL, and EL for $t = 3$, $n = 3$, and true $\mathbf{p}' = (1/3, 1/3, 1/3)$. Two observations based on this figure are of interest here. First, the loss rises more quickly for RSEL than SEL as the action $\mathbf{a}$ "moves toward the boundary." Second, EL behaves like SEL for $\mathbf{a}$ near $\mathbf{p}$ but it imposes a very heavy penalty for $\mathbf{a}$ near the boundary.

Consider the performance of the MLE with respect to $L_S(\cdot, \cdot)$ and $L_R(\cdot, \cdot)$; estimation with respect to $L_E(\cdot, \cdot)$ is examined in Problem 2.18. The risk of the MLE $\hat{\mathbf{p}}$ under SEL is easily calculated as:

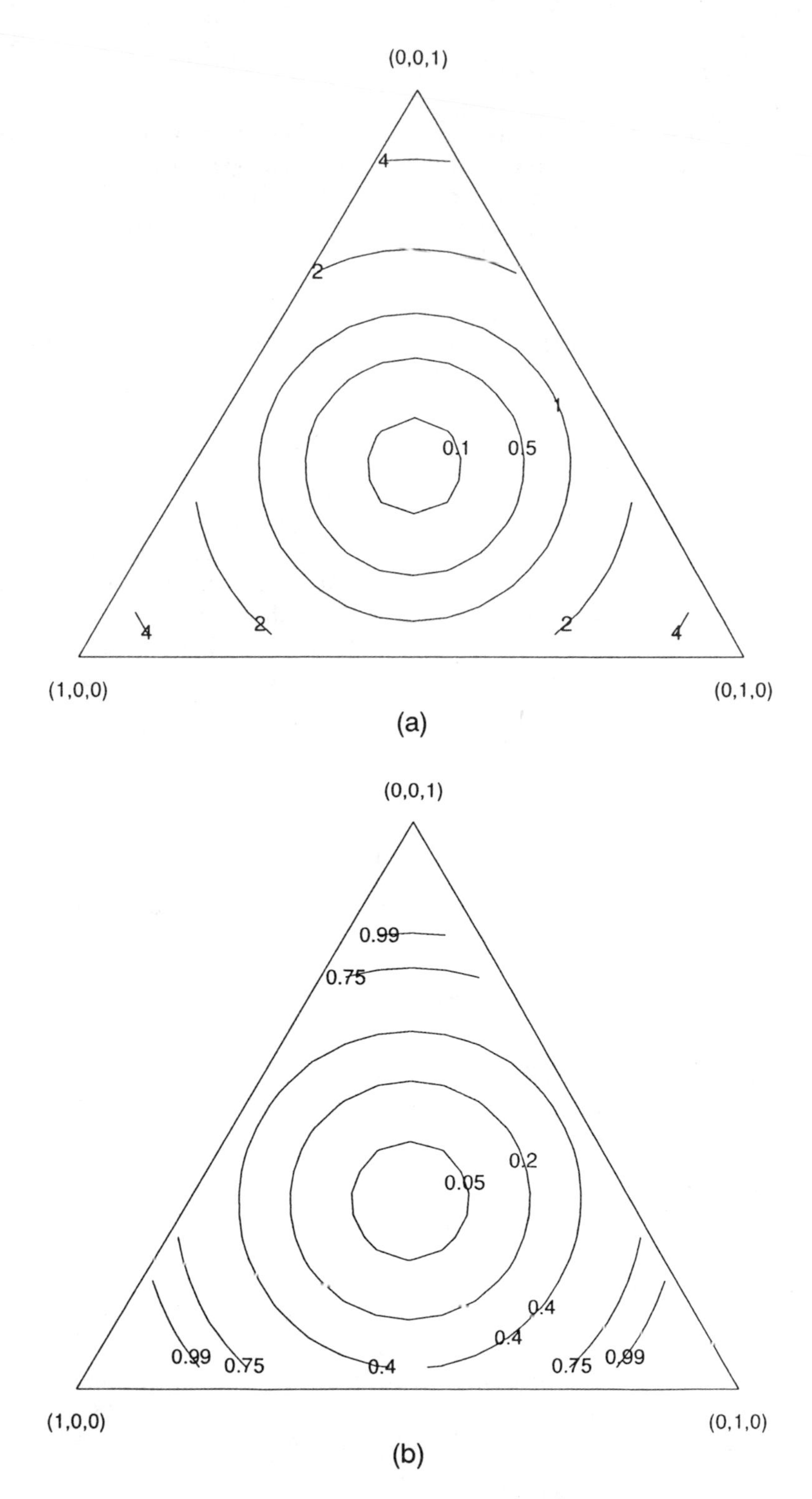

Figure 2.2.1. Iso-loss curves under SEL (a), RSEL (b), and EL (c) for $t = 3 = n$, and true $\mathbf{p}' = (1/3, 1/3, 1/3)$.

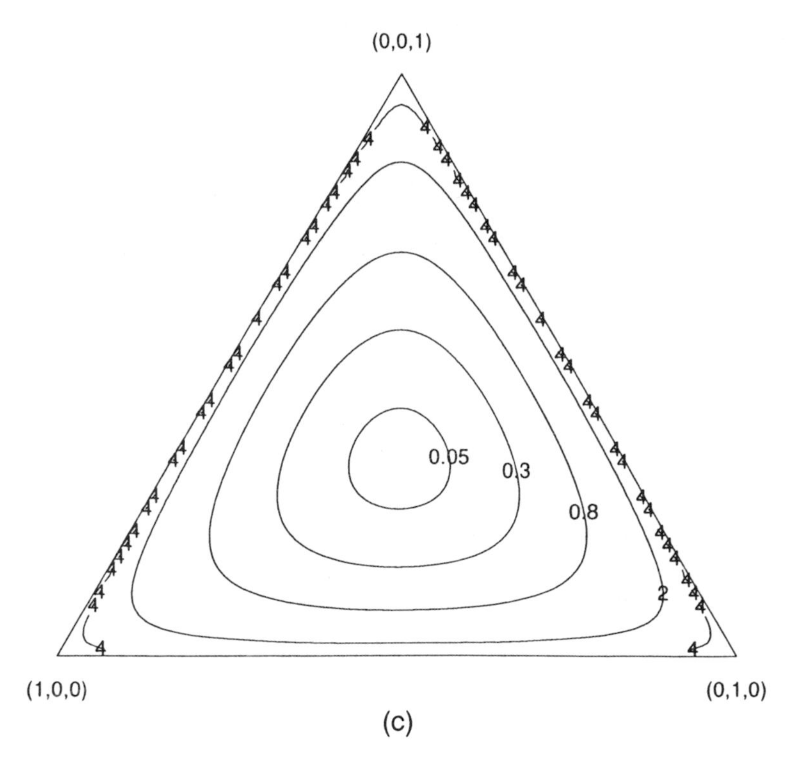

Figure 2.2.1. (cont.)

$$R_S(\mathbf{p}, \hat{\mathbf{p}}) = nE\left[\sum_{i=1}^{t}(\hat{p}_i - p_i)^2\right] = n\sum_{i=1}^{t}\text{Var}(\hat{p}_i)$$
$$= n\sum_{i=1}^{t}\frac{p_i(1-p_i)}{n}$$
$$= 1 - ||\mathbf{p}||^2$$

where $||\mathbf{x}|| = (\sum_i x_i^2)^{1/2}$. Thus $\hat{\mathbf{p}}$ performs well only at (near) vertices where it has (near) zero risk. Conversely, $\hat{\mathbf{p}}$ performs *worst* at the center $\mathbf{c}' := (1/t, \ldots, 1/t)$ of $\mathcal{S}$ since

$$\max_{\mathbf{p}\in\mathcal{S}} R_S(\mathbf{p}, \hat{\mathbf{p}}) = R_S(\mathbf{c}, \hat{\mathbf{p}}) = 1 - \frac{1}{t}.$$

Johnson (1971) uses the phrase "tyranny of the boundary" to describe behavior such as this in which an estimator excels only on (part of) the boundary of the parameter space. That the MLE does so well at vertices suggests it may be admissible; Johnson (1971) and Alam (1979) prove this is true.

Under RSEL and for $\mathbf{p} > 0$ (i.e., $p_i > 0$ for $i = 1(1)t$),

$$\begin{aligned} R_R(\mathbf{p}, \hat{\mathbf{p}}) &= nE_{\mathbf{p}}\left[\sum_{i=1}^{t} \frac{(Y_i/n - p_i)^2}{p_i}\right] \\ &= \frac{1}{n}\sum_{i=1}^{t} \frac{E_{\mathbf{p}}(Y_i - np_i)^2}{p_i} = \frac{1}{n}\sum_{i=1}^{t} \frac{np_i(1-p_i)}{p_i} \\ &= \sum_{i=1}^{t}(1 - p_i) = t - 1. \end{aligned} \tag{2.2.6}$$

Separate consideration of the case where one or more $p_i = 0$, shows that for arbitrary $\mathbf{p} \in \mathcal{S}$:

$$R_R(\mathbf{p}, \hat{\mathbf{p}}) = (\# \text{ of } i \text{ such that } p_i > 0) - 1. \tag{2.2.7}$$

In particular, equation (2.2.7) reduces to (2.2.6) for $\mathbf{p}$ in the interior of $\mathcal{S}$; equation (2.2.7) also shows $R_R(\mathbf{p}, \hat{\mathbf{p}}) = 0$ at vertices $\mathbf{p}$ of $\mathcal{S}$. As in the binomial case, $\hat{\mathbf{p}}$ is admissible and minimax under RSEL. A proof of this is given later as a consequence of Proposition 2.2.2 which shows that the MLE is Bayes with respect to RSEL for the Dirichlet prior.

The admissibility of the MLE $\hat{\mathbf{p}}$ under SEL and RSEL for both the binomial and multinomial problems is a special case of a general result due to Gutmann (1982). Gutmann shows that the Stein effect (ensemble estimators formed from *admissible* component estimators can be *inadmissible* when the loss for the ensemble problem is the sum of the component losses, Berger (1985), p. 10) is impossible for problems with finite sample spaces. However the poor risk properties of $\hat{\mathbf{p}}$ under SEL away from vertices and its minimax character under RSEL motivate the development of alternative estimators with better MSE properties than $\hat{\mathbf{p}}$.

Example 2.2.1. The data in Columns 1 and 2 of Table 2.2.1 are from Knoke (1976). Based on a 1972 survey, 834 individuals are classified according to occupational group. The MLE for $\mathbf{p} = (p_1, \ldots, p_9)'$, the vector of cell probabilities, is given in Column 3.

Bayes and Related Estimators

Both Bayesian and smoothing techniques have been applied to suggest alternative estimators to the MLE. These two methods are based on the premise that additional information about the problem is available. As Section 2.1 illustrated, Bayes techniques assume prior information about $\mathbf{p}$, and possibly also loss information evaluating each potential guess of $\mathbf{p}$

Table 2.2.1. Eight Hundred and Thirty-Four Individuals Classified According to Occupational Group (Reprinted with permission from *Change and Continuity in American Politics* by David Knoke. The Johns Hopkins Univ. Press, Baltimore, MD, 1976.)

Occupational Group	Observed Number	$\hat{p}_i$	$\hat{p}_i^{B\,1}$	$\hat{p}_i^{B\,2}$	$\hat{p}_i^{O}$	$\hat{p}_i^{T}$	$\hat{p}_i^{U}$	$\hat{p}_i^{G\,3}$	$\hat{p}_i^{S}$
Professionals	159	.191	.190	.188	.190	.190	.189	.189	.189
Business, Managers	148	.177	.177	.175	.176	.176	.176	.176	.176
Clerical	53	.064	.064	.065	.065	.064	.065	.065	.065
Sales	41	.049	.049	.051	.050	.050	.050	.050	.050
Craftsman, Foreman	199	.239	.238	.235	.237	.237	.237	.237	.237
Operatives	146	.175	.175	.173	.174	.174	.174	.174	.174
Service Worker	39	.047	.047	.049	.048	.048	.048	.048	.048
Farmer	21	.025	.026	.028	.026	.027	.027	.027	.027
Laborer	28	.033	.034	.036	.034	.034	.034	.034	.034
Total	834								

[1] $(\boldsymbol{\mu}, K) = (\mathbf{c}, 4.5)$.

[2] $(\boldsymbol{\mu}, K) = (\mathbf{c}, 28.29)$.

[3] $X^2 > 400$.

relative to the true cell probabilities $\mathbf{p}$; smoothing techniques assume prior information about relationships among the components p_i. As will be seen below, both Bayes and smoothing methods lead to estimators of the form

$$\alpha\boldsymbol{\mu} + (1-\alpha)\hat{\mathbf{p}} \tag{2.2.8}$$

where $0 \le \alpha \le 1$ and $\boldsymbol{\mu} \in \mathcal{S}$. First Bayesian methods will be discussed and then smoothing techniques.

The Bayes method and its variants introduced in Section 2.1 for the problem of estimating a binomial p have been generalized to allow for multinomial responses. The simplest Bayes estimator to compute is that with respect to the (conjugate) Dirichlet prior under SEL. The vector $\mathbf{W}' = (W_1, \ldots, W_t)$ with $W_t = 1 - \sum_{i=1}^{t-1} W_i$ has the *Dirichlet distribution* with parameter $\boldsymbol{\beta}' = (\beta_1, \ldots, \beta_t) > 0$ and is denoted by $\mathbf{W} \sim \mathcal{D}_t(\boldsymbol{\beta})$ if $(W_1, \ldots, W_{t-1})$ has the density function:

$$\frac{\Gamma(\sum_{i=1}^t \beta_i)}{\prod_{i=1}^t \Gamma(\beta_i)} \prod_{i=1}^{t-1} w_i^{\beta_i - 1} \left(1 - \sum_{i=1}^{t-1} w_i\right)^{\beta_t - 1},$$

$$\mathbf{w} \in \left\{\mathbf{x} \in \mathrm{I\!R}^{t-1} : \sum_{i=1}^{t-1} x_i < 1, x_i > 0\right\}.$$

The Dirichlet distribution reduces to the beta distribution when $t = 2$. Calculation shows that the mean vector of $\mathbf{W} \sim \mathcal{D}_t(\boldsymbol{\beta})$ is

$$E[\mathbf{W}] = \left(\ldots, \frac{\beta_i}{\sum_{j=1}^t \beta_j}, \ldots\right).$$

Reparametrizing to $\mu_i := \beta_i / \sum_{j=1}^t \beta_j$, $1 \le i \le t$, and $K := \sum_{j=1}^t \beta_j$ gives $E[\mathbf{W}] = \boldsymbol{\mu}' := (\mu_1, \ldots, \mu_t)$ and $\mathrm{Var}(W_i) = \{\mu_i(1-\mu_i)\}/(K+1)$. Thus the $\mathcal{D}_t(K\boldsymbol{\mu})$ prior has the interpretation that $\boldsymbol{\mu}$ is the prior mean of $\mathbf{p}$ and K expresses the prior certainty about the mean.

A straightforward calculation shows that the posterior of $\mathbf{p}$ given $\mathbf{Y} = \mathbf{y}$ is the $\mathcal{D}_t(\boldsymbol{\mu}K + \mathbf{y})$ distribution. The Bayes estimator of $\mathbf{p}$ under SEL is the mean of the posterior distribution of $\mathbf{p}$ which is

$$\begin{aligned} E(\mathbf{p} \mid \mathbf{Y}) &= \frac{\mathbf{Y} + \boldsymbol{\mu}K}{\sum_{i=1}^t (Y_i + \mu_i K)} = (\mathbf{Y} + \boldsymbol{\mu}K)/(n+K) \\ &= \left(\frac{n}{n+K}\right)\hat{\mathbf{p}} + \left(\frac{K}{n+K}\right)\boldsymbol{\mu} =: \hat{\mathbf{p}}^B(K, \boldsymbol{\mu}) = \hat{\mathbf{p}}^B, \text{ say,} \end{aligned} \tag{2.2.9}$$

and is of the form (2.2.8). As in the binomial case, the Bayes estimator is a convex combination of $\hat{\mathbf{p}}$ and the prior mean $\boldsymbol{\mu}$ with weights depending on the sample size n and the "confidence" K associated with the prior mean.

For any $\boldsymbol{\mu} \in \mathcal{S}$ with $\boldsymbol{\mu} > 0$ and $K > 0$, $\hat{\mathbf{p}}^B(K, \boldsymbol{\mu})$ is admissible with respect to $L_S(\cdot, \cdot)$ as it is a unique positive Bayes estimator. Denoting $\omega = n/(n + K)$, the MSE of $\hat{\mathbf{p}}^B$ is

$$\begin{aligned} R_S(\mathbf{p}, \hat{\mathbf{p}}^B) &= nE_{\mathbf{p}}\left[\sum_{i=1}^{t}\{[\omega\hat{p}_i + (1-\omega)\mu_i] - p_i\}^2\right] \\ &= nE_{\mathbf{p}}\left[\sum_{i=1}^{t}\{[\omega\hat{p}_i - \omega p_i] + [(1-\omega)\mu_i - (1-\omega)p_i]\}^2\right] \\ &= \omega^2(1 - \|\mathbf{p}\|^2) + n(1-\omega)^2\|\mathbf{p} - \boldsymbol{\mu}\|^2. \end{aligned} \tag{2.2.10}$$

Comparing $R_S(\mathbf{p}, \hat{\mathbf{p}}^B)$ and $R_S(\mathbf{p}, \hat{\mathbf{p}})$, it is easy to see that $\hat{\mathbf{p}}^B$ is superior to $\hat{\mathbf{p}}$ for $\mathbf{p}$ sufficiently near $\boldsymbol{\mu}$ since

$$R_S(\mathbf{p}, \hat{\mathbf{p}}^B) < 1 - \|\mathbf{p}\|^2 = R_S(\mathbf{p}, \hat{\mathbf{p}})$$

in this case.

Two special choices for K and $\boldsymbol{\mu}$ have additional properties.

Case 1: For $\boldsymbol{\mu} = \mathbf{c} := (1/t, \ldots, 1/t)'$ and $K = t/2$

$$\hat{p}_i^B = \frac{Y_i + 1/2}{n + t/2}.$$

This estimator adds 1/2 to each observed cell count. Berkson (1955) and others have advocated the same modification based on frequentist arguments.

Case 2: The risk of the Bayes estimator with $\boldsymbol{\mu} = \mathbf{c}$ and $K = \sqrt{n}$ is

$$\begin{aligned} R_S(\mathbf{p}, \hat{\mathbf{p}}^B(\mathbf{c}, \sqrt{n})) &= \left(\frac{n}{n + \sqrt{n}}\right)^2 (1 - \|\mathbf{p}\|^2) + \left(\frac{\sqrt{n}}{n + \sqrt{n}}\right)^2 n \left(\|\mathbf{p}\|^2 - \frac{1}{t}\right) \\ &= \frac{n^2 - n^2/t}{(n + \sqrt{n})^2} \end{aligned}$$

which is independent of p. Thus $\hat{\mathbf{p}}^B(\sqrt{n}, \mathbf{c})$ is a minimax estimator of $\mathbf{p}$ under SEL as it has constant risk.

Of course, the Bayes estimator changes with the prior and loss function. To illustrate the modifications in the Bayes procedure caused by changing loss functions, considering estimating $\mathbf{p}$ under the Dirichlet distribution $\mathcal{D}_t(K\boldsymbol{\mu})$ for RSEL. The $\mathcal{D}_t(K\boldsymbol{\mu})$ prior remains conjugate for RSEL if the prior parameters $K\mu_i$ are all sufficiently large.

Proposition 2.2.2. *If $K\mu_i + y_i > 1$ for all i, then the Bayes estimator with respect to RSEL is*

$$\hat{\mathbf{p}}^{BR} = \hat{\mathbf{p}}^{BR}(K, \boldsymbol{\mu}) := \left(\frac{n}{n + K - t}\right)\hat{\mathbf{p}} + \left(\frac{K - t}{n + K - t}\right)(K\boldsymbol{\mu} - 1)/K - t)$$

where

$$(K\boldsymbol{\mu}-1)/(K-t) := \left(\frac{K\mu_1-1}{K-t},\ldots,\frac{K\mu_t-1}{K-t}\right)' \in \mathcal{S}$$

has all positive components.

Proof. The estimator $\hat{\mathbf{p}}^{BR}$ minimizes the posterior expected risk, $E[L_R(\mathbf{p},\mathbf{a})\,|\,\mathbf{y}]$, where expectation is with respect to the conditional distribution of $\mathbf{p}$ given $\mathbf{Y}=\mathbf{y}$; i.e., the $\mathcal{D}_t(K\boldsymbol{\mu}+\mathbf{y})$ distribution. Calculation gives

$$\begin{aligned}
E[L_R(\mathbf{p},\mathbf{a})\,|\,\mathbf{y}] &= n\sum_{i=1}^{t} E\left[\frac{(p_i-a_i(\mathbf{y}))^2}{p_i}\,\middle|\,\mathbf{y}\right]\\
&= n\sum_{i=1}^{t} E\left[p_i - 2a_i(\mathbf{y}) + \frac{a_i^2(\mathbf{y})}{p_i}\,\middle|\,\mathbf{y}\right]\\
&= n\sum_{i=1}^{t}\left\{\frac{K\mu_i+y_i}{K+n} - 2a_i(\mathbf{y}) + a_i^2(\mathbf{y})E\left[\frac{1}{p_i}\,\middle|\,\mathbf{y}\right]\right\}\\
&= n\sum_{i=1}^{t}\left\{\frac{K\mu_i+y_i}{K+n} - 2a_i(\mathbf{y}) + a_i^2(\mathbf{y})\frac{K+n-1}{K\mu_i+y_i-1}\right\},
\end{aligned}$$

where the last equality relies on the assumption $K\mu_i + y_i > 1$ for $i = 1(1)t$ in the computation of $E[\frac{1}{p_i}\,|\,\mathbf{y}]$. Simplification yields

$$E[L_R(\mathbf{p},\mathbf{a})\,|\,\mathbf{y}] = n\sum_{i=1}^{t} a_i^2(\mathbf{y})\frac{K+n-1}{K\mu_i+y_i-1} - 1. \tag{2.2.11}$$

The Bayes estimator is the minimum of (2.2.11) over $(a_1(\mathbf{y}),\ldots,a_t(\mathbf{y}))$ in $\mathcal{S}$. A straightforward application of Lagrange multipliers gives the solution

$$\hat{p}_i^{BR} = \frac{K\mu_i+y_i-1}{K+n-t}. \qquad \square$$

As an application, when $\boldsymbol{\mu}=\mathbf{c}$, $K=t$, and $\mathbf{y}>0$ then $K\mu_i+y_i>1$ for $i=1(1)t$ and so $\hat{\mathbf{p}}^{BR}(\mathbf{y}) = \hat{\mathbf{p}}$. Ighodaro (1980) shows $\hat{\mathbf{p}}$ is also the Bayes estimator for RSEL when $\min_{1\le i\le t}\{y_i\}=0$. Hence, $\hat{\mathbf{p}}$ is the unique Bayes estimator for RSEL with respect to this prior and is minimax and admissible.

Example 2.2.1 (continued). Columns 4 and 5 of Table 2.2.1 illustrate two conjugate Bayes estimates for the occupation data; Column 4 is the "add 1/2" estimator, $\hat{\mathbf{p}}^B$ with $(\boldsymbol{\mu},K)=(\mathbf{c},9/2)$, and Column 5 is the minimax estimator with respect to SEL, $\hat{\mathbf{p}}^B$ with $(\boldsymbol{\mu},K)=(\mathbf{c},\sqrt{834})=(\mathbf{c},28.89)$.

Columns 4 and 5 are very close to $\hat{\mathbf{p}}$. The primary reason is that the sample size $n = 834$ is large relative to the values of $K = 9/2$ or 28.29.

Consider now the case when there is inadequate information to construct a loss function but there is prior information regarding the parameter. This might be the case in basic scientific or engineering studies. Many papers implicitly adopt this perspective; their goal is the construction of meaningful priors and the determination of the posterior distribution (or a summary of it such as its mean, mode, or a $100(1-\alpha)\%$ highest posterior credible region). For example, Lindley (1964) presents an argument in favor of the "noninformative" prior proportional to $\prod_{i=1}^{t} p_i^{-1}$. Leonard (1973) is another example; he generalizes his earlier work on estimating a binomial p to estimating the multinomial $\mathbf{p}$ by using a multivariate normal prior distribution on the vector of logits of the p_i. By an appropriate choice of covariance matrix this prior permits adjacent cells (or other relevant sets of cells) to be highly associated. Recently, Lenk (1987) has studied general exchangeable priors which are closed to posterior analysis. He considers priors of the form $(W_1/S, \ldots, W_t/S)$ where $S := \sum_{i=1}^{t} W_i$ and $(W_1, \ldots, W_t)$ is a vector of positive random variables. Other examples which adopt the perspective that only knowledge of the prior is available will be given in Sections 3.4 and 5.4.

When the prior (or the parameters defining it) are unknown, the gamma minimax, hierarchical Bayes, pseudo Bayes, or empirical Bayes techniques are possible alternatives. The last three will be discussed below.

Hierarchical Bayes rules have been studied by Good (1965, 1967, 1983), Leonard (1977a) and the references therein. For example, consider a two-stage prior which mixes Dirichlet distributions each having the same known $\boldsymbol{\mu} \in \mathcal{S}$ with $\boldsymbol{\mu} > 0$ by choosing $K \in (0, \infty)$ according to a prior $q(K)$ in a first stage, and then $\mathbf{p}$ according to the $\mathcal{D}_t(K\boldsymbol{\mu})$ distribution in a second stage. The resulting estimators have the intuitively attractive form

$$\frac{n}{K(\mathbf{Y})+n}\mathbf{p} + \frac{K(\mathbf{Y})}{K(\mathbf{Y})+n}\boldsymbol{\mu}$$

where $K(\cdot)$ is a function of $\mathbf{Y}$ which is difficult (impossible) to calculate in closed form for most problems. If both K and $\boldsymbol{\mu}$ are unknown then the first stage chooses $(K, \boldsymbol{\mu})$ according to a prior $q(K, \boldsymbol{\mu})$. Approximations to hierarchical estimators based on two-stage Dirichlet priors are of interest given the difficulty of computing them exactly. One such approximation is given by Leonard (1977a) who considers the case of known $\boldsymbol{\mu}$ and $q(K)$ of a certain form; the resulting estimator is exactly the cross-validation smoothing estimator of Stone (1974) which will be described later.

As discussed in Section 2.1, if the prior mean $\boldsymbol{\mu}$ is known and K is unknown, the pseudo Bayes technique first determines that the risk minimizing choice of K is:

$$K = K(\mathbf{p}, \boldsymbol{\mu}) = (1 - ||\mathbf{p}||^2)/||\mathbf{p} - \boldsymbol{\mu}||^2$$

(Problem 2.11). The formula $K(\mathbf{p},\boldsymbol{\mu})$ is a generalization of (2.1.9) for $t \geq 3$ with the same intuition as (2.1.9) for the cases (i) $\mathbf{p} = \mathbf{p}^v$ a vertex, and (ii) $\mathbf{p} \to \boldsymbol{\mu}$. Fienberg and Holland (1973) propose estimating K by the maximum likelihood estimator based on $\mathbf{Y} \,|\, \mathbf{p}$ which is

$$\hat{K}^O := K(\hat{\mathbf{p}},\boldsymbol{\mu}) = (1 - \|\hat{\mathbf{p}}\|^2)/\|\hat{\mathbf{p}} - \boldsymbol{\mu}\|^2. \tag{2.2.12}$$

The corresponding PB estimator of $\mathbf{p}$ is

$$\hat{\mathbf{p}}^O := \hat{\mathbf{p}}^B(\hat{K}^O,\boldsymbol{\mu}) = \frac{n}{n+\hat{K}^O}\hat{\mathbf{p}} + \frac{\hat{K}^O}{n+\hat{K}^O}\boldsymbol{\mu}.$$

Ighodaro and Santner (1982) study an estimator of $K(\mathbf{p},\boldsymbol{\mu})$ which is motivated by the observation that if $\hat{\mathbf{p}}^O$ has smaller risk than $\hat{\mathbf{p}}$ over a substantial portion of the parameter space near $\boldsymbol{\mu}$, then intuitively $K(\hat{\mathbf{p}}^O,\boldsymbol{\mu})$ should be a better estimate of $K(\mathbf{p},\boldsymbol{\mu})$ than $K(\hat{\mathbf{p}},\boldsymbol{\mu})$. Now use $K(\hat{\mathbf{p}}^O,\boldsymbol{\mu})$ in the Bayes formula (2.2.9) to reestimate p. The reasoning is diagramed as follows

$$\begin{aligned} \hat{\mathbf{p}} &\to \hat{K}^O := K(\hat{\mathbf{p}},\boldsymbol{\mu}) \to \hat{\mathbf{p}}^O := \hat{\mathbf{p}}^B(\hat{K}^O,\boldsymbol{\mu}) \\ &\to \hat{K}^T := K(\hat{\mathbf{p}}^O,\boldsymbol{\mu}) \to \hat{\mathbf{p}}^T := \hat{\mathbf{p}}^B(\hat{K}^T,\boldsymbol{\mu}). \end{aligned} \tag{2.2.13}$$

The notations $\hat{\mathbf{p}}^O$ and $\hat{\mathbf{p}}^T$ are mnemonics for "one-step" and "two-step" estimators, respectively. There is a temptation to continue iterating. Ighodaro and Santner (1982) show that iterating to infinity yields an estimator of $\mathbf{p}$ which has inferior risk compared with $\hat{\mathbf{p}}^T$.

Sutherland, Feinberg, and Holland (1974) proposed a "ratio unbiased" PB estimator of $K(\mathbf{p},\boldsymbol{\mu})$. No unbiased estimator of $K(\mathbf{p},\boldsymbol{\mu})$ exists, but separate unbiased estimators of the numerator and denominator of $K(\mathbf{p},\boldsymbol{\mu})$ do exist. Sutherland et al. estimate K by the ratio of these unbiased estimators:

$$\begin{aligned} \hat{K}^U &= \frac{1 - \frac{\Sigma Y_i(Y_i-1)}{n(n-1)}}{\Sigma\frac{Y_i(Y_i-1)}{n(n-1)} - \frac{2\Sigma Y_i\mu_i}{n} + \Sigma\mu_i^2} \\ &= \frac{n^2 - \sum Y_i^2}{\sum Y_i^2 - n - 2(n-1)\sum Y_i\mu_i + n(n-1)\sum\mu_i^2}. \end{aligned} \tag{2.2.14}$$

The corresponding PB estimator of $\mathbf{p}$ is $\hat{\mathbf{p}}^U = \hat{\mathbf{p}}^B(\hat{K}^U,\boldsymbol{\mu})$. Bishop et al. (1975) recommend the following modification of $\hat{K}^U$: if equation (2.2.14) is less than zero, then set $\hat{K}^U = 0$ (and hence estimate $\mathbf{p}$ by $\hat{\mathbf{p}}$). However, (2.2.14) is less than zero if and only if the denominator is less than zero which indicates the true $\mathbf{p}$ is near $\boldsymbol{\mu}$. Thus the appropriate modification is to set $\hat{K}^U = \infty$ and estimate $\mathbf{p}$ by $\boldsymbol{\mu}$ when (2.2.14) is negative.

Empirical Bayes estimators of $\mathbf{p}$ can be derived by maximizing the marginal distribution of $\mathbf{Y}$ given the unknown prior parameters. This requires numerical evaluation even in the case of the conjugate $\mathcal{D}_t(K\boldsymbol{\mu})$ prior with known $\boldsymbol{\mu}$. However, Good (1965) states that his numerical studies show the EB estimator is often quite close to $\hat{\mathbf{p}}^O$.

Smoothing Estimators

Some smoothing techniques assume that the multinomial counts $\mathbf{Y}$ are obtained by grouping the values of a continuous variable. The adjacent cell probabilities are then adjacent areas under a density function and are expected to be related (especially if the number of cells is large). Simonoff (1983) studies maximum penalized likelihood estimation (MPLE) which is based on such an assumption. The MPLE of $\mathbf{p}$ maximizes

$$\sum_{i=1}^{t} Y_i \ln(p_i) - \beta \sum_{i=1}^{t-1} \{\ln(p_i/p_{i+1})\}^2 \tag{2.2.15}$$

where $\beta > 0$ is a given smoothness penalty; the estimator forces adjacent cell probabilities to be close by penalizing disparate estimates of p_i and p_{i+1} based on $\{\ln(p_i/p_{i+1})\}^2$. Formally the MPLE is equivalent to Leonard's (1973) Bayes estimator when the covariance is chosen so that the prior is proportional to

$$\sum_{i=1}^{t-1} \{\ln(p_i/p_{i+1})\}^2.$$

Kernel estimators are also based on the idea of smoothing the raw counts to reflect an assumed relationship between the cells. Aitchison and Aitken (1976), Titterington (1980), Brown and Rundell (1985), and the references therein study the case of related adjacent cells. They consider estimators of the form

$$\hat{p}_i^K = \sum_{j=1}^{t} Y_j K(j \mid i, \lambda) \tag{2.2.16}$$

where $K(\cdot \mid i, \lambda)$ is called the "kernel function" and λ is the "bandwidth" or "smoothing" parameter. The function $K(\cdot \mid i, \lambda)$ is typically selected to be a probability mass function with mode at i. Thus the kernel estimator of p_i is based on a weighted average of the cell counts with greatest weight placed on Y_i and the adjacent cells. For example, Aitchison and Aitken (1976) propose

$$K(j \mid i, \lambda) = \begin{cases} \lambda, & j = i \\ (1-\lambda)/(t-1), & j \neq i \end{cases}$$

with $t^{-1} \leq \lambda \leq 1$. In this case (2.2.16) yields the estimator

$$\hat{\mathbf{p}}^K = \hat{\mathbf{p}}^K(\lambda) = \alpha \mathbf{c} + (1-\alpha)\hat{\mathbf{p}} \tag{2.2.17}$$

where $\alpha = t(1-\lambda)/(t-1)$. If λ is unknown, as will typically be the case, Aitchison and Aitken suggest estimating it by maximizing

$$\prod_{i=1}^{t} \{\hat{p}_i^K(\lambda; \mathbf{Y} - \boldsymbol{\delta}^i)\}^{Y_i} \tag{2.2.18}$$

where $\boldsymbol{\delta}^i$ is the vector of $(t-1)$ 0's and a single 1 in the ith cell, and $\hat{p}_i^K(\lambda; \mathbf{Y} - \boldsymbol{\delta}^i)$ is the kernel estimate of the probability of producing a count in the ith cell based on data $\mathbf{Y} - \boldsymbol{\delta}^i$. Equation (2.2.18) is the jackknife likelihood of producing the observed data; it is the product of the estimated probabilities of producing the n observed outcomes when the probability of generating the jth multinomial trial result for $1 \le j \le n$ is estimated based on all the data *except* itself.

Another approach to smoothing starts directly with the class of estimators $\{\alpha\boldsymbol{\mu} + (1-\alpha)\hat{\mathbf{p}}\colon 0 \le \alpha \le 1\}$ for some given $\boldsymbol{\mu}$ and focuses on the problem of choosing α. The paragraphs above show several applications of the Bayes principle which lead to this class, as does kernel estimation. Formally this problem is identical to that of choosing K in pseudo Bayes estimation since $\alpha = n/(n+K)$. For example, Fienberg and Holland's estimator $\hat{\mathbf{p}}^O$ corresponds to

$$\hat{\alpha}^O = (n - Z)/(n + [n-1]Z) \tag{2.2.19}$$

where $Z := X^2/(t-1)$ and $X^2 = \sum_{i=1}^t \{(Y_i - n\mu_i)^2/n\mu_i\}$ is the chi-squared statistic for testing $H_0\colon \mathbf{p} = \boldsymbol{\mu}$ versus $H_A\colon \mathbf{p} \ne \boldsymbol{\mu}$.

Starting with this class, perhaps the earliest proposal for estimating α is due to Good (1965) who suggested the "testimator"

$$\hat{\alpha}^G := \begin{Bmatrix} 1, & X^2 \le t-1 \\ Z^{-1}, & X^2 > t-1 \end{Bmatrix} = \min\{1, Z^{-1}\} \tag{2.2.20}$$

where Z is as in (2.2.19). Thus the associated estimator $\hat{p}^G = \hat{\alpha}^G\boldsymbol{\mu} + (1 - \hat{\alpha}^G)\hat{\mathbf{p}}$ guesses $\boldsymbol{\mu}$ when the test statistic X^2 of $H_0\colon \mathbf{p} = \boldsymbol{\mu}$ falls on or below its null mean $t-1$ and otherwise guesses a value between $\boldsymbol{\mu}$ and $\hat{\mathbf{p}}$.

Stone (1974) adopts a prediction perspective for choosing α based on cross validation. Let

$$Q(\alpha) := \frac{1}{n}\sum_{i=1}^{t} Y_i L_S\left(\alpha\boldsymbol{\mu} + \{1-\alpha\}\hat{\mathbf{p}}(\mathbf{Y} - \boldsymbol{\delta}^i), \boldsymbol{\delta}^i\right)$$

where $\boldsymbol{\delta}^i$ is as in (2.2.18) and $\hat{\mathbf{p}}(\mathbf{Y} - \boldsymbol{\delta}^i)$ is the MLE of $\mathbf{p}$ computed using all the data except for a single count in the ith cell. Intuitively $Q(\alpha)$ is the average SEL over the n outcomes using $\alpha\boldsymbol{\mu} + (1-\alpha)\hat{\mathbf{p}}$ to predict each outcome based on all the data except itself. Stone proposes choosing α to minimize $Q(\alpha)$; he derives a closed form expression for the optimal α.

When $\boldsymbol{\mu} = \mathbf{c}$ the minimizing α is

$$\hat{\alpha}^S := \begin{cases} 1, & X^2 \leq t-1 \\ \frac{n-Z}{1+(n-2)Z}, & X^2 > t-1 \end{cases}$$

$$= \min\{1, (n-Z)/(1+(n-2)Z)\} \tag{2.2.21}$$

where Z is defined below equation (2.2.19) with $\boldsymbol{\mu} = \mathbf{c}$ (Problem 2.19). Similar to Good's proposal, the result $\hat{\mathbf{p}}^S = \hat{\alpha}^S \mathbf{c} + (1-\hat{\alpha}^S)\hat{\mathbf{p}}$ is a testimator. It guesses $\mathbf{c}$ when X^2 falls on or below the "cutoff point" $t-1$, $\hat{\mathbf{p}}$ when $\mathbf{Y}$ is a sample vertex, and a value in between otherwise. As noted earlier, Leonard (1977a) shows that $\hat{\mathbf{p}}^S$ is an approximate Bayes estimator. There is also overlap with Aitchison and Aitken (1976) as their proposal for choosing λ can be viewed as cross validation with entropy loss (Problem 2.19).

Example 2.2.1 (continued). Columns 6, 7, and 8 of Table 2.2.1 list the estimators $\hat{\mathbf{p}}^O$, $\hat{\mathbf{p}}^T$, and $\hat{\mathbf{p}}^U$ for Knoke's occupational group data; Columns 9 and 10 list $\hat{\mathbf{p}}^G$ and $\hat{\mathbf{p}}^S$. The homogeneity hypothesis is strongly rejected and consequently Good's estimator does not shrink the $\{\hat{\mathbf{p}}_i\}$ very much. The very large sample sizes swamp the external information used in all five estimators and they are virtually identical to each other, to the MLE, and to the previously computed Bayes estimators. (Problem 2.12 illustrates a small sample application in which the prior has greater influence.)

Comparison of Estimators

While estimators of $\mathbf{p}$ abound, there have not been a great many comparisons among the various proposals. One method of comparison is by exact small-sample risk curves; Bishop, Fienberg, and Holland (1975) present the risk curves for several problems in which $\hat{\mathbf{p}}^O$ and $\hat{\mathbf{p}}^U$ both have smaller MSE than $\hat{\mathbf{p}}$. Similarly Titterington (1980) compares $\hat{\mathbf{p}}^S$ with $\hat{\mathbf{p}}^O$ and other kernel based methods; $\hat{\mathbf{p}}^S$ performed well in his study.

A second method of comparison is asymptotic. Two types of large sample comparisons have been considered in the literature. The first is (standard) large-strata asymptotics in which t, $\mathbf{p}$, and $\boldsymbol{\mu}$ are fixed and $n \to \infty$. For example, most of the sequences of estimators described earlier are known to be consistent as $n \to \infty$. Sutherland et al. (1974) gives another example of this type in proving

$$\lim_{n\to\infty} \frac{R_S(\mathbf{p}, \hat{\mathbf{p}}^O)}{R_S(\mathbf{p}, \hat{\mathbf{p}})} = \lim_{n\to\infty} \frac{R_S(\mathbf{p}, \hat{\mathbf{p}}^U)}{R_S(\mathbf{p}, \hat{\mathbf{p}})} = \begin{cases} 1 & \text{if } \mathbf{p} \neq \boldsymbol{\mu} \\ \frac{1}{4} & \text{if } \mathbf{p} = \boldsymbol{\mu}. \end{cases}$$

Thus for any true $\mathbf{p} \neq \boldsymbol{\mu}$, the three estimators $\hat{\mathbf{p}}$, $\hat{\mathbf{p}}^O$, and $\hat{\mathbf{p}}^U$ are essentially equivalent when all cells have large sample sizes. More recently Hall (1981) and Bowman, Hall and Titterington (1984) have studied optimally smoothed kernel estimators in the sense of minimizing the risk under squared error loss. In general, the large-strata asymptotic results that have been established do *not* distinguish between the MLE and its competitors.

The second type of asymptotic analysis, called "large sparse" (LS) asymptotics, does discriminate between estimators. In this case both n and t approach infinity so that $n/t \to \delta \in (0, \infty)$ as $n \to \infty$. Thus not only does the number of observations become large, but also the number of cells and in such a way that the average proportion of observations in each cell approaches a constant. Note that LS asymptotics are not possible in the binomial case. The new technical requirement for LS asymptotics is to smoothly vary $\mathbf{p}$ as t increases. One way to do this is to fix a probability density function $p(\cdot)$ on $[0,1]$ and define

$$p_i := \int_{(i-1)/t}^{i/t} p(w)dw$$

for $i = 1(1)t$. Geometrically, $\mathbf{p}$ is the area under $p(\cdot)$ based on a t-cell equipartition of $[0,1]$.

One class of results under such LS models concerns consistency; consistency of a generic sequence of estimators $\{\hat{\mathbf{p}}^*\}_{n\geq 1}$ is usually taken to mean that $\sup_i |\hat{p}_i^*/p_i - 1| \to 0$ as $n \to \infty$. Neither the MLE nor the Bayes estimator (2.2.9) is consistent in this sense. Simonoff (1983) proves that under smoothness conditions on the sequence of true underlying probabilities (which hold if the density $p(\cdot)$ above has bounded second derivative), the sequence of maximum penalized likelihood estimators is consistent.

A second class of results concerns approximations for the risk of the sequence $\{\hat{\mathbf{p}}^*\}_{n\geq 1}$ under $L_S(\cdot,\cdot)$. For example, Sutherland et al. (1974) derive the first-order risk expansions (as $t \to \infty$) for a class of estimators containing $\hat{\mathbf{p}}$, $\hat{\mathbf{p}}^O$, and $\hat{\mathbf{p}}^U$. Define a sequence of smoothly varying $\boldsymbol{\mu}$ vectors for increasing t by

$$\mu_i := \int_{(i-1)/t}^{i/t} \mu(w)dw$$

for $i = 1(1)t$ where $\mu(\cdot)$ is a fixed probability density function $\mu(\cdot)$ on $[0,1]$. In terms of $D := \delta \int_0^1 (p(z) - \mu(z))^2 dz$ the Sutherland et al. expansions are:

$$R_S(p(\cdot), \hat{\mathbf{p}}) = 1 - \frac{1}{t}\int_0^1 p^2(z)dz + o\left(\frac{1}{t}\right),$$

$$R_S(p(\cdot), \hat{\mathbf{p}}^O) = \frac{D^2 + 3D + 1}{(D+2)^2} + o(1), \text{ and}$$

$$R_S(p(\cdot), \hat{\mathbf{p}}^U) = \frac{D}{D+1} + o(1).$$

Thus to first-order $R_S(p(\cdot), \hat{\mathbf{p}}^U) \leq R_S(p(\cdot), \hat{\mathbf{p}}^O) \leq R_S(p(\cdot), \hat{\mathbf{p}})$ for any $p(\cdot)$ and $\mu(\cdot)$ since

$$\frac{D}{D+1} \leq \frac{D^2 + 3D + 1}{(D+2)^2} \leq 1 \quad \text{for} \quad D \geq 0. \tag{2.2.22}$$

For example, when $D = 0$ (i.e., $\mu(\cdot) = p(\cdot)$ and shrinkage is toward the true state of nature) inequality (2.2.22) gives the first-order risk comparison $0 \leq 1/4 \leq 1$. As a second example, Table 2.2.2 lists the first-order comparisons among $\hat{\mathbf{p}}$, $\hat{\mathbf{p}}^U$, and $\hat{\mathbf{p}}^O$ when $\delta = 5$, shrinkage is toward the center of simplex (i.e., $\mu(w) = 1$ for $0 \leq w \leq 1$), and the true $\mathbf{p}$ is specified by

$$p(w) = \alpha w^{\alpha - 1}, \quad 0 \leq w \leq 1 \tag{2.2.23}$$

for $\alpha = 1, 2, 3$, and 5. The true state $\mathbf{p}$ is also the center of the simplex when $\alpha = 1$ ($D = 0$), but it approaches a vertex as $\alpha \to \infty$. To first-order, $\hat{\mathbf{p}}$ is dominated by both $\hat{\mathbf{p}}^O$ and $\hat{\mathbf{p}}^U$ for all four values of α with $\hat{\mathbf{p}}^U$ having the smallest first-order values.

Table 2.2.2. Values of First-Order Terms of $R_S(\cdot, \cdot)$ for $\hat{\mathbf{p}}$, $\hat{\mathbf{p}}^O$, and $\hat{\mathbf{p}}^U$ when $\delta = 5 = n/t$

α	$\hat{\mathbf{p}}$	$\hat{\mathbf{p}}^O$	$\hat{\mathbf{p}}^U$
1	1	.25	0
2	1	.65	.62
3	1	.81	.80
5	1	.90	.90

Although more difficult, asymptotic risk expressions have been developed for some of the iterative estimators described above. Simonoff (1983) gives the asymptotic risk expression for the MPLE. Burman (1987) and Hall and Titterington (1987) study kernel estimators; their techniques are similar to those that occur in kernel estimation of densities. Both papers consider the problem of optimal choice of bandwidth (for fixed kernel); Hall and Titterington derive a data-dependent asymptotically optimal kernel estimator based on cross-validation.

Figure 2.2.2 summarizes the results of a simulation study of the MSE of $\hat{\mathbf{p}}$, $\hat{\mathbf{p}}^O$, $\hat{\mathbf{p}}^U$, $\hat{\mathbf{p}}^T$, and $\hat{\mathbf{p}}^S$ in the LS asymptotic framework. Two questions are addressed.

(1) For $\hat{\mathbf{p}}$, $\hat{\mathbf{p}}^O$, and $\hat{\mathbf{p}}^U$, how large must n and t be for the first-order terms to give approximately correct risk expressions?

(2) How large must t be for the large-sample risk improvements over $\hat{\mathbf{p}}$ of $\hat{\mathbf{p}}^O$, $\hat{\mathbf{p}}^U$ (and presumably $\hat{\mathbf{p}}^T$, $\hat{\mathbf{p}}^G$, and $\hat{\mathbf{p}}^S$) to carry over to small samples?

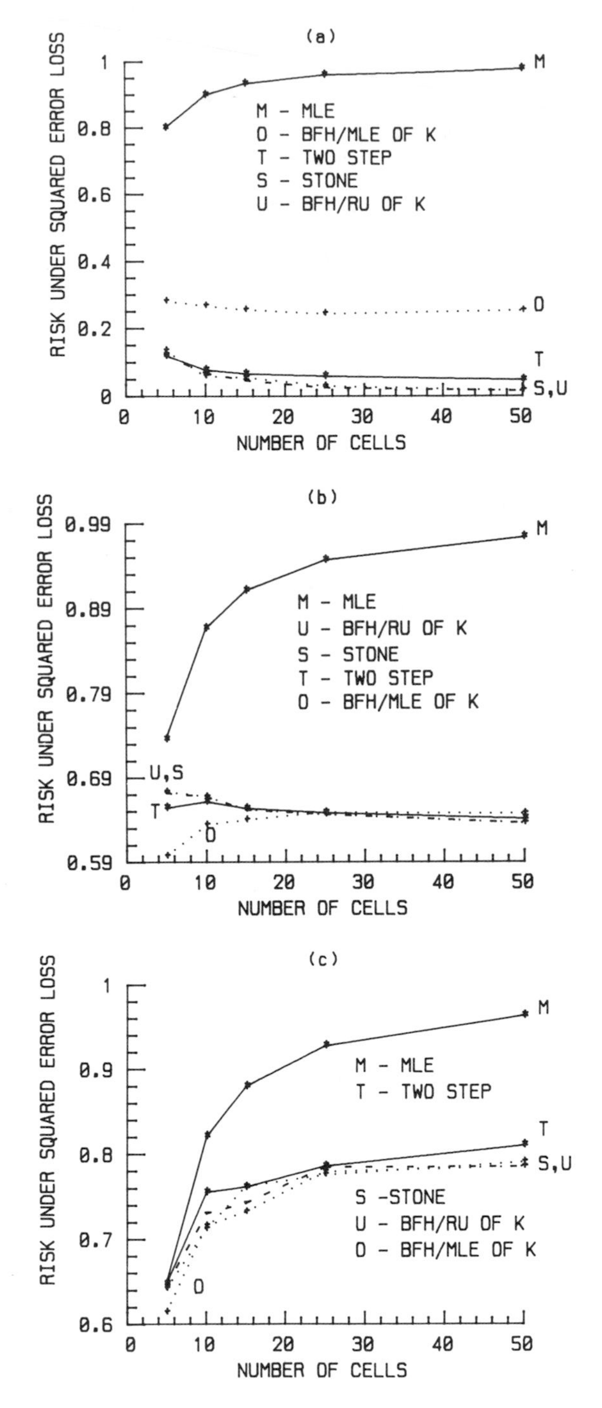

Figure 2.2.2. Risk values under squared error loss for $\hat{\mathbf{p}}(M)$, $\hat{\mathbf{p}}^O(O)$, $\hat{\mathbf{p}}^U(U)$, $\hat{\mathbf{p}}^T(T)$, and $\hat{\mathbf{p}}^S(S)$ for $\delta = 5 = n/t$ and $\alpha = 1$ (a), $\alpha = 2$ (b), $\alpha = 3$ (c), and $\alpha = 5$ (d).

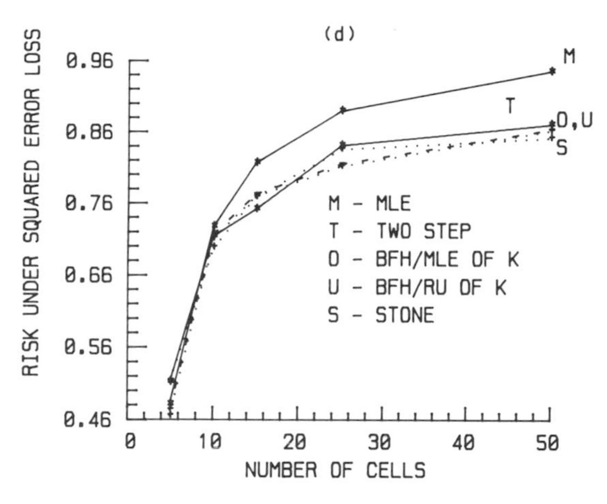

Figure 2.2.2. (cont.)

The study shrinks all alternative estimators toward the center of the simplex (i.e., $\mu(w) = 1$) and uses true $\mathbf{p}$ given by (2.2.23) for $\alpha = 1, 2, 3, 5$. Five cases of $t = 5$, 10, 15, 25, 50 were used, and for each value of t the risks were simulated for $n = t\delta$ with $\delta = 1$, 5, 10. Since the alternative estimators shrink toward $\mathbf{c}$, their greatest improvement over the MLE $\hat{\mathbf{p}}$ should be when the true $\mathbf{p}$ is specified by $\alpha = 1$. Conversely, the setup becomes more favorable for $\hat{\mathbf{p}}$ as $\alpha \to \infty$. Recall that $\hat{\mathbf{p}}$ has risk $R_S(\mathbf{p}, \hat{\mathbf{p}}) = 1 - ||\mathbf{p}||^2$ which is zero at vertices. To illustrate, when $t = 5 = \alpha$ then $\mathbf{p} = (3 \times 10^{-4}, 10^{-2}, .067, .250, .672)$ which is already fairly close to a vertex so that the risk of the MLE will be $1 - ||\mathbf{p}||^2 = .48$ in this case. Figure 2.2.2 (a)–(d) plots the *exact* risk of $\hat{\mathbf{p}}$ and the *simulated* risks of $\hat{\mathbf{p}}^O$, $\hat{\mathbf{p}}^T$, $\hat{\mathbf{p}}^U$, and $\hat{\mathbf{p}}^S$ for $n = 5t$ ($\delta = 5$) and $\alpha = 1$, 2, 3, 5.

The following conclusions are based on Figure 2.2.2, Table 2.2.2, and the risk plots for $\delta = 1$ and $\delta = 10$ (not given). All the estimators are competitive with the MLE $\hat{\mathbf{p}}$, even in the case most favorable for the MLE, $(\alpha, \delta) = (5, 10)$. The alternatives do considerably better when α is small. For $\hat{\mathbf{p}}$, $\hat{\mathbf{p}}^U$, and $\hat{\mathbf{p}}^O$ the first-order risk terms are approximately correct for $t \geq 25$, $\delta \geq 1$, and all $\alpha \geq 1$. The asymptotic superiority of $\hat{\mathbf{p}}^U$ over $\hat{\mathbf{p}}^O$ is only manifested when $\alpha = 1$; for $\alpha = 2$, 3, and 5, $\hat{\mathbf{p}}^U$ and $\hat{\mathbf{p}}^O$ are essentially indistinguishable. Lastly, the risks of the Stone estimator $\hat{\mathbf{p}}^S$ and the two step estimator $\hat{\mathbf{p}}^T$ are very similar to that of the ratio unbiased estimator $\hat{\mathbf{p}}^U$.

These LS asymptotic comparisons have implications for other multinomial problems such as cross-classified data in which it is easy to imagine sparse tables with large numbers of cells. Since the usual method of estimation is maximum likelihood based on some loglinear model, the comparisons above suggest the possibility of significant improvements by alternative estimators.

B. Hypothesis Tests about $\mathbf{p}$

This section considers three classes of testing problems for multinomial data: (i) simple null versus global alternative, (ii) simple null versus restricted alternative, and (iii) composite null versus global alternative. The discussion assumes the reader is familiar with the likelihood ratio, score, and Wald principles for constructing hypothesis tests. A brief review of these methods is given in Appendix 4.

Tests of Simple Null versus the Global Alternative

Let $\mathbf{Y} \sim M_t(n, \mathbf{p})$ where $\mathbf{p} \in \mathcal{S}$ is unknown. Suppose it is desired to test the simple null hypothesis $H_0\colon \mathbf{p} = \mathbf{p}^0$ versus the global alternative $H_A\colon \mathbf{p} \neq \mathbf{p}^0$. Example A.4.1 shows the likelihood ratio, score, and Wald tests of H_0 versus H_A are

$$G^2 = 2\sum_{i=1}^{t} Y_i \ln\left(\frac{Y_i}{np_i^0}\right) \geq \chi^2_{\alpha,t-1}, \tag{2.2.24}$$

$$X^2 = \sum_{i=1}^{t} \frac{[Y_i - np_i^0]^2}{np_i^0} \geq \chi^2_{\alpha,t-1}, \quad \text{and} \tag{2.2.25}$$

$$W = \sum_{i=1}^{t} \frac{[Y_i - np_i^0]^2}{Y_i} \geq \chi^2_{\alpha,t-1}, \tag{2.2.26}$$

respectively.

Cressie and Read (1984) provided a unified way of viewing (2.2.24)–(2.2.26) by introducing the power divergence family of test statistics, denoted by $\{I^\lambda : \lambda \in \mathbb{R}\}$, which contains the G^2, X^2, and W test statistics as special cases. The statistic I^λ is defined by

$$I^\lambda := \frac{2}{\lambda(\lambda+1)} \sum_{i=1}^{t} Y_i \left\{ \left(\frac{Y_i}{n\hat{p}_i^0}\right)^\lambda - 1 \right\} \tag{2.2.27}$$

for $\lambda \in \mathbb{R}$ where I^0 and I^{-1} are defined by continuity. The following are special cases of I^λ : $I^0 = G^2$, $I^1 = X^2$, $I^{-2} = W$, and $I^{-1/2} = 4\sum_{i=1}^{t}[(Y_i)^{1/2} - (np_i^0)^{1/2}]^2$ which is called the Freeman–Tukey statistic. Their recent book (Read and Cressie, 1988) summarizes most of the work on I^λ and its extensions to date. For example, they prove that for any $\lambda \in \mathbb{R}$

$$I^\lambda = X^2 + o_p(1) \tag{2.2.28}$$

under H_0 so that

$$I^\lambda \geq \chi^2_{\alpha,t-1}$$

is an asymptotic size α test of H_0 (Problem 2.13).

Which of the tests X^2, G^2, W, or some other member of the Cressie Read family is to be preferred for testing H_0? Simulation studies, exact probability calculations, and theoretical studies have been performed to answer this question (Slakter, 1966; Good, Gover and Mitchell, 1970; Yarnold, 1970; Larntz, 1978; Read and Cressie, 1988). In sum, I^λ for $\lambda \in [1/3, 3/2]$ and in particular $X^2 = I^1$ approaches the limiting χ^2_{t-1} distribution more rapidly than the other I^λ statistics. In the case of X^2, one theoretical fact which suggests the accuracy of the χ^2_{t-1} limiting distribution is the conformity of the moments of the two distributions. When $\mathbf{p} = \mathbf{p}^0$,

$$\begin{aligned} E[X^2] &= \sum_{i=1}^{t} \frac{E[Y_i - np_i^0]}{np_i^0} \\ &= t - 1 \\ &= E[\chi^2_{t-1}] \end{aligned}$$

for all n. Thus the first moment of X^2 agrees with that of χ^2_{t-1}. Similar but more tedious calculations under $\mathbf{p} = \mathbf{p}^0$ show that $\text{Var}(X^2) = 2(t-1)(1-1/n)$ while $\text{Var}(\chi^2_{t-1}) = 2(t-1)$. Thus the second moments agree to order $1/n$. Cressie and Read recommend using $I^{2/3}$ as an omnibus test.

Given that the score statistic X^2, $I^{2/3}$, or some other I^λ with $\lambda \in [1/3, 3/2]$ is preferred over G^2 or W based on rapidity with which their actual size approaches the nominal size, one must still determine when the sample size n is sufficiently large that the asymptotic chi-squared critical point can safely be used. The traditional advice is due to Cochran (1952); he recommends that if $np_i^0 \geq 5$ for all i then the approximation for X^2 is adequate. More recent studies show that if $np_i^0 \geq 1$ for all cells with $np_i^0 \geq 5$ for 80% of the cells, then the χ^2 approximation is very good. Before discussing the power of these tests, their use is illustrated.

Example 2.2.2. G. Mendel postulated that the resulting offspring from crossing round (R) yellow (Y) peas with wrinkled (W) green (G) peas would be RY–RG–WY–WG in the proportions 9:3:3:1. If $\mathbf{p}$ is the true probability vector for these four outcomes then it is of interest to test $H_0\colon \mathbf{p} = (\frac{9}{16}, \frac{3}{16}, \frac{3}{16}, \frac{1}{16})' =: \mathbf{p}^0$ versus $H_A\colon \mathbf{p} \neq \mathbf{p}^0$. Table 2.2.3 lists the number of offspring of each of the four combinations from 556 crossings and the expected number of offspring under H_0. The X_3^2 $(t-1=3)$ distribution is an adequate approximation to the null distribution of X^2 since all $np_i^0 \geq 30$. Calculation gives $X^2 = .47 = I^{2/3}$. The P-value for this test is .93 which is not only significant, but also suspiciously high since the probability of observing a chi-squared value as low as .47 is only .07. Some speculate that the data may have been tampered.

Table 2.2.3. Mendel's Pea Data and Null Expected Frequencies

Type	Y_i	p_i^0	np_i^0
RY	315	9/16	312.75
RG	108	3/16	104.25
WY	101	3/16	104.25
WG	32	1/16	34.75
Total	556	1	556.00

In designed experiements with categorical data, questions of sample size determination arise for which it is necessary to know the power function of the test being used. The easiest case to study is that of contiguous alternatives. Suppose that $\mathbf{p}^n := \mathbf{p}^0 + \mathbf{u}/\sqrt{n}$ is a sequence of probability vectors in $\mathcal{S}$ with $\Sigma u_i = 0$ and suppose that $\mathbf{Y}^n \sim M_t(n, \mathbf{p}^n)$ for $n \geq 1$. Example A.4.7 proves that $X^2 \xrightarrow{\mathcal{L}} \chi^2_{t-1}(\sum_1^t u_i^2/p_i^0)$ as $n \to \infty$; Problem 2.13 shows that $I^\lambda = X^2 + o_p(1)$ holds in this case for any $\lambda \in \mathbb{R}$ and thus I^λ has the same limiting distribution as X^2. To apply this result suppose that there is a specific alternative $\mathbf{p}^*$ ($\neq \mathbf{p}^0$) of interest. Write $\mathbf{p}^* = \mathbf{p}^0 + \mathbf{u}/\sqrt{n}$ where $\mathbf{u} := \sqrt{n}(\mathbf{p}^* - \mathbf{p}^0)$. An approximate expression for any I^λ is:

$$P_{\mathbf{p}^*}(I^\lambda \geq \chi^2_{\alpha,t-1}) \simeq P\left[\chi^2_{t-1}\left(\frac{n\Sigma[p_i^* - p_i^0]^2}{p_i^0}\right) \geq \chi^2_{\alpha,t-1}\right]. \qquad (2.2.29)$$

Empirical work by Drost et al. (1987a) shows that this approximation can be improved by tailoring the formula for the noncentrality parameter to λ. They recommend using expression (2.2.27), the formula for I^λ, with np_i^* substituted for Y_i. This gives (2.2.29) for the power of X^2 but for G^2 gives

$$P_{\mathbf{p}^*}(G^2 \geq \chi^2_{\alpha,t-1}) \simeq P\left[\chi^2_{t-1}\left(2n\sum_{i=1}^t p_i^* \ln(p_i^*/p_i^0)\right) \geq \chi^2_{\alpha,t-1}\right].$$

Tables of the noncentral chi-square distribution function (Hayam, Govindarajulu, and Leone, 1973) can be used to choose sample sizes to approximately achieve a given power at $\mathbf{p}^*$ (Problem 2.14). Theoretical study of the power of I^λ for noncontiguous alternatives is more limited. See Drost et al. (1989).

For completeness it should be noted that for testing simple null hypotheses with large sparse multinomial data, a number of alternative null approximations to the test statistics described above have been suggested as well as new test statistics (Koehler and Larntz, 1980; Simonoff, 1985, 1986; Zelterman, 1987; Read and Cressie, 1988). Such tests are of greater practical importance in the analysis of cross-classified data and composite null hypotheses where sparse data often occur.

Tests of Simple Null versus Restricted Alternatives

This section begins with two motivating examples, the first of which illustrates "smooth" alternatives and the second "isotonic" alternatives. Then tests for these two problems will be derived.

Example 2.2.3 (Circadian Rhythms). Fix, Hodges and Lehmann (1959) consider the problem of testing whether events occur uniformly in time versus the alternative that they occur in cyclic patterns (called circadian rhythms). For example, the events could be diagnoses of a particular type of cancer or suicide attempts.

Suppose time is divided into t equal periods, n is the fixed total number of events to be observed, and Y_i is the number of events in time period i for $i = 1(1)t$. Then $\mathbf{Y} \sim M_t(n, \mathbf{p})$. For $\eta_1' > 0$, $\eta_2' \in \mathbb{R}$, and $i = 1(1)t$ define

$$p_i(\eta_1', \eta_2') = \frac{1}{t} + \int_{2\pi(i-1)/t}^{2\pi i/t} \eta_1' \sin(u - \eta_2')du.$$

Let $\mathcal{T} = \{\mathbf{p} \in \mathcal{S} : p_i = p_i(\eta_1', \eta_2')$ for some $\eta_1' \geq 0$, $\eta_2' \in \mathbb{R}\}$. Then $\mathcal{T} \subset \mathcal{S}$ strictly and the problem of testing uniformity versus the occurrence of circadian rhythms can be formulated as testing $H_0: \mathbf{p} = \mathbf{c}$ versus $H_A: \mathbf{p} = \mathcal{T}/\{\mathbf{c}\}$. Such alternatives are termed "smooth" because the parameters p_i vary smoothly according to a function of (η_1', η_2').

Example 2.2.2 (continued). Recall Mendel's experiment in which peas are round (R) or wrinked (W) and yellow (Y) or green (G). Round and yellow genes are dominant. Suppose α_1 = relative frequency of gene R in the population, and α_2 = relative frequency of gene Y in the population. Then the proportion of round peas in the population is $\beta_1 = \alpha_1^2 + 2\alpha_1(1 - \alpha_1)$ and the proportion of yellow peas is $\beta_2 = \alpha_2^2 + 2\alpha_2(1 - \alpha_2)$. Assuming that these two characteristics are independently determined, the pea population will be divided in four groups according to the proportions in Table 2.2.4. As a special case, if $\alpha_1 = 1/2 = \alpha_2$ then $\beta_1 = \beta_2 = 3/4$ and $\mathbf{p} = (9/16, 3/16, 3/16, 1/16)'$. However, as long as α_1 and α_2 are greater than or equal to $1 - \sqrt{2}/2 \approx .29$, β_1 and β_2 will be greater than or equal to 1/2 which is equivalent to the vector $\mathbf{p}$ satisfying $p_1 \geq p_2, p_3 \geq p_4$. This theory suggests that it is more pertinent to test $H_0: \mathbf{p} = (9/16, 3/16, 3/16, 1/16)'$ versus $H_A: p_1 \geq p_2, p_3 \geq p_4$ rather than the global alternative. Hypotheses such as H_A are called "isotonic."

Table 2.2.4. Theoretical Proportion of Peas with Color and Shape Characteristics

Paired Characteristics	Proportion
RY	$p_1 = \beta_1\beta_2$
RG	$p_2 = \beta_1(1-\beta_2)$
WY	$p_3 = (1-\beta_1)\beta_2$
WG	$p_4 = (1-\beta_1)(1-\beta_2)$

Restricted Alternatives 1: Smooth Case

First a generic model which generalizes Example 2.2.2 is formulated. The observed data is $\mathbf{Y} \sim M_t(n, \mathbf{p})$; it is assumed that $\mathbf{p}$ belongs to

$$\mathcal{T} := \{\mathbf{p}(\boldsymbol{\eta}) \in \mathcal{S} : \boldsymbol{\eta} = (\eta_1, \dots, \eta_r)' \in \mathcal{N} \subset \mathrm{I\!R}^r\}$$

and that $\mathcal{N}$ is identifiable meaning that if $\mathbf{p}(\boldsymbol{\eta}) = \mathbf{p}(\boldsymbol{\eta}^*)$ then $\boldsymbol{\eta} = \boldsymbol{\eta}^*$. Fix $\boldsymbol{\eta}^0 \in \mathcal{N}$ and consider testing $H_0{:}\, \mathbf{p} = \mathbf{p}^0 := \mathbf{p}(\boldsymbol{\eta}^0)$ versus $H_A{:}\, \mathbf{p} \in \mathcal{T} \setminus \{\mathbf{p}^0\}$. The asymptotic normality of the standardized multinomial data will serve to reduce the problem to that of testing a linear hypothesis with known covariance.

To review the relevant linear model theory, suppose an $m \times 1$ data vector $\mathbf{Z}$ has the $N_m[\boldsymbol{\mu}, \Sigma]$ distribution where Σ is *known* and $\boldsymbol{\mu} \in \mathcal{C} := \mathcal{C}(\mathbf{X}) = \{\mathbf{X}\boldsymbol{\beta} : \boldsymbol{\beta} \in \mathrm{I\!R}^r\}$, the column space of a known $m \times r$ matrix $\mathbf{X}$ having rank r ($\leq m$); hence the dimension of $\mathcal{C}$ is r. Consider testing $H_0 : \boldsymbol{\mu} \in V$ versus $H_A : \boldsymbol{\mu} \in \mathcal{C} \setminus V$, where V is a q ($< r$) dimensional subspace of $\mathcal{C}$. The likelihood ratio test rejects H_0 if and only if

$$(\mathbf{Z} - \hat{\boldsymbol{\mu}}_V)'\boldsymbol{\Sigma}^{-1}(\mathbf{Z} - \hat{\boldsymbol{\mu}}_V) - (\mathbf{Z} - \hat{\boldsymbol{\mu}}_{\mathcal{C}})'\boldsymbol{\Sigma}^{-1}(\mathbf{Z} - \hat{\boldsymbol{\mu}}_{\mathcal{C}})$$
$$= (\hat{\boldsymbol{\mu}}_{\mathcal{C}} - \hat{\boldsymbol{\mu}}_V)'\boldsymbol{\Sigma}^{-1}(\hat{\boldsymbol{\mu}}_{\mathcal{C}} - \hat{\boldsymbol{\mu}}_V) \geq \chi^2_{\alpha, r-q},$$

where

$$\begin{aligned} \hat{\boldsymbol{\mu}}_V &:= \text{weighted least squares estimate of } \boldsymbol{\mu} \text{ based on } V, \text{ and} \\ \hat{\boldsymbol{\mu}}_{\mathcal{C}} &:= \text{weighted least squares estimate of } \hat{\boldsymbol{\mu}} \text{ based on } \mathcal{C}, \\ &= \mathbf{X}\hat{\boldsymbol{\beta}} \text{ where } \hat{\boldsymbol{\beta}} \text{ minimizes } (\mathbf{Z} - \mathbf{X}\boldsymbol{\beta})'\boldsymbol{\Sigma}^{-1}(\mathbf{Z} - \mathbf{X}\boldsymbol{\beta}). \end{aligned}$$

Returning to the multinomial problem, let

$$\mathbf{Z}^n = \sqrt{n}\,\left((\hat{p}_1 - p_1^0), \dots, (\hat{p}_{t-1} - p_{t-1}^0)\right)'$$

where $\hat{\mathbf{p}}$ is the MLE of $\mathbf{p}$ in $\mathcal{T}$. Fix $\boldsymbol{\delta} = (\delta_1, \dots, \delta_r)'$ and suppose the sequence of vectors $\boldsymbol{\eta}^n = \boldsymbol{\eta}^0 + \boldsymbol{\delta}/\sqrt{n}$ is in $\mathcal{N}$; let $\mathbf{p}^n = \mathbf{p}(\boldsymbol{\eta}^n)$. Lehmann

(1986) applies the delta method to prove the following result showing that asymptotically $\mathbf{Z}^n$ follows a linear model.

Proposition 2.2.3. *If* $(\partial p_i(\boldsymbol{\eta}))/(\partial \eta_j)$ *exists and is continuous at* $\boldsymbol{\eta}^0$ *for* $i = 1(1)t$ *and* $j = 1(1)r$, *then* $\mathbf{Z}^n \xrightarrow{\mathcal{L}} N_{t-1}[\boldsymbol{\mu}, \boldsymbol{\Sigma}]$ *as* $n \to \infty$ *where* $\boldsymbol{\mu} = [\nabla \mathbf{p}(\boldsymbol{\eta}^0)] \times \boldsymbol{\delta}$, $\nabla \mathbf{p}(\boldsymbol{\eta}^0)$ *is the* $(t-1)\times r$ *Jacobian matrix*, $\boldsymbol{\Sigma} = \text{Diag}(\mathbf{p}^0_-) - (\mathbf{p}^0_-)(\mathbf{p}^0_-)'$, *and* $\mathbf{p}^0_- = (p^0_1, \ldots, p^0_{t-1})'$.

Thus asymptotically $\mathbf{Z}^n$ follows a linear model with known covariance, design matrix defined by

$$(\mathbf{X})_{ij} = \frac{\partial p_i(\boldsymbol{\mu}^0)}{\partial \eta_j}, \quad \text{for } 1 \le i < t, 1 \le j \le r$$

and parameter vector $\boldsymbol{\delta}$. The hypothesis $H_0: \mathbf{p} = \mathbf{p}^0$ corresponds to $V = \{\mathbf{0}_{t-1}\}$ and $H_A: \mathbf{p} \in \mathcal{T} \setminus \{\mathbf{p}^0\}$ corresponds to $H'_A: \boldsymbol{\mu} \in \mathcal{C}(\mathbf{X}) \setminus \{\mathbf{0}_{t-1}\}$. Therefore the asymptotic likelihood ratio test of H_0 versus H_A rejects H_0 if and only if

$$(\hat{\boldsymbol{\mu}}_V - \hat{\boldsymbol{\mu}}_C)'\boldsymbol{\Sigma}^{-1}(\hat{\boldsymbol{\mu}}_V - \hat{\boldsymbol{\mu}}_C) \ge \chi^2_{\alpha,r}$$

where $\hat{\boldsymbol{\mu}}_V = \mathbf{0}_{t-1}$, $\hat{\boldsymbol{\mu}}_C = [\nabla \mathbf{p}(\boldsymbol{\mu}^0)] \times \hat{\boldsymbol{\delta}}$, and $\hat{\boldsymbol{\delta}}$ minimizes

$$(\mathbf{Z} - \nabla \mathbf{p}(\boldsymbol{\eta}^0)\boldsymbol{\delta})'\boldsymbol{\Sigma}^{-1}(\mathbf{Z} - \nabla \mathbf{p}(\boldsymbol{\eta}^0)\boldsymbol{\delta}) = \sum_{i=1}^{t} \frac{1}{p^0_i}\left(Z^n_i - \sum_{j=1}^{r} \frac{\partial p_i(\boldsymbol{\eta}^0)}{\partial \eta_j}\delta_j\right)^2.$$

Example 2.2.3 (continued). The asymptotic likelihood ratio test following Proposition 2.2.3 will be derived for this example and compared to the general-purpose chi-squared test. Recall that

$$p_i(\eta'_1, \eta'_2) = \frac{1}{t} + \eta'_1 \int_{2\pi(i-1)/t}^{2\pi i/t} \sin(u - \eta'_2)du, \quad i = 1(1)t.$$

Applying standard trigonometric identities shows that

$$p_i(\eta'_1, \eta'_2) = \frac{1}{t} + x_{i1}\eta_1 + x_{i2}\eta_2$$

where $(\eta_1, \eta_2) := (\eta'_1 \times \cos \eta'_2, \eta'_1 \times \sin \eta'_2)$ and

$$(x_{i1}, x_{i2}) := \left(2\left(\sin\left(\frac{\pi}{t}\right)\right)\sin\left((2i-1)\frac{\pi}{t}\right), -2\sin\left(\frac{\pi}{t}\right)\cos\left((2i-1)\frac{\pi}{t}\right)\right)$$

for $i = 1(1)t$. Thus $r = 2$. To test $H_0: \mathbf{p} = \mathbf{c}$ versus $H_A: \mathbf{p} \in \mathcal{T} \setminus \{\mathbf{c}\}$, first observe that H_0 holds if and only if $\boldsymbol{\eta}^0 = (0, 0)'$. Calculation gives $(\partial p_i(\boldsymbol{\eta}))/(\partial \eta_j) = x_{ij}$, which implies $\nabla \mathbf{p}(\boldsymbol{\eta}) = \mathbf{X}$, where $\mathbf{X} = (x_{ij})$ is of order $(t-1) \times 2$. The estimated mean is $\hat{\boldsymbol{\mu}}_C = \mathbf{X}\hat{\boldsymbol{\delta}}$ where $\hat{\boldsymbol{\delta}} = (\hat{\delta}_1, \hat{\delta}_2)'$ minimizes

$$\sum_{i=1}^{t} \frac{1}{t}\left(\sqrt{n}\left(\hat{p}_i - \frac{1}{t}\right) - x_{i1}\delta_1 - x_{i2}\delta_2\right)^2 .$$

Taking partial derivatives and using $\sum_{i=1}^{t} x_{i1} = 0 = \sum_{i=1}^{t} x_{i2} = \sum_{i=1}^{t} x_{i1}x_{i2}$ yields

$$(\hat{\delta}_1, \hat{\delta}_2) = \left(\sqrt{n}\sum_{i=1}^{t}\hat{p}_i x_{i1} \Big/ \sum_{i=1}^{t} x_{i1}^2, \sqrt{n}\sum_{i=1}^{t}\hat{p}_i x_{i2} \Big/ \sum_{i=1}^{t} x_{i2}^2\right).$$

Thus H_0 is rejected if and only if

$$\sum_{i=1}^{t} \frac{1}{t}\left(\sqrt{n}\left(\hat{p}_i - \frac{1}{t}\right) - x_{i1}\hat{\delta}_1 - x_{i2}\hat{\delta}_2\right)^2 \geq \chi^2_{\alpha,2},$$

or, equivalently, if and only if

$$2n\left(\sum_{i=1}^{t}\hat{p}_i \sin\left(\frac{(2i-1)\pi}{t}\right)\right)^2 + 2n\left(\sum_{i=1}^{t}\hat{p}_i \cos\left(\frac{(2i-1)\pi}{t}\right)\right)^2 \geq \chi^2_{\alpha,2}.$$

Next consider the powers of the unrestricted X^2 test and the restricted LR test. Fix an alternative $\mathbf{p}(\eta_1', \eta_2')$ in H_A, where η_1' and η_2' denote the phase shift and magnitude of the cyclic disturbance of the alternative. The asymptotic power of the *restricted* test is

$$P\left[\chi_2^2\left(n\eta_1'^2 t^2 \sin^2\left(\frac{\pi}{t}\right)\right) \geq \chi^2_{\alpha,2}\right],$$

while the asymptotic power of the *unrestricted* X^2 test is

$$P\left[\chi_{t-1}^2\left(2n\eta_1'^2 t^2 \sin^2\left(\frac{\pi}{t}\right)\right) \geq \chi^2_{\alpha,t-1}\right].$$

Both tests have power independent of the phase shift η_2' of the sine wave. For fixed degrees of freedom, the power of either test increases as η_1' increases or n increases. Rewriting $t^2\sin^2(\pi/t) = [\frac{\sin(\pi/t)}{\pi/t}]^2\pi^2$, shows $t^2\sin^2(\pi/t)$ increases to π^2 as $t \to \infty$; thus the power of the restricted test increases as the number of cells t increases. However the increase in power is not substantial as t grows large because $t^2\sin^2(\pi/t)$ approaches its limit very quickly ($t^2\sin^2(\pi/t) > .95\pi^2$ for $t \geq 6$ and $> .98\pi^2$ for $t \geq 12$). Analyzing the behavior of the unrestricted test as t increases is more difficult as both the degrees of freedom and the noncentrality parameter change. If the noncentrality parameter were held constant; e.g., by changing α as t increases, then as t increased the power would decrease. Although the overall effect is unclear, due to the speed at which $t^2\sin^2(\pi/t)$ approaches

π^2, it seems likely that numerical work would show that the noncentrality parameter stays relatively constant and thus the power of the unrestricted test *decreases* in t.

Restricted Alternatives 2: Isotonic Case

Perhaps the simplest isotonic alternative to the null hypothesis of equiprobable cells is $p_1 \leq p_2 \leq \ldots \leq p_t$ leading to the so-called test for trend. Chacko (1966) derives the LRT of $H_0: \mathbf{p} = \mathbf{c}$ versus $H_A: p_1 \leq p_2 \leq \ldots \leq p_t$ (not all p_i's equal). This test requires the calculation of the MLE of $\mathbf{p}$ under H_A. The limiting distribution of the test statistic is a chi-bar-squared distribution, which is a linear combination of chi-squared distributions. The only known optimality property of this test is consistency; i.e., the power of the test converges to one for all $\mathbf{p}$ in H_A.

Y.J. Lee (1977) proposes using the test of H_0 versus H_A which maximizes the minimum power over a portion of the parameter space in H_A (a minimax principle applied with a special 0–1 loss). To apply this criterion one must restrict attention to a proper subset of the alternative since $\min\{P_{\mathbf{p}}[\text{Reject } H_0] : \mathbf{p} \text{ in } H_A\} = \alpha$ for all tests with continuous power functions; this shows all such tests are equivalent if the entire alternative region is considered. Instead Lee determines the test which maximizes the minimum power over the alternative,

$$H_A^*: \mathbf{p} \in \mathcal{S}(\boldsymbol{\delta}^*) := \{\mathbf{p} \in \mathcal{S} : p_{i+1} - p_i \geq \delta_i^*,\ 1 \leq i \leq t-1\}$$

where $\boldsymbol{\delta}^*$ is a fixed vector of positive constants satisfying $\sum_{j=1}^{t-1}(t-j)\delta_j^* \leq 1$. The latter requirement insures $\mathcal{S}(\boldsymbol{\delta}^*) \neq \phi$. The maximin test of H_0 versus H_A^* is randomized; it rejects H_0 with probability 1 when

$$\sum_{i=1}^{t} Y_i \log p_i^* > K \tag{2.2.30}$$

and with probability γ if equality holds in (2.2.30). The vector $\mathbf{p}^*$ has components $p_2^* = p_1^* + \delta_1^*, \ldots, p_t^* = p_{t-1}^* + \delta_{t-1}^*$ with p_1^* chosen so that $\sum_{i=1}^{t} p_i^* = 1$. The constants K and γ are chosen so that $P_{\mathbf{p}=\mathbf{c}}[\text{Reject } H_0] = \alpha$.

Likelihood ratio tests have been derived for many other isotonic alternatives. Robertson (1978) gives results for alternatives in a general cone; this formulation includes many previously considered problems as special cases. The following definition is required to state his results.

Definition. A set $K \subset \mathbb{R}^t$ is a *cone* means that $\beta\mathbf{x} \in K$ for all $\mathbf{x} \in K$ and all $\beta \geq 0$.

For a given cone $K \subset \mathbb{R}^t$ and a given $\mathbf{p}^0 \in \mathcal{O} := K \cap \mathcal{S}$, Robertson (1978) considers testing $H_0: \mathbf{p} = \mathbf{p}^0$ versus $H_A: \mathbf{p} \in \mathcal{O} \setminus \{\mathbf{p}^0\}$. For example, if $K = \{\mathbf{x} \in \mathbb{R}^t : x_t \geq x_{t-1} \geq \ldots \geq x_1 \geq 0\}$ and $\mathbf{p}^0 = \mathbf{c}$ then $\mathcal{O} = \{\mathbf{p} \in \mathcal{S} : p_1 \leq p_2 \leq \ldots \leq p_t\}$ which yields Chacko's alternative hypothesis. As a second example, if $K = \{\mathbf{x} \in \mathbb{R}^t : x_1 \geq x_2,\, x_3 \geq x_4 \geq 0\}$, and $\mathbf{p}^0 = (9/16, 3/16, 3/16, 1/16)'$ then $\mathcal{O} = \{\mathbf{p} \in \mathcal{S} : p_1 \geq p_2,\, p_3 \geq p_4\}$ which gives the null and alternative considered earlier in Mendel's experiment.

Robertson derives the limiting null distribution of the likelihood ratio test of H_0 versus H_A and shows the test is consistent. C.C. Lee (1987) has studied the Wald and Pearson chi-squared tests for the same null and alternative hypotheses. All three tests are based on the same asymptotic null distribution. While presumably isotonic tests are more powerful than general omnibus tests, the quantification of these gains has not been studied extensively. C.C. Lee (1987) presents the results of a simulation study of the size and power of the LRT, Wald and Pearson tests for trend and provides some guidelines for sample size selection. Another interesting area for additional research is to determine the form of the maximin test for Robertson's null and alternative.

Tests of Composite Null versus the Global Alternative

Suppose it is desired to test

$$H_0 : \mathbf{p} \in \{\mathbf{p}(\boldsymbol{\eta}) : \boldsymbol{\eta} \in \mathcal{N}\} \text{ versus } H_A\text{: not } H_0 \tag{2.2.31}$$

based on $\mathbf{Y} \sim M_t(n, \mathbf{p})$ where $\mathcal{N} \subset \mathbb{R}^r$. The null hypothesis is composite and parameterized by $\boldsymbol{\eta}$ while the alternative is global. The following example illustrates the general formulation.

Example 2.2.4. Edwards and Fraccaro (1960) display the data in Table 2.2.5 on the number of boys among the first four children in 3343 Swedish families having at least four children. If the number of families with i boys is denoted by Y_i for $i = 0(1)4$ then $\mathbf{Y}$ is distributed $M_5(3343, \mathbf{p})$ where $\mathbf{p} \in \mathcal{S}$, the four dimensional simplex. Consider the model which postulates (i) a constant probability η, $0 < \eta < 1$, of each birth being a boy and (ii) that the determinations of the sexes of a family's children are mutually independent events. Under (i) and (ii)

$$p_j(\eta) = \binom{4}{j} \eta^j (1-\eta)^{4-j}$$

is the probability that the number of boys in a family is j for $j = 0(1)4$. A test of the model is $H_0 : \mathbf{p} \in \omega$ versus $H_A : \mathbf{p} \in \mathcal{S} \setminus \omega$ where $\omega = \{\mathbf{p} = \mathbf{p}(\eta) : 0 < \eta < 1\}$.

Table 2.2.5. Number of Boys Among the First Four Children in 3343 Swedish Families (Reprinted with permission from A.W.F. Edwards and M. Fraccaro: "Distribution and Sequences of Sexes in a Selected Sample of Swedish Families," *Annals of Human Genetics*. Cambridge Univ. Press, 1960, vol. 24, pg. 246.)

	0	1	2	3	4	Total
Number of Families	183	789	1250	875	246	3343

Returning to the general composite null versus global alternative testing problem formulated in (2.2.31), both the score test and the likelihood ratio test of H_0 versus H_A are derived next. Let $\mathbf{p}_- := (p_1, \ldots, p_{t-1})'$, $\Omega := \{\mathbf{u} = (u_1, \ldots, u_{t-1})' : u_i \geq 0,\ \sum_{i=1}^{t-1} u_i \leq 1\}$ and $\omega := \{(p_1, \ldots, p_{t-1}) \in \Omega : \mathbf{p}_- = \mathbf{p}_-(\boldsymbol{\eta}) \text{ for some } \boldsymbol{\eta} \in \mathcal{N}\}$. Let $\hat{\boldsymbol{\eta}}$ be the maximum likelihood estimate of $\boldsymbol{\eta}$ based on $\mathbf{Y}$; i.e., $\hat{\boldsymbol{\eta}}$ maximizes $\prod_{i=1}^{t}(p_i(\boldsymbol{\eta}))^{Y_i}$. Thus the maximum likelihood estimate of $\mathbf{p}_-$ restricted to ω is $\hat{\mathbf{p}}_-^{\omega} := (p_1(\hat{\boldsymbol{\eta}}), \ldots, p_{t-1}(\hat{\boldsymbol{\eta}}))'$ by invariance. The score vector is the $(t-1) \times 1$ column vector

$$\mathbf{S}(\mathbf{p}_-) = \left(\ldots, \frac{Y_i}{p_i} - \frac{Y_t}{p_t}, \ldots\right)'$$

and the information matrix is

$$\mathbf{I}(\mathbf{p}_-) = n\left[\text{Diag}\left(\frac{1}{p_1}, \ldots, \frac{1}{p_{t-1}}\right) - \frac{1}{p_t}\mathbf{1}_{t-1}\mathbf{1}_{t-1}'\right]$$

(Appendix A.4). The score test statistic is

$$(\mathbf{S}(\hat{\mathbf{p}}_-^{\omega}))'\mathbf{I}^{-1}(\hat{\mathbf{p}}_-^{\omega})\mathbf{S}(\hat{\mathbf{p}}_-^{\omega}) = \sum_{i=1}^{t} \frac{[Y_i - np_i(\hat{\boldsymbol{\eta}})]^2}{np_i(\hat{\boldsymbol{\eta}})} =: X^2,$$

which is the Pearson chi-squared statistic for testing H_0 versus H_A. The likelihood ratio test statistic is

$$G^2 = 2\sum_{i=1}^{t} Y_i \ln\left(\frac{Y_i}{np_i(\hat{\boldsymbol{\eta}})}\right).$$

As in the simple null hypothesis testing case, these tests are members of the Cressie Read family which is defined for this problem by

$$I^{\lambda} := \frac{2}{\lambda(\lambda+1)} \sum_{i=1}^{t} Y_i \left\{\left(\frac{Y_i}{np_i(\hat{\eta})}\right)^{\lambda} - 1\right\}.$$

Read and Cressie (1988) apply the delta method to prove the following result which gives conditions under which χ^2_{t-1-r} is the asymptotic null distribution of any I^{λ}, $\lambda \in \mathbb{R}$ (see also Rao, 1965, p. 325).

Proposition 2.2.4. *Suppose* $\mathbf{Y} \sim M_t(n, \mathbf{p}(\boldsymbol{\eta}^0))$ *where* $\boldsymbol{\eta}^0 \in \mathcal{N}$,

(i) $p_i(\boldsymbol{\eta})$ *has continuous first partials with respect to* η_j *at* $\boldsymbol{\eta}^0$ *for every* $1 \le i < t$ *and* $1 \le j \le r$, *and*

(ii) *the* $(t-1) \times r$ *matrix* $M(\boldsymbol{\eta}) := \{(\partial p_i(\boldsymbol{\eta}^0))/(\partial \eta_j)\}$ *has rank* r.

Then every I^λ *converges in law to the* χ^2_{t-1-r} *distribution as* $n \to \infty$ *for every* $\lambda \in \mathbb{R}$.

Proposition 2.2.4 shows that if hypotheses (i) and (ii) hold for all $\mathbf{p} \in \omega$ then asymptotically the effect of $\boldsymbol{\eta}$ "washes out" and the null distribution is the same for any $\mathbf{p}$ in H_0. (However note that the exact small sample distribution of I^λ depends on $\mathbf{p} = \mathbf{p}(\boldsymbol{\eta})$ and λ.)

Work by Odoroff (1970), Larntz (1978), Lawal (1984), and Read and Cressie (1988) in specific composite H_0 cases shows that, as in the case of simple null hypotheses, I^λ converges to χ^2_{t-1-r} more rapidly for $\lambda \in [1/3, 3/2]$ than for other choices of λ. In particular, the test based on X^2 has achieved size closer to nominal than G^2. There has also been recent work on deriving *normal* null approximations to the distribution of I^λ, in particular for X^2 and G^2 (Simonoff, 1985; Koehler 1986; Zelterman, 1987; Read and Cressie, 1988). In the case of multinomial distributions based on large sparse contingency tables, normal approximations can yield tests with more nearly nominal size. Weiss (1975) studies the asymptotic null distributions of X^2, G^2 and related test statistics in "non-regular" cases not covered by Proposition 2.2.4. The asymptotic power of the X^2 and G^2 tests is less well understood when H_0 is composite than when H_0 is simple although Feder (1968) does study the asymptotic behavior of these tests under H_A. The use of X^2, G^2, and $I^{2/3}$ are illustrated in Example 2.2.4.

Example 2.2.4 (continued). Regarded as a function of η, the likelihood is proportional to

$$\prod_{j=0}^{t} \left\{ \binom{4}{j} \eta^j (1-\eta)^{4-j} \right\}^{Y_j},$$

and so, apart from constants, the log likelihood is

$$\ln L(\eta) = \ln \eta \sum_{j=0}^{4} jY_j + \ln(1-\eta) \sum_{j=0}^{4} (4-j)Y_j.$$

Differentiating and solving gives the maximizing value $\hat{\eta} = \frac{1}{4(3343)} \sum_{j=0}^{4} jY_j = .515$ which is the proportion of boy births. Due to the large sample size, $X^2 = G^2 = I^{2/3} = .627$ to three decimal places. The limiting null distribution of X^2, G^2, and $I^{2/3}$ is χ^2_3; thus the P-value of the test for this data is $P[\chi^2_3 \ge .627] \approx .89$ and there is no evidence against the hypothesized model.

One interesting application of the composite null testing procedure is testing goodness-to-fit to a given parametric family of continuous distributions, such as the normal family. The test is conducted by grouping the observed (continuous) data into cells to form multinomial data. In this application, it is often considerably easier to estimate the parameters $\boldsymbol{\eta}$ of the null hypothesis using the original continuous data rather than the grouped data (Problem 2.17). Chernoff and Lehmann (1954) prove that, under mild conditions, the asymptotic null distribution of the X^2 test statistic is no longer χ^2_{t-1} and worse changes with the true null distribution of the continuous data. Moore (1971), Rao and Robson (1974), Moore and Spruill (1975), Weiss (1975) and other authors have considered modifications of X^2 which allow the use of the MLE of $\boldsymbol{\eta}$ based on the continuous data and have the *same* limiting null distribution throughout the null hypothesis.

C. Interval Estimation

The object of this subsection is to derive and compare several sets of simultaneous confidence intervals for families of functions of $\mathbf{p}$. Attention will be restricted to the families $\{p_i : 1 \le i \le t\}$ and $\{p_i - p_j : 1 \le i < j \le t\}$ and to asymptotic methods.

Consider first the case of simultaneous confidence intervals for $\{p_i : 1 \le i \le t\}$. For ease of exposition, the notation $\mathbf{p}_- := (p_1, \ldots, p_{t-1})'$ and

$$\hat{\mathbf{p}}_- := \left(\frac{Y_1}{n}, \ldots, \frac{Y_{t-1}}{n}\right)'$$

will be used. Gold (1963) defines the confidence region

$$R_G(\mathbf{Y}) := \left\{ \mathbf{w} \in \mathrm{I\!R}^{t-1} : (\hat{\mathbf{p}}_- - \mathbf{w})'\hat{\boldsymbol{\Sigma}}^{-1}(\hat{\mathbf{p}}_- - \mathbf{w}) \le \frac{\chi^2_{\alpha,t-1}}{n} \right\}$$

where $\hat{\boldsymbol{\Sigma}} = \text{Diag}(\hat{\mathbf{p}}_-) - \hat{\mathbf{p}}_-(\hat{\mathbf{p}}_-)'$. Then $P_{\mathbf{p}}[\mathbf{p}_- \in R_G(\mathbf{Y})] \to 1 - \alpha$ as $n \to \infty$ since $n^{1/2}(\hat{\mathbf{p}}_- - \mathbf{p}_-) \xrightarrow{\mathcal{L}} N_{t-1}[0, \boldsymbol{\Sigma} := (\text{Diag}(\mathbf{p}_-) - \mathbf{p}_-\mathbf{p}_-')]$. Applying Scheffe's projection method (Scheffe (1959), Sec. 3.4) to $R_G(\mathbf{Y})$ gives,

$$\{\mathbf{p}_- \in R_G(\mathbf{Y})\} = \bigcap_{\boldsymbol{\ell} \in \mathrm{I\!R}^{t-1}} \left\{ \boldsymbol{\ell}'\mathbf{p}_- \in \boldsymbol{\ell}'\hat{\mathbf{p}}_- \pm \left(\frac{\chi^2_{\alpha,t-1}}{n}\right)^{1/2} (\boldsymbol{\ell}'\hat{\boldsymbol{\Sigma}}\boldsymbol{\ell})^{1/2} \right\}.$$

Geometrically, $\mathbf{p}_-$ is in the ellipse $R_G(\mathbf{Y})$ if and only if for every direction $\boldsymbol{\ell}$, $\boldsymbol{\ell}'\mathbf{p}_-$ falls in the projection of $R_G(\mathbf{Y})$ on the line through the origin in the direction of $\boldsymbol{\ell}$. Gold (1963) proposes using the projections on the right hand side to give confidence intervals for the family of all linear combinations of the p_i or any subfamily of these linear combinations. As a special case, the subfamily $\{(1,0,\ldots,0),\ldots,(0,\ldots,0,1),\ (-1,\ldots,-1)\}$ of $\boldsymbol{\ell}$ produces the asymptotic simultaneous confidence intervals

$$I_i^G = \hat{p}_i \pm \left(\frac{\hat{p}_i(1-\hat{p}_i)}{n}\right)^{1/2} (\chi^2_{\alpha,t-1})^{1/2}, \quad i = 1(1)t \tag{2.2.32}$$

for $\{p_i : 1 \leq i \leq t\}$.

Quesenberry and Hurst (1964) propose confidence intervals based on a projection method similar to Gold's, but starting from the ellipse

$$R_{QH}(\mathbf{Y}) = R_{QH} := \{\mathbf{w} \in \mathbb{R}^{t-1} : (\hat{\mathbf{p}}_- - \mathbf{w})'[\boldsymbol{\Sigma}(\mathbf{w})]^{-1}(\hat{\mathbf{p}}_- - \mathbf{w}) \leq \frac{\chi^2_{\alpha,t-1}}{n}\}$$

where $\boldsymbol{\Sigma}(\mathbf{w}) = \text{Diag}(\mathbf{w}) - \mathbf{w}\mathbf{w}'$. The ellipse R_{QH} is asymptotically equivalent to R_G as they differ only in that R_G uses an estimated variance–covariance matrix while R_{QH} uses the population quantity; R_G and R_{QH} are analogs of the binomial success probability confidence intervals I_N and I_S, respectively, which are discussed in Section 2.1. The Quesenberry and Hurst simultaneous confidence interval for p_i is

$$\begin{aligned} I_i^{QH} &= \left\{ w_i \in \mathbb{R}^1 : \frac{(\hat{p}_i - w_i)^2}{w_i(1-w_i)} \leq \frac{\chi^2_{\alpha,t-1}}{n} \right\} \\ &= \frac{2Y_i + c \pm \left(c[c + \frac{4Y_i(n-Y_i)}{n}]\right)^{1/2}}{2(n+c)} \end{aligned} \tag{2.2.33}$$

where $c = \chi^2_{\alpha,t-1}$.

Neither family of intervals $\{I_i^G\}_{i=1}^t$ or $\{I_i^{QH}\}_{i=1}^t$ is satisfactory. First, the calculations in Ghosh (1979) comparing I_N and I_S make the distribution theory of the $\{I_i^G\}_{i=1}^t$ intervals suspect. Second, even if the asymptotic distribution theory were exact, *both* families are conservative because they start with an ellipse having asymptotic probability $(1-\alpha)$ and extract a finite subset of projections.

The concerns of the previous paragraph prompted Goodman (1965) to study simultaneous confidence intervals derived from Bonferroni's inequality. Recall that if for $j = 1(1)t$, $P[E_j] \geq 1-\alpha/t$, then Bonferroni's inequality states that

$$\begin{aligned} P\left[\bigcap_{j=1}^t E_j\right] &= 1 - P\left[\bigcup_{j=1}^t (\sim E_j)\right] \\ &\geq 1 - \sum_{j=1}^t P[\sim E_j] \geq 1 - \sum_{j=1}^t (\alpha/t) = 1 - \alpha \end{aligned}$$

where $\sim E_j$ denotes the complement of E_j. The lower bound is relatively sharp for small α. Goodman applies the Bonferroni inequality to the events,

$$E_j := \left\{ \frac{n(\hat{p}_j - p_j)^2}{p_j(1-p_j)} \le \chi^2_{\alpha/t,1} \right\}, \quad j = 1(1)t,$$

for which $P_{\mathbf{p}}[E_j] \to 1 - \alpha/t$ as $n \to \infty$, to yield asymptotic confidence intervals $\{I_i^{GM}\}_{i=1}^t$. These intervals have the same form as $\{I_i^{QH}\}_{i=1}^t$ in equation (2.2.33) with the modification $c = \chi^2_{\alpha/t,1}$. The following example compares $\{I_i^{GM}\}_{i=1}^t$ and $\{I_i^{QH}\}_{i=1}^t$.

Example 2.2.5. Table 2.2.6 from Quesenberry and Hurst (1964) classifies 870 machine failures according to one of 10 failure modes. Table 2.2.7 lists 90% simultaneous QH and GM confidence intervals for $\{p_i\}_{i=1}^{10}$. Perhaps surprisingly, the Bonferroni intervals are *uniformly* shorter.

In general Goodman shows,

$$\frac{\text{length of } GM \text{ interval}}{\text{length of } QH \text{ interval}} \xrightarrow{p} \left(\frac{\varsigma^2_{\alpha/t,1}}{\chi^2_{\alpha,t-1}} \right)^2 \quad \text{as } n \to \infty. \qquad (2.3.34)$$

The Bonferroni intervals are asymptotically superior when the right-hand side of (2.2.34) is less than unity; Goodman found it to be *less* than unity for $\alpha \le .10$ and a wide range of t. In Example 2.2.5, $(\alpha, t) = (.10, 10)$ and the limiting ratio is .67.

Table 2.2.6. Failure Modes for 870 Machines (Reprinted with permission from C.P. Quesenberry and D.C. Hurst: "Large Sample Simultaneous Confidence Intervals for Multinomial," *Technometrics,* vol. 6, no. 2, 1964. American Society for Quality Control and the American Statistical Association.)

	Failure Mode									
	1	2	3	4	5	6	7	8	9	10
Number of Machines	5	11	19	30	58	67	92	118	123	217

Table 2.2.7. 90% $\{I_i^{QH}\}_{i=1}^{10}$ and $\{I_i^{GM}\}_{i=1}^{10}$ Intervals for Machine Failure Data

i	$\hat{p}_i$	Q–H		GM	
1	.006	.001	.027	.002	.017*
2	.013	.044	.037	.006	.027*
3	.022	.009	.060	.013	.039*
4	.034	.017	.067	.022	.054*
5	.067	.041	.107	.048	.092*
6	.077	.049	.119	.057	.104*
7	.106	.072	.152	.082	.136*
8	.136	.097	.186	.108	.168*
9	.199	.152	.256	.166	.236*
10	.341	.283	.405	.301	.384*

*Shorter interval.

In principle, for *small sample* cases one can use individual Blyth/Still confidence intervals for each p_i with a Bonferroni constant to obtain simultaneous confidence intervals for $\{p_i\}_{i=1}^t$ with probability $1-\alpha$. However in practice this is not feasible as only 95% and 99% Blyth/Still intervals have been tabulated. A computationally feasible alternative is to use the Bonferroni constant with (conservative) tail intervals since they can be constructed from (2.1.13) for any α.

Projections of Gold's ellipse R_G can be used to obtain simultaneous confidence intervals for the differences between cell proportions, $\{p_i - p_j : 1 \le i < j \le t\}$. The resulting intervals are

$$\hat{p}_i - \hat{p}_j \pm (\chi^2_{\alpha,t-1})^{1/2}\left\{\frac{\hat{p}_i + \hat{p}_j - (\hat{p}_i - \hat{p}_j)^2}{n}\right\}^{1/2}. \tag{2.2.35}$$

The factor for the standard deviation on the right comes from estimating

$$\begin{aligned}\mathrm{Var}(\hat{p}_i - \hat{p}_j) &= \mathrm{Var}(\hat{p}_i) + \mathrm{Var}(\hat{p}_j) - 2\,\mathrm{Cov}(\hat{p}_i, \hat{p}_j)\\ &= \frac{p_i(1-p_i)}{n} + \frac{p_j(1-p_j)}{n} - \frac{2(-p_i p_j)}{n}\\ &= \frac{p_i + p_j - (p_i - p_j)^2}{n}.\end{aligned}$$

Projections based on the Quesenberry–Hurst region R_{QH} appear impossible to derive since the variance of $(\hat{p}_i - \hat{p}_j)$ does not depend on p_i and p_j only through $p_i - p_j$. Goodman (1965) proposes using Bonferroni intervals by replacing $\chi^2_{\alpha,t-1}$ in (2.2.35) by $\chi^2_{z,1}$ with $z = \alpha/\binom{t}{2}$. The extra conservatism of the projection method suggests that Goodman's system of intervals are the method of choice among the three discussed above. In closing, it should be noted that a number of papers in the literature have addressed the problem of sample size determination when interval estimation is the goal (Angers, 1974; Thompson, 1987; and the references therein). These papers primarily use the Bonferroni method based on the asymptotic normality of $\hat{\mathbf{p}}$.

D. Selection and Ranking

Let $\mathbf{Y} \sim M_t(n, \mathbf{p})$ and denote the ordered p_i's by

$$p_{[1]} \leq \cdots \leq p_{[t]}.$$

Consider the problem of selecting the cell associated with the largest cell probability $p_{[t]}$. This is a multiple decision problem. The two classical formulations of this problem are (i) the *indifference zone formulation* due to Bechhofer (1954) and (ii) the *subset selection formulation* due to Gupta (1956, 1965). The former is primarily concerned with experimental design while the latter can be thought of as an analysis technique useful no matter what the sample size.

The goal of the indifference zone approach is to select a *single* cell and claim that it has cell probability $p_{[t]}$. Consider the intuitive procedure which selects the cell associated with $\hat{p}_{[t]} := \max_{1 \leq i \leq t}\{\hat{p}_i\}$. The statistical issue is to determine the minimum sample size n to be used with the above procedure to guarantee a prespecified probability of correctly identifying the cell associated with $p_{[t]}$ (a "Correct Selection" for this formulation). Bechhofer, Elmagraby and Morse (1959) propose the following design criterion. Given $\delta > 0$ and $1 > \alpha > 0$, choose the smallest n such that

$$P_{\mathbf{p}}[\text{Correct Selection}] \geq 1 - \alpha$$

for all $\mathbf{p} \in \mathcal{S}(\delta) := \{\mathbf{p} \in \mathcal{S} : p_{[t]} \geq \delta p_{[t-1]}\}$. This formulation derives its name from the fact that the experimenter is regarded as indifferent to which cell is selected when $\mathbf{p} \in \mathcal{S} \setminus \mathcal{S}(\delta)$. Equivalently, the sample size n is chosen so that

$$\inf_{\mathbf{p} \in \mathcal{S}(\delta)} P_{\mathbf{p}}[\text{Corrected Selection}] \geq 1 - \alpha. \tag{2.2.36}$$

Intuitively, one suspects that the infimum on the left-hand side of (2.2.36) will occur when $p_{[t]} = \delta p_{[t-1]}$ and $p_{[1]} = \ldots = p_{[t-1]}$, or equivalently when $p_{[1]} = \ldots = p_{[t-1]} = \frac{1}{\delta+t-1}$ and $p_{[t]} = \frac{\delta}{\delta+t-1}$. Kesten and Morse (1959) prove this intuitive conjecture is correct. Tables of n which guarantee (2.2.36) for various choices of (δ, t, α) can be found in Bechhofer, Elmagraby and Morse (1959), and Gibbons, Olkin and Sobel (1977).

It is interesting to note that although it might appear that the problem of selecting the cell associated with $p_{[1]}$ should be symmetric to that of selecting the cell associated with $p_{[t]}$, this is not the case. Alam and

Thompson (1972) prove that there does not exist an n which guarantees a given probability of correct selection for all $\mathbf{p} \in \mathcal{S}$ such that $p_{[1]} \leq \delta p_{[2]}$.

A second formulation of ranking and selection problems is the subset selection approach. The goal in subset selection is to select a *subset* of the cells $1, \ldots, t$ which contains the cell associated with $p_{[t]}$ (a "Correct Selection" in this case). Gupta and Nagel (1967) study the intuitive procedure which places cell i in the selected subset whenever

$$\hat{p}_i \geq \hat{p}_{[t]} - D. \tag{2.2.37}$$

They propose choosing $D = D(\alpha, n, t)$ to satisfy the following probability requirement

$$P_{\mathbf{p}}[\text{Correct Selection}] \geq 1 - \alpha \quad \text{for all} \quad \mathbf{p} \in \mathcal{S}. \tag{2.2.38}$$

The fact that (2.2.38) is required for *all* $\mathbf{p} \in \mathcal{S}$ rather than $\mathbf{p}$ outside some indifference zone, forces the selected subset to be of *random size.*

Philosophically, subset selection procedures are similar in intention, if not operating characteristics, to confidence intervals. For example, consider forming a 95% confidence interval for a normal mean when the true mean and variance are both unknown. The usual t-interval adjusts both its center and length so that it contains the true mean in roughly 95% of independent applications. Similarly subset procedures adjust the number of populations they select to guarantee the selected subset contains the cell associated with $p_{[t]}$ no matter what the true $\mathbf{p} \in \mathcal{S}$. However, the 95% normal mean t-interval achieves exactly .95 coverage probability no matter what the true mean and variance, whereas the probability of correct selection depends on $\mathbf{p}$; it is at least $(1 - \alpha)$, but can be extremely close to unity if $p_{[t]}$ is "far" from $p_{[t-1]}$.

Much additional work has been done on both the indifference zone and subset selection approaches to this problem. Some of the topics which have been explored are (i) inverse sampling and other sequential procedures (Panchapakesan, 1971; Cacoullos and Sobel, 1966; Ramey and Alam, 1979; Bechhofer and Goldsman, 1985, 1986; and the references therein), (ii) comparisons with a control (Chen, 1987), and (iii) curtailment (Bechhofer and Kulkarni, 1984). Gupta and Panchapakesan (1979) contains a detailed survey of the selection procedures.

2.3 Poisson Responses

Consider mutually independent Poisson random variables $Y_1, \ldots, Y_t$ with $E[Y_i] = T_i \lambda_i$, where $T_i > 0$ is known and $\lambda_i > 0$ is unknown for $i = 1(1)t$. Recall from Section 1.3 the following two examples which lead to this model. If $Y_{i1}^*, \ldots, Y_{im_i}^*$ is a random sample from the $P(\lambda_i)$ distribution, then sufficiency reduces the data to $Y_i := \sum_{j=1}^{m_i} Y_{ij}^* \sim P(m_i \lambda_i)$, $i = 1(1)t$.

A second example is that of t independent homogeneous Poisson processes which are observed for varying amounts of time. The count for the ith process, Y_i, is distributed $P(T_i\lambda_i)$ where λ_i is its hazard rate and T_i is the time the process is observed. The Poisson process model holds under mild conditions (Karlin and Taylor, 1975, Chapter 4) for phenomena occurring in time or space such as printing errors in newspapers, industrial accidents, cases of rare diseases in a given region, and fire or crime incidence.

The vector of counts $\mathbf{Y} = (Y_1, \ldots, Y_t)'$ can be thought of as a Poisson one-way layout. The generalization to the case of $E[\mathbf{Y}]$ following an arbitrary linear model is considered in Sections 3.2 and 3.4. The conditional distribution of $(Y_1, \ldots, Y_t)$ given $S := \sum_{i=1}^{t} Y_i$ is the multinomial distribution with S trials, t cells, and vector of "cell probabilities" $\mathbf{p} := (\ldots, T_i\lambda_i / \sum_{j=1}^{t} T_j\lambda_j, \ldots)'$. This relationship might make one suspect that the results from Section 2.2 for the multinomial case can be adapted to Poisson responses. While this is partially true, especially for testing hypotheses, there are also important differences between the two models particularly for the problem of point estimation of $\boldsymbol{\lambda}$.

Throughout this section, interest focuses on the vector $\boldsymbol{\lambda}' = (\lambda_1, \ldots, \lambda_t)$. As in Section 2.2 point estimation of $\boldsymbol{\lambda}$ is discussed first, followed by results on tests for $\boldsymbol{\lambda}$, and a brief introduction to the problems of interval estimation and selection and ranking.

A. Point Estimation of $\boldsymbol{\lambda}$

Much of the early literature on the problem of point estimation of $\boldsymbol{\lambda}$ assumes

$$T_1 = \ldots = T_t = 1. \tag{2.3.1}$$

Results which require the restriction (2.3.1) will be explicitly noted. First the MLE of $\boldsymbol{\lambda}$ is derived and its properties in large and small samples are studied. For the latter purpose, several loss functions that have been proposed in the Poisson literature are examined. Then alternative estimators are developed.

Maximum Likelihood Estimation

By the independence of $Y_1, \ldots, Y_t$, it suffices to consider the MLE for a single component. The likelihood function at λ_i based on Y_i is

$$L(\lambda_i) = \frac{\exp(-\lambda_i T_i)(\lambda_i T_i)^{y_i}}{y_i!}$$

for $y_i \in Z := \{0, 1, \ldots\}$, the set of nonnegative integers. The log likelihood is proportional to

$$-\lambda_i T_i + y_i \ln(\lambda_i T_i). \tag{2.3.2}$$

For $y_i \geq 1$, (2.3.2) is strictly concave with unique maximum at y_i/T_i. Thus $\hat{\boldsymbol{\lambda}} := (Y_1/T_1, \ldots, Y_t/T_t)'$ is the MLE of $\boldsymbol{\lambda}$ (where the estimation space is taken as $\times_{i=1}^{t}[0, \infty)$ to make the statement correct for $y_i = 0$).

Besides its intuitive appeal as a moment estimator, $\hat{\boldsymbol{\lambda}}$ has important classical optimality properties. First, for the asymptotic setup in which $\lambda_i > 0$ is fixed and $T_i \to \infty$, $\hat{\lambda}_i = Y_i/T_i$ is an efficient estimator of λ_i. Second, $\hat{\lambda}_i$ is the (small sample) UMVUE of λ_i.

Loss Functions

An alternative small sample analysis of $\hat{\boldsymbol{\lambda}}$ arises by considering various loss functions. Perhaps the simplest loss functions are squared error (L_S) and relative squared error (L_R) loss:

$$L_S(\boldsymbol{\lambda}, \mathbf{a}) = \sum_{i=1}^{t} (\lambda_i - a_i)^2 \quad \text{and} \tag{2.3.3}$$

$$L_R(\boldsymbol{\lambda}, \mathbf{a}) = \sum_{i=1}^{t} (\lambda_i - a_i)^2/\lambda_i. \tag{2.3.4}$$

The factor λ_i in the denominator of $L_R(\cdot, \cdot)$ is the variance of Y_i; it serves to penalize guesses a_i which do not perform well for small λ_i values. Several generalizations of $L_S(\cdot, \cdot)$ and $L_R(\cdot, \cdot)$ have been considered in the literature. The *k-normalized squared error* (k-NSEL) is

$$L_k(\boldsymbol{\lambda}, \mathbf{a}) := \sum_{i=1}^{t} (\lambda_i - a_i)^2/\lambda_i^k \tag{2.3.5}$$

for integer $k \geq 0$. Note that $L_S(\cdot, \cdot) = L_0(\cdot, \cdot)$ and $L_R(\cdot, \cdot) = L_1(\cdot, \cdot)$. For $k \geq 2$, L_k places an even higher premium on precise estimation of small λ_i's than $L_R(\cdot, \cdot)$. A more flexible version of $L_k(\cdot, \cdot)$ is **d**-*normalized squared error loss* (**d**-NSEL)

$$L_{\mathbf{d}}(\boldsymbol{\lambda}, \mathbf{a}) := \sum_{i=1}^{t} (\lambda_i - a_i)^2/\lambda_i^{d_i} \tag{2.3.6}$$

with $\mathbf{d} = (d_1, \ldots, d_t)' \geq \mathbf{0}$ which allows individual components to be weighted differentially. A final generalization of $L_{\mathbf{d}}(\cdot, \cdot)$ is *weighted* **d**-*NSEL*

$$L_{\mathbf{w},\mathbf{d}}(\boldsymbol{\lambda}, \mathbf{a}) := \sum_{i=1}^{t} w_i(\lambda_i - a_i)^2/\lambda_i^{d_i} \tag{2.3.7}$$

where $\mathbf{w} = (w_1, \ldots, w_t)' > \mathbf{0}$ and $\mathbf{d} = (d_1, \ldots, d_t)' \geq \mathbf{0}$. The reason for the importance of weighted **d**-NSEL is that studies of the MLE $\hat{\boldsymbol{\lambda}} =$

$(Y_1/T_1, \ldots, Y_t/T_t)'$ assuming **d**-NSEL are mathematically equivalent to studies of the MLE $\hat{\boldsymbol{\lambda}} = \mathbf{Y}$ under (2.3.1) assuming weighted **d**-NSEL. This holds because

$$\begin{aligned} R_{\mathbf{d}}(\boldsymbol{\lambda}, \hat{\boldsymbol{\lambda}}) &:= \sum_{i=1}^{t} E_{\boldsymbol{\lambda}}\left((\lambda_i - Y_i/T_i)^2\right)/\lambda_i^{d_i} \\ &= \sum_{i=1}^{t} T_i^{d_i-2} E_{\boldsymbol{\lambda}}\left((T_i\lambda_i - Y_i)^2\right)/(T_i\lambda_i)^{d_i} \\ &= R_{\mathbf{w},\mathbf{d}}(\mathbf{T}\boldsymbol{\lambda}, \hat{\boldsymbol{\lambda}}) \end{aligned}$$

where $\mathbf{w} = (T_1^{d_1-2}, \ldots, T_t^{d_t-2})'$, and $\mathbf{T}\boldsymbol{\lambda} = (T_1\lambda_1, \ldots, T_t\lambda_t)'$.

As in Section 2.2 on multinomial observations, none of the loss functions introduced above distinguishes between positive and zero estimates of λ_i but entropy does make this distinction. The *entropy loss* of estimating the true $\boldsymbol{\lambda}$ by $\mathbf{a}$ is the Kullback–Liebler distance between the Poisson distributions $P(\boldsymbol{\lambda})$ and $P(\mathbf{a})$:

$$L_E(\boldsymbol{\lambda}, \mathbf{a}) = E_{\boldsymbol{\lambda}}\left[\ln\left(\frac{f(\mathbf{y} \mid \boldsymbol{\lambda})}{f(\mathbf{y} \mid \mathbf{a})}\right)\right] = \sum_{i=1}^{n}[a_iT_i - \lambda_iT_i + y_i \ln(\lambda_i/a_i)] \quad (2.3.8)$$

where $f(\cdot \mid \cdot)$ denotes the point probability mass function of $\mathbf{Y}$, $\ln(0/0) := 0$, and $\ln(\lambda_iT_i/0) := \infty$ for $\lambda_i > 0$. If the parameter space for $\boldsymbol{\lambda}$ is $X_{i=1}^{t}(0, \infty)$ then only estimators which are positive in *all* components can have finite risk under $L_E(\cdot, \cdot)$. Ghosh and Yang (1988) discuss estimation with respect to entropy loss under (2.3.1).

The risk of the MLE $\hat{\boldsymbol{\lambda}} = (Y_1/T_1, \ldots, Y_t/T_t)'$ with respect to $L_{\mathbf{d}}(\cdot, \cdot)$ is easily calculated to be

$$R_{\mathbf{d}}(\boldsymbol{\lambda}, \hat{\boldsymbol{\lambda}}) = \sum_{i=1}^{t} T_i^{-2}\lambda_i^{-d_i+1}. \quad (2.3.9)$$

In particular under (2.3.1), the risk under squared error loss is

$$R_S(\boldsymbol{\lambda}, \hat{\boldsymbol{\lambda}}) = E\left[\sum_{i=1}^{t}(\lambda_i - Y_i)^2\right] = \sum_{i=1}^{t}\lambda_i,$$

and the risk for relative squared error loss is

$$R_R(\boldsymbol{\lambda}, \hat{\boldsymbol{\lambda}}) = t. \quad (2.3.10)$$

Thus under (2.3.1) the MLE has low MSE when the true λ_i's are all small but has constant risk under $L_R(\cdot, \cdot)$ (and is minimax, Berger (1985), p. 360, Section 5.4.4). In general when $Y_i \sim P(\lambda_iT_i)$, $\hat{\boldsymbol{\lambda}} = (Y_1/T_1, \ldots, Y_t/T_t)'$ has small k-NSEL risk

$$R_k(\boldsymbol{\lambda}, \hat{\boldsymbol{\lambda}}) = \sum_{i=1}^{t} T_i^{-2}\lambda_i^{-k+1}$$

when the true λ_i's are all large (small) for $k > 1$ ($k < 1$).

The most important difference between the problems of simultaneously estimating a vector of multinomial cell probabilities and estimating a vector of Poisson rates is that the latter exhibits a Stein effect. Recall from Section 2.2 that the Stein effect is the inadmissibility of a vector estimator formed from admissible components. Thus combining independent Poisson problems and accepting some bias in exchange for reduced variability leads to estimators with uniformly (in $\boldsymbol{\lambda}$) lower risk than $\hat{\boldsymbol{\lambda}}$ despite the fact that the individual $\hat{\lambda}_i$ are admissible for estimating λ_i. The exact conditions under which inadmissibility occurs depend on the loss function and the dimension of $\hat{\boldsymbol{\lambda}}$. The impact of this result is that theoretical research on alternative estimators to the MLE has focused on the development of estimators which *dominate* $\hat{\boldsymbol{\lambda}}$ and to a lesser extent on the development of Bayes (and related) estimators. In particular, smoothing techniques have received little attention in the Poisson case. For convenience, the discussion below first considers SEL, and then RSEL and other loss functions. Of course, general results for k-NSEL and $\mathbf{d}$-NSEL apply to SEL and RSEL, and, where useful in the discussions of the latter, specific results will be mentioned.

Alternative Estimators Under Squared Error Loss

Consider first alternative estimators to $\hat{\boldsymbol{\lambda}}$ suggested by Bayesian considerations. The mathematically simplest of these is the Bayes estimator with respect to the (conjugate) gamma prior. Specifically assume that $\lambda_1, \ldots, \lambda_t$ have independent prior distributions with $\lambda_i \sim \Gamma(\alpha_i, \beta_i)$, $\alpha_i > 0$, $\beta_i > 0$, $1 \leq i \leq t$. Thus $\boldsymbol{\lambda}$ has joint prior density

$$g(\boldsymbol{\lambda}) = \prod_{i=1}^{t} \frac{\exp(-\lambda_i/\beta_i)\lambda_i^{\alpha_i - 1}}{\beta_i^{\alpha_i}\Gamma(\alpha_i)}$$

for $\boldsymbol{\lambda} \in \times_{i=1}^{t}(0, \infty)$. The prior mean and variance of λ_i are $\mu_i = E[\lambda_i] = \alpha_i\beta_i$ and $\mathrm{Var}(\lambda_i) = \alpha_i\beta_i^2 = \beta_i\mu_i$, respectively. Thus this two parameter family allows the independent prior specification of a point estimate of λ_i via μ_i and a prior degree of belief in μ_i via β_i. The posterior distribution of λ_i given $Y_i = y_i$ is $\Gamma(\alpha_i + y_i, \beta_i/(T_i + \beta_i))$. Hence the Bayes estimator of $\boldsymbol{\lambda}$ with respect to SEL is

$$\hat{\lambda}_i^B = E[\lambda_i \mid \mathbf{Y}] = \left(\frac{1}{1 + T_i\beta_i}\right)\mu_i + \left(\frac{T_i\beta_i}{1 + T_i\beta_i}\right)\hat{\lambda}_i. \tag{2.3.11}$$

The form of (2.3.11) shows that $\hat{\boldsymbol{\lambda}}^B = (\hat{\lambda}_1^B, \ldots, \hat{\lambda}_t^B)'$ pulls the MLE toward the prior mean $\boldsymbol{\mu} = (\mu_1, \ldots, \mu_t)'$. For fixed mean μ_i, the degree of translation is greater for smaller β_i (i.e., for smaller prior variance). In the limit,

$\hat{\lambda}_i^B \to \mu_i$ as $\beta_i \to 0$. The MSE of $\hat{\boldsymbol{\lambda}}^B$ is

$$R_S(\boldsymbol{\lambda}, \hat{\boldsymbol{\lambda}}^B) = \sum_{i=1}^{t} E_{\boldsymbol{\lambda}} \left\{ \left(\frac{T_i \beta_i}{1 + T_i \beta_i} \right) [\hat{\lambda}_i - \lambda_i] + \left(\frac{1}{1 + T_i \beta_i} \right) [\mu_i - \lambda_i] \right\}^2$$

$$= \sum_{i=1}^{t} \left\{ \left(\frac{T_i \beta_i}{1 + T_i \beta_i} \right)^2 \frac{\lambda_i}{T_i} + \left(\frac{1}{1 + T_i \beta_i} \right)^2 (\mu_i - \lambda_i)^2 \right\}. \tag{2.3.12}$$

One implication of equation (2.3.12) is that $\hat{\boldsymbol{\lambda}}^B$ *does not dominate* $\hat{\boldsymbol{\lambda}}$ even if t is large; $\hat{\boldsymbol{\lambda}}^B$ has smaller risk for $\boldsymbol{\lambda}$ near $\boldsymbol{\mu}$ but larger risk for $\boldsymbol{\lambda}$ far from $\boldsymbol{\mu}$, particularly for $\boldsymbol{\lambda}$ near the origin. This emphasizes the difference in attitude of Bayesian estimation and the frequentist school. For the Bayesian, assuming the selected gamma prior expresses one's prior belief, then $\hat{\boldsymbol{\lambda}}^B$ is of interest in its own right independent of its possible domination of $\hat{\boldsymbol{\lambda}}$. However, if one is a frequentist primarily interested in using Bayes estimators to search for dominating estimators, then the class of Bayes estimators with respect to gamma priors is not of interest.

Albert (1981b) proposed a class of estimators similar in spirit, but not philosophy, to the pseudo-Bayes estimators of Sections 2.1 and 2.2. Albert's starting premise is that important practical gains in risk are only possible by using prior information. Thus his estimator requires the specification of both prior parameters μ_i and β_i. He considers the class $\hat{\boldsymbol{\lambda}}^c$, $c \in \mathbb{R}$, defined by

$$\hat{\lambda}_i^c = \left(\frac{c}{1 + T_i \beta_i} \right) \mu_i + \left(1 - \frac{c}{1 + T_i \beta_i} \right) \hat{\lambda}_i$$

which is a generalization of the Bayes estimator (2.3.11) for which $c = 1$. Albert calculates the $c = c(\boldsymbol{\lambda})$ which minimizes the risk $R_S(\boldsymbol{\lambda}, \hat{\boldsymbol{\lambda}}^c)$ as

$$c = \frac{\sum_{i=1}^{t} \lambda_i T_i / (1 + \beta_i)}{\sum_{i=1}^{t} T_i \lambda_i / (1 + T_i \beta_i)^2 + \sum_{i=1}^{t} (T_i (\lambda_i - \mu_i) / (1 + T_i \beta_i))^2} \tag{2.3.13}$$

(Problem 2.21). Substituting $\hat{\lambda}_i$ for λ_i in (2.3.13), Albert proposes using the data-based value $c^A = \min\{c(\hat{\boldsymbol{\lambda}}), 1\}$ where the minimum prevents greater shrinkage toward $\boldsymbol{\mu}$ than $\hat{\boldsymbol{\lambda}}^B$. Albert shows $\hat{\boldsymbol{\lambda}}^A := (\hat{\lambda}_1^A, \ldots, \hat{\lambda}_t^A)'$ with

$$\hat{\lambda}_i^A = \left(\frac{c^A}{1 + T_i \beta_i} \right) \mu_i + \left(1 - \frac{c^A}{1 + T_i \beta_i} \right) \hat{\lambda}_i$$

has similar risk improvements as $\hat{\boldsymbol{\lambda}}^B$ for $\boldsymbol{\lambda}$ "near" $\boldsymbol{\mu}$ while its risk is not much larger than $\hat{\boldsymbol{\lambda}}$ when $\boldsymbol{\lambda}$ is "far" from $\boldsymbol{\mu}$.

Two other approaches which rely on prior information will be discussed: empirical Bayes and hierarchical Bayes estimation. However papers adopting these approaches differ from their multinomial counterparts in that they deemphasize the loss function and emphasize the calculation of the posterior distribution (or summaries of it such as the posterior mean, mode, or a $100(1-\alpha)\%$ credible region).

Maritz (1969) studies empirical Bayes estimation based on the following model. Given $\boldsymbol{\lambda}$ the data $Y_1, \ldots, Y_t$ are mutually independent with $Y_i \sim P(\lambda_i)$, $1 \leq i \leq t$; $\lambda_1, \ldots, \lambda_t$ are iid with unknown prior $G(\cdot)$. Maritz develops estimators of the posterior mean $E[\lambda_t \mid Y_t = y_t]$ for the cases of gamma $G(\cdot)$ and completely nonparametric $G(\cdot)$. (The related problem of estimating the prior $G(\cdot)$ in the nonparametric case has been studied by Simar (1976) and Laird (1978b).) Gavor and O'Muircheartaigh (1987) consider the general case in which given $\boldsymbol{\lambda}$, $Y_1, \ldots, Y_t$ are independent with $Y_i \sim P(T_i\lambda_i)$, $1 \leq i \leq t$ while the prior $G(\cdot)$ has either a gamma or log-Student's t-distribution with *unknown* location and scale parameters and *known* degrees of freedom. Another interesting application of the empirical Bayes model with $Y_i \sim P(T_i\lambda_i)$ and gamma $G(\cdot)$ is Hoadley (1981) who used it to devise a quality management plan.

Hierarchical Bayes estimation is an alternative to empirical Bayes estimation when the prior is not completely known. In a series of applied papers, Kaplan and his coworkers use the hierarchical Bayes approach to assess failure data in the nuclear industry. For example, Kaplan (1983) uses the model: (i) given $\boldsymbol{\lambda}$, $Y_1, \ldots, Y_t$ are mutually independent with $Y_i \sim P(T_i\lambda_i)$, $1 \leq i \leq t$, (ii) given (μ, σ), the first-stage prior assumes $\ln(\lambda_1), \ldots, \ln(\lambda_t)$ iid $N(\mu, \sigma^2)$, and (iii) (μ, σ) are given an (improper) uniform second-stage prior. An important contribution of this paper is the development of numerical techniques to calculate point and interval estimates of the relevant Bayesian quantities. A second example of hierarchical Bayes methodology is the work of Ghosh and Parsian (1981) and Ghosh (1983). Among other results they derive closed form hierarchical Bayes estimators with respect to SEL (as well as the more general k-NSEL and $\mathbf{d}$-NSEL) for the model: (i) given $\boldsymbol{\lambda}$, $Y_1, \ldots, Y_t$ are mutually independent with $Y_i \sim P(\lambda_i)$, $1 \leq i \leq t$, (ii) given $(\boldsymbol{\alpha}, \beta)$, $\lambda_1, \ldots, \lambda_t$ are iid $\Gamma(\alpha_i, \beta)$, and (iii) given α^*, $\beta^* > 0$, $(1+\beta)^{-1} \sim \text{Be}(\alpha^*, \beta^*)$ is the second stage prior. Two important features of their estimators are that they are *both* automatically *admissible* (they are unique proper Bayes) and they *dominate* $\hat{\boldsymbol{\lambda}}$ (in the k-NSEL case).

As noted above, the motivation for developing several of the hierarchical Bayes estimators in the previous paragraph was research on estimators dominating the MLE. Historically, the first author to specifically focus attention on the domination problem was Peng (1975). For the case $T_1 = \ldots = T_t = 1$, he prove that $\hat{\boldsymbol{\lambda}}$ is admissible for $t = 1$ or 2 dimensions and is inadmissible for 3 or more dimensions. To show inadmissibility, he demonstrated that the estimator with components

$$\hat{\lambda}_i^P = Y_i - \frac{(t - N_0 - 2)^+ h(Y_i)}{\sum_{i=1}^t h^2(Y_i)} \tag{2.3.14}$$

uniformly dominated $\hat{\boldsymbol{\lambda}}$ where $N_0 :=$ (the number of i with $Y_i = 0$), $h(0) = 0$, $h(u) = \sum_{j=1}^{u}(1/j)$ for positive integers u, and $z^+ = \max(0, z)$. Estimator (2.3.14) has an interesting intuitive explanation. The correction term to $\hat{\lambda}_i = Y_i$ is based on the transformed variable $h(Y_i)$ which is close to $\ln(Y_i)$ when Y_i is large and close to $(Y_i)^{1/2}$ when Y_i is small. Thus $h(Y_i)$ has roughly a normal distribution and the correction is that of the James–Stein estimator for Gaussian data. The analytic method used to derive $\hat{\boldsymbol{\lambda}}^p$ was to express the difference between the risks of the estimators $\hat{\boldsymbol{\lambda}} + Q(\mathbf{Y})$ and $\hat{\boldsymbol{\lambda}}$ as $R_S(\boldsymbol{\lambda}, \hat{\boldsymbol{\lambda}}) - R_S(\boldsymbol{\lambda}, \hat{\boldsymbol{\lambda}} + Q(\mathbf{Y})) = E_{\boldsymbol{\lambda}}[\mathcal{D}(\mathbf{Y}, Q)]$ by applying integration-by-parts. Then the (difference) inequality $\mathcal{D}(\mathbf{Y}, Q) < 0$ is solved for $Q(\cdot)$ yielding an estimator which dominates $\hat{\boldsymbol{\lambda}}$ (Stein (1973)). Note that if $N_0 < t - 2$ and $Y_i > 0$ then $\hat{\lambda}_i^P$ translates $\hat{\lambda}_i = Y_i$ toward the origin. Hwang (1982) extended this result by providing a large *class* of estimators which dominate $\hat{\boldsymbol{\lambda}}$ under $L_S(\cdot, \cdot)$ and shrink towards the origin.

The dominating estimators discussed thus far show significant risk decreases over the MLE only for $\boldsymbol{\lambda}$ near the origin. In cases where it is not reasonable to suppose that the true $\boldsymbol{\lambda}$ is near the origin, they offer only minimal risk improvement. Tsui (1981), Hudson and Tsui (1981) and Ghosh, Hwang and Tsui (1983) have developed estimators which either shrink towards an arbitrary non-negative integer or to certain data-defined points (the minimum of the Y_i's or the jth smallest Y_i, for example). These estimators are also derived from algebraic expressions for risk reduction using Stein's (1973) method and are similar in form to (2.3.14). They uniformly dominate $\hat{\boldsymbol{\lambda}}$ for t large ($t \geq 3, 4$ or 5, depending on the particular estimator). Simulation studies comparing their risks to the MLE and to Peng's estimator show that considerable improvement is possible over a large set of $\boldsymbol{\lambda}$ (up to a 40% reduction in relative risk, $100\% \times [R_S(\boldsymbol{\lambda}, \hat{\boldsymbol{\lambda}}) - R_S(\boldsymbol{\lambda}, \hat{\boldsymbol{\lambda}}^*)]/R_S(\boldsymbol{\lambda}, \hat{\boldsymbol{\lambda}})$, for various competitors $\boldsymbol{\lambda}^*$ compared to the MLE). The improvement is particularly impressive when shrinkage is towards a data-determined point.

Tsui (1979) and Hwang (1982) explicitly discuss estimation under $L_S(\cdot, \cdot)$ for the general case $Y_i \sim P(T_i\lambda_i)$, $1 \leq i \leq t$. Roughly, their results give estimators which dominate the MLE as long as no one T_i greatly exceeds the others.

Finally, there is work by Brown and Farrell (1985a, 1985b) when $Y_i \sim P(\lambda_i)$, $1 \leq i \leq t$, which derives *linear* (in $\mathbf{Y}$) estimators of $\boldsymbol{\lambda}$. The simple Bayes estimate (2.3.11) and the MLE are linear; all other estimators described above are not. Brown and Farrell provide some motivation for considering linear estimators and characterize the class of admissible linear estimators under SEL.

Example 2.3.1 (Albert, 1981b). The following estimation problem has a known answer and thus allows the computation of actual loss values for various estimators. Consider simultaneous estimation of the mean number of fires in New York City during May 1920 for the 7 days of the week

Sunday through Saturday which are denoted $\lambda_1, \ldots, \lambda_7$, respectively. The available data are the average number of fires for the months of March and April 1920 for each of the 7 days. Assume that the monthly distribution of fires over days of the week is constant during this three month period so that the March and April data provide information from which to estimate fires in May. Consider the model which postulates that the *total* number Y_i of fires for the ith day, $1 \leq i \leq 7$, during April 1920 satisfies $Y_i \sim P(T_i \lambda_i)$ where T_i is the number of such days during the month (typically $T_i = 4$). The MLE of the vector of rates is $\hat{\boldsymbol{\lambda}} = (Y_1/T_1, \ldots, Y_7/T_7)'$. To use either the Bayes or Albert estimates, the prior parameters $\{\mu_i\}$ and $\{\beta_i\}$ must be known. Suppose these quantities are based on the data for March 1920 in the following way. For $1 \leq i \leq 7$ and $1 \leq j \leq n_i$ let X_{ij} denote the number of fires on the jth day of type i during March 1920. Then marginally $E[X_{ij}] = E(E[X_{ij} \mid \lambda_i]) = E[\lambda_i] = \mu_i$ and similarly $\text{Var}(X_{ij}) = \mu_i(\beta_i + 1)$. Therefore $\hat{\mu}_i = \overline{X}_i = n_i^{-1} \sum_{j=1}^{n_i} X_{ij}$ and $\hat{\beta}_i = \max\{0, V_i^2/\overline{X}_i - 1\}$ for $V_i^2 = n_i^{-1} \sum_{j=1}^{n_i} (X_{ij} - \overline{X}_i)^2$ are moment estimators of the required prior and variance parameters. Table 2.3.1 lists $\hat{\boldsymbol{\mu}}$ and $\hat{\boldsymbol{\beta}}$ based on the March data, as well as the estimates $\hat{\boldsymbol{\lambda}}$, $\hat{\boldsymbol{\lambda}}^B$, and $\hat{\boldsymbol{\lambda}}^A$. Note that $\hat{\boldsymbol{\lambda}}$ is the vector of observed mean numbers of fires during April 1920. In addition, the actual mean number of fires for each day during May 1920 is given in Column 8 and denoted $\boldsymbol{\lambda}$. The summed square errors for the three estimators are:

$$\sum_{i=1}^{7} (\hat{\lambda}_i - \lambda_i)^2 = 137.73,$$

$$\sum_{i=1}^{7} (\hat{\lambda}_i^A - \lambda_i)^2 = 108.04, \quad \text{and}$$

$$\sum_{i=1}^{7} (\hat{\lambda}_i^B - \lambda_i)^2 = 59.10$$

so that $\hat{\boldsymbol{\lambda}}^A$ gives a 22% reduction and $\hat{\boldsymbol{\lambda}}^B$ a 57% reduction in summed squared error over $\hat{\boldsymbol{\lambda}}$.

Table 2.3.1. Three Estimators of the Mean Numbers of Fires in New York City during May 1920 for Each of the 7 Days of the Week, Sunday through Saturday (Reprinted with permission from James H. Albert: "Simultaneous Estimation of Poisson Means," *Journal of Multivariate Analysis,* vol. 11, Academic Press, 1981.)

i	Day	$\hat{\mu}_i$	$\hat{\beta}_i$	$\hat{\lambda}_i$	$\hat{\lambda}_i^B$	$\hat{\lambda}_i^A$	λ_i
1	Sunday	20.25	.81	15.75	16.64	15.89	17.00
2	Monday	17.80	0	16.75	17.80	16.89	17.60
3	Tuesday	17.20	0	17.50	17.20	17.46	13.50
4	Wednesday	23.40	0	13.25	23.40	14.61	19.50
5	Thursday	19.75	1.97	19.40	19.44	19.40	15.00
6	Friday	19.25	0	20.80	19.25	20.59	16.25
7	Saturday	14.25	0	21.75	14.25	20.74	15.40

Alternative Estimators Under Other Loss Functions

The following paragraphs primarily develop alternative estimators which dominate $\hat{\boldsymbol{\lambda}} = (Y_1, \ldots, Y_t)'$ with respect to $L_R(\cdot,\cdot)$ under (2.3.1). Recall that for $L_R(\cdot,\cdot)$, estimators which dominate $\hat{\boldsymbol{\lambda}}$ must be minimax since $\hat{\boldsymbol{\lambda}}$ is minimax. Clevenson and Zidek (1975) were the first to propose a class of dominating estimators when $t \geq 2$ ($\hat{\lambda}_1 = Y_1$ is admissible when $t = 1$). They study

$$\hat{\boldsymbol{\lambda}}^{CZ} = \left(1 - \frac{\varphi(\Sigma Y_i)}{\Sigma Y_i + t - 1}\right)^+ \mathbf{Y} \qquad (2.3.15)$$

where $[x]^+ = \max\{0, x\}$ and $\varphi(\cdot)$ is nondecreasing, not identically zero, and bounded by $0 \leq \varphi(w) \leq 2(t-1)$. Clevenson and Zidek's estimators translate $\hat{\boldsymbol{\lambda}}$ toward the origin; they can also be shown to have uniformly lower risk than the MLE when $t \geq 2$ under $L_k(\cdot,\cdot)$ for all $k \geq 2$. Although $\hat{\boldsymbol{\lambda}}^{CZ}$ dominates $\hat{\boldsymbol{\lambda}}$, Hwang (1982) showed that if the function $\varphi(\Sigma Y_i)$ equals a constant φ^*, then $\hat{\boldsymbol{\lambda}}^{CZ}$ is inadmissible for all $\varphi^* < t - 1$. Whether $\hat{\boldsymbol{\lambda}}^{CZ}$ is admissible for non-constant $\varphi(\cdot)$ functions is an open question.

Extensions of Clevenson–Zidek type estimators have been investigated by Tsui and Press (1982), Hwang (1982) and Ghosh, Hwang and Tsui (1983). Tsui and Press (1982) and Hwang (1982) describe classes of minimax estimators which translate toward the origin, dominate the MLE when $t \geq 2$, and include $\hat{\boldsymbol{\lambda}}^{CZ}$. In addition, both papers generalize the results to provide dominating estimators under $L_k(\cdot,\cdot)$, $k \geq 2$. Tsui and Press also discuss the general model $Y_i \sim P(T_i \lambda_i)$, $i = 1(1)t$.

Ghosh, Hwang, and Tsui (1983) develop dominating estimators which shrink $\hat{\boldsymbol{\lambda}} = (Y_1, \ldots, Y_t)'$ towards a *fixed nonnegative point* or certain *data-defined points.* They also derive dominating estimators for $L_k(\cdot,\cdot)$, $k \geq 2$, and $L_{\mathrm{d}}(\cdot,\cdot)$ as well as for general discrete exponential distributions. Particular emphasis is placed on the Poisson and negative binomial classes.

Bayes estimators and their relatives have been studied for $L_R(\cdot,\cdot)$ as noted in the previous subsection. The following interesting example shows that the Clevenson–Zidek estimator is empirical Bayes. Suppose $Y_1, \ldots, Y_t$ are mutually independent Poisson random variables with means $\lambda_1, \ldots, \lambda_t$, respectively. In addition, suppose $\lambda_1, \ldots, \lambda_t$ are iid with exponential prior

$$\prod_{i=1}^{t} \mu^{-1} \exp\{-\lambda_i/\mu\}$$

so that each component has common mean μ. The posterior distribution of λ_i given $Y_i = y_i$ is $\Gamma(y_i + 1, \mu/(\mu+1))$. The Bayes estimator with respect to $L_R(\cdot,\cdot)$ at $\mathbf{y}$ is that action $\mathbf{a}$ minimizing

$$E\left[\sum_{i=1}^{t} \lambda_i^{-1}(a_i - \lambda_i)^2 \,|\, Y_i = y_i\right]$$

$$= \sum_{i=1}^{t} \int_0^\infty (a_i - \lambda_i)^2 \lambda_i^{y_i - 1} \exp[-\lambda_i(1+\mu)/\mu] d\lambda_i. \quad (2.3.16)$$

The form of (2.3.16) shows that a_i is the mean of the $\Gamma(y_i, \mu/(1+\mu))$ distribution which is

$$\left(\frac{\mu}{1+\mu}\right) Y_i, \quad 1 \le i \le t. \quad (2.3.17)$$

Intuitively, the closer the prior mean μ is to zero, the more (2.3.17) pulls $\mathbf{Y}$ toward zero. The derivation above obviously extends to any $\Gamma(\alpha, \beta)$ prior with $\alpha > 1$. Now suppose the prior parameter μ in (2.3.17) is unknown. Empirical Bayes estimation is based on the marginal distribution of Y_i under the Bayes model with exponential prior which is

$$P[Y_i = y] = \int_0^\infty \frac{e^{-\lambda}\lambda^y}{y!} \lambda^y \mu^{-1} e^{-\lambda/\mu} d\lambda$$

$$= \left(\frac{1}{1+\mu}\right)\left(\frac{\mu}{1+\mu}\right)^y \quad (2.3.18)$$

for $y \in Z$. Thus Y_i has a geometric distribution with "success" probability $\mu/(1+\mu)$; in particular $E[Y_i] = \mu$. Consider estimation of $\mu/(1+\mu)$ based on $Y_1, \ldots, Y_t$ iid with distribution (2.3.18). Substituting the UMVUE of $\mu/(1+\mu)$, $(t-1)/(S+t-1)$ with $S := \sum_{i=1}^{t} Y_i$, into (2.3.17) leads to

$$\lambda_i^E = \left(1 - \frac{S}{S+t-1}\right) Y_i$$

which is precisely $\hat{\boldsymbol{\lambda}}^{CZ}$ for $\varphi(s) = s$. (See Tamura and Young, 1987 for other methods of estimating geometric distribution parameters.)

An alternate approach in the case of unknown prior parameters is the hierarchical Bayes method. Ghosh and Parsian (1981) use a hierarchical two-stage prior to develop a class of proper Bayes (hence admissible) minimax estimators which dominate the MLE for $t \geq 3$. In the first stage, $W = w$ is observed with density $g(\cdot)$ having support in $(0, \infty)$ and satisfying mild conditions while in the second stage, $\lambda_1, \ldots, \lambda_t$ are iid exponential with parameter w. For example, if $g(\cdot)$ is the beta density with parameters $\alpha > 0$ and $0 < \beta \leq t-2$, then the Bayes estimator is

$$\hat{\lambda}_i^{GP} = \left(\frac{\sum_{i=1}^t Y_i + \alpha - 1}{\sum_{i=1}^t Y_i + \alpha + \beta + t - 1} \right) Y_i. \tag{2.3.19}$$

Like $\hat{\boldsymbol{\lambda}}^{CZ}$, $\hat{\boldsymbol{\lambda}}^{GP}$ shows substantial risk improvement when the true $\boldsymbol{\lambda}$ is near the origin.

Recently, Tsui (1984, 1986) has explored the distributional robustness of Clevenson–Zidek type estimators. Specifically, he has shown that $\hat{\boldsymbol{\lambda}}^{CZ}$ continues to dominate the MLE when the true distribution of $\mathbf{Y}$ falls in a class which includes all mixtures of Poisson distributions. His results further support the "safety" of such estimators: not only is their risk everywhere less than that of $\hat{\boldsymbol{\lambda}}$, but this risk domination holds even if the underlying distribution is not precisely Poisson. Note also that Brown and Farrell's (1985a, 1985b) investigation of the admissibility of linear estimators for Poisson means considers $L_R(\cdot, \cdot)$ as well as $L_S(\cdot, \cdot)$.

Example 2.3.2 (Clevenson and Zidek, 1975). As in Example 2.3.1, this example compares several estimators for a case where the actual losses can be computed. Consider the 36 half-year periods between 1953 and 1970. Suppose it is desired to estimate the number of oil well discoveries in Alberta, Canada obtained by wildcat exploration during the third months of each of these half-year periods; i.e., during March and September. The data used are the average monthly number of discoveries for the remaining five months of the half-year period. Table 2.3.2 lists the data $\mathbf{Y} = \hat{\boldsymbol{\lambda}}$, the true $\boldsymbol{\lambda}$, and $\hat{\boldsymbol{\lambda}}^{CZ}$. The actual relative squared errors for $\hat{\boldsymbol{\lambda}}$ and $\hat{\boldsymbol{\lambda}}^{CZ}$ are

$$\sum_{i=1}^{36} (\lambda_i - \hat{\lambda}_i)^2 / \lambda_i = 39.12 \quad \text{and}$$

$$\sum_{i=1}^{36} (\lambda_i - \hat{\lambda}_i^{CZ}) / \lambda_i = 14.26$$

the latter of which is a 72% reduction in the RSEL of $\hat{\boldsymbol{\lambda}}$. As in Example 2.3.1, this calculation suggests the practical importance of alternative estimators to $\hat{\boldsymbol{\lambda}}$.

Table 2.3.2. Comparison of the MLE and Clevenson and Zidek Estimator of the Number of Oilwell Discoveries in Alberta, Canada for 36 Months during the Period 1953–1970 (Reprinted with permission from M.L. Clevenson and J.V. Zidek: "Simultaneous Estimation of the Means of Independent Poisson Laws," *Journal of the American Statistical Association,* vol. 70, 1975.)

i	$Y_i = \hat{\lambda}_i$	$\hat{\lambda}_i^{CZ}$	λ_i
1	0	0	1.17
2	0	0	.83
3	0	0	.50
4	1	.45	1.00
5	2	.89	.83
6	1	.45	.83
7	0	0	1.17
8	2	.89	.83
9	0	0	.67
10	0	0	.17
11	0	0	.00
12	1	.45	.33
13	3	1.34	1.50
14	0	0	.50
15	0	0	1.17
16	3	1.34	1.33
17	0	0	.50
18	2	.89	1.17
19	1	.45	.50
20	2	.89	.50
21	0	0	1.33
22	0	0	.83
23	0	0	.33
24	1	.45	1.50
25	5	2.23	1.33
26	0	0	.67
27	1	.45	.67
28	0	0	.33
29	0	0	.33
30	1	.45	.33
31	0	0	.50
32	1	.45	.83
33	0	0	.67
34	1	.45	.33
35	0	0	.00
36	1	.45	.50

B. Hypothesis Tests

Both graphical and analytic procedures will be introduced for several testing problems. First the problem of testing fit to the Poisson assumption itself will be considered, and then testing homogeneity of $\lambda_1, \ldots, \lambda_t$ will be discussed. Lastly, analogs of some of the multinomial tests for isotonic alternatives will be given.

Testing Fit to the Poisson Model

Consider the single sample problem of testing goodness-of-fit to the Poisson model based on a random sample $Y_1, \ldots, Y_t$ of count data. Hoaglin (1980) proposes a graphical technique of assessing whether the Y_i are iid $P(\lambda)$ for some $\lambda > 0$. For each count k observed in the data set, plot

$$(k, \ln(k!) + \ln(F_k)) \tag{2.3.20}$$

where $F_k := \sum_{j=1}^{t} [Y_j = k]$ is the number of data values Y_j equal to k. The motivation for (2.3.20) is that

$$E[F_k] = E_\lambda \left(\sum_{j=1}^{t} [Y_j = k] \right) = te^{-\lambda} \lambda^k / k!$$

so that taking the natural logarithm of both sides suggests $\ln(F_k)$ approximately satisfies

$$\ln(F_k) \simeq k \ln(\lambda) - \lambda + \ln(t) - \ln(k!).$$

Only the data set must be large enough that a sufficient number of distinct counts be observed.

If the fit to the Poisson model is adequate, then the points should lie on a straight line with slope approximately $\ln(\lambda)$ and intercept approximately $(\ln(t) - \lambda)$. In this case a crude estimate of λ can be read from the plot. Non-linearities in the plot indicate deviations from the Poisson model. For example, if a few scattered points deviate then they might be checked for errors; if the point corresponding to $k = 0$ is unusually large then an added-zeros model may be appropriate (Problem 2.25); if many of the values $\ln(k!) + \ln(F_k)$ corresponding to large k are unusually small then there is evidence that the observed counts come from a distribution with a lighter (right) tail than the Poisson. Problem 2.23 illustrates this plot for data on radioactive decay. Ord (1967) and Gart (1970) suggest alternative graphical methods to assess fit to the Poisson model.

One analytic test for the Poisson distribution is the chi-square test for multinomial data applied as in Example 2.2.3. However this is an omnibus test and thus can lack power against interesting alternatives. Perhaps the oldest analytic test of fit, devised specifically for the Poisson model, is

Fisher's *dispersion index* test which rejects the null hypothesis of the Poisson model when

$$\sum_{i=1}^{t}(Y_i - \overline{Y})^2/\overline{Y} \geq \chi^2_{\alpha,t-1} \tag{2.3.21}$$

where $\overline{Y} := t^{-1}\sum_{j=1}^{t} Y_j$. The intuition behind (2.3.21) is that it rejects when the sample variance of the data is large compared to its sample mean thereby exploiting the relationship $\text{Var}[Y_i] \geq E[Y_i]$, which holds for Poisson mixtures (see Section 1.3). The dispersion test statistic is the score statistic for testing the Poisson null model against the negative binomial alternative (1.3.7) (Problem 2.29). Moran (1970) showed that the dispersion test is asymptotically equivalent to the likelihood ratio test for a wide class of mixed Poisson alternatives.

There are several generalizations of the dispersion test based on gamma mixture alternatives. In all cases the null hypothesis is that $Y_1, \ldots, Y_t$ are independent Poisson observations with mean vector satisfying one of the assumptions listed below. Collins and Margolin (1985) consider the cases (i) $E[\lambda_i] = T_i\lambda$, $1 \leq i \leq t$, where the T_i are known and $\lambda > 0$ is unknown, and (ii) the one-way layout in which the Y_i belong to one of k groups where all the observations in the kth group have a common mean. In the latter, multiple observations are required from each of the groups. Cameron and Trivedi (1986) and Lee (1986) allow general regression models $E[Y_i] = \lambda(\mathbf{x}_i, \boldsymbol{\beta})$.

Tests of Homogeneity

Suppose $Y_1, \ldots, Y_t$ are independent Poisson with $E[Y_i] = T_i\lambda_i$, $1 \leq i \leq t$. Consider first the problem of testing the ANOVA-like hypothesis of homogeneity of the rates, i.e.,

$$H_0\colon \lambda_1 = \ldots = \lambda_t \text{ versus } H_A\colon (\text{not } H_0). \tag{2.3.22}$$

The hypothesis H_0 is the analog of testing a simple null versus global alternative for the multinomial model because the conditional distribution of $\mathbf{Y}$ under H_0 given $S = \sum_{i=1}^{t} Y_i$ is $M_t(S, \mathbf{p})$ where $\mathbf{p} = (\ldots, T_i/\sum_{j=1}^{t} T_j, \ldots)'$. After considering an example, the likelihood ratio and score tests for this problem are derived.

Example 2.3.3. Table 2.3.3 from Lee (1963) lists all cases of acute lymphatic leukemia recorded by the British Cancer Registry from 1946 to 1960 classified according to month of clinical detection. The multinomial model is not appropriate since the total number of observed cases is not fixed ahead of time. Let Y_i be the number of cases reported in month i for $i = 1(1)12$ and assume that $Y_1, \ldots, Y_{12}$ are independent Poisson processes with Y_i having rate equal to (14 years $\times \lambda_i$ occurrences per year); i.e., $Y_i \sim P(14\lambda_i)$, $i = 1(1)12$. Consider the problem of assessing the evidence against the hypothesis that monthly incidence rates are identical.

Table 2.3.3. Number of Cases of Acute Lymphatic Leukemia Recorded by the British Cancer Registry from 1946 to 1960 (Reprinted with permission from J.A. Lee: "Seasonal Variation in Leukemia Incidence," *British Medical Journal,* vol. 2, pg. 623, British Medical Association.)

	J	F	M	A	M	J	J	A	S	O	N	D
Number of Cases	40	34	30	44	39	58	51	56	36	48	33	38

The likelihood based on $Y_i \sim P(T_i\lambda_i)$ is

$$L(\boldsymbol{\lambda}) = \exp\left(-\sum_{i}^{t} \lambda_i T_i\right) \prod_{i=1}^{t} \left(\frac{(\lambda_i T_i)^{y_i}}{y_i!}\right). \tag{2.3.23}$$

Under H_0, the MLE of the common $\lambda = \lambda_1 = \ldots = \lambda_t$ is

$$\hat{\lambda} = \sum_{j=1}^{t} y_j / \sum_{j=1}^{t} T_j = \overline{y}/\overline{T}$$

where $\overline{y} := t^{-1} \sum_{j=1}^{t} y_j$ and $\overline{T} = t^{-1} \sum_{j=1}^{t} T_j$. When $T_1 = \ldots = T_t = T$, $\hat{\lambda}$ reduces to $\overline{y}/T$. Thus the numerator of the LRT test statistic is

$$\exp(-t\overline{y})[\overline{T}]^{-t\overline{y}} \prod_{i=1}^{t} ((T_i\overline{y})^{y_i}/y_i!)$$

and similarly its denominator is $\exp(-t\overline{y}) \prod_{i=1}^{t} (y_i^{y_i}/y_i!)$ yielding the test statistic

$$\prod_{i=1}^{t} (T_i\overline{y}/\overline{T}y_i)^{y_i}.$$

When $T_1 = \ldots = T_t = T$, the likelihood ratio statistic reduces to $\prod_{i=1}^{t} (\overline{y}/y_i)^{y_i}$. In terms of the general description of the likelihood ratio test in Appendix 4, $\theta = (\lambda_1, \ldots, \lambda_t)'$, $\Omega = X_{i=1}^{t}(0, \infty)$ and $\omega = \{(\lambda, \ldots, \lambda) : \lambda \in (0, \infty)\}$. The likelihood ratio test rejects H_0 if and only if

$$G^2 := 2 \sum_{i=1}^{t} y_i \ln\left(\frac{\overline{T}y_i}{T_i\overline{y}}\right) \geq C. \tag{2.3.24}$$

The choice of critical constant will be considered after the construction of the score test.

Apart from constants the loglikelihood based on (2.3.23) is

$$\ln(L) = \sum_{i=1}^{t} y_i \ln(\lambda_i T_i) - \sum_{i=1}^{t} \lambda_i T_i.$$

It is straightforward to calculate that the score vector has ith element $S_i(\boldsymbol{\lambda}) = (y_i - T_i\lambda_i)/\lambda_i$ and the information matrix is $\mathbf{I}(\boldsymbol{\lambda}) = \text{Diag}(\lambda_1/T_1, \ldots, \lambda_t/T_t)$. Since $\hat{\boldsymbol{\theta}}_\omega = (\overline{y}/\overline{T}, \ldots, \overline{y}/\overline{T})'$, the score statistic is

$$X^2 = \{S(\hat{\boldsymbol{\theta}}_\omega\}'\mathbf{I}^{-1}(\hat{\boldsymbol{\theta}}_\omega)\{S(\hat{\boldsymbol{\theta}}_\omega)\}$$
$$= \sum_{i=1}^{t} \frac{[Y_i - T_i\overline{y}/\overline{T}]^2}{T_i\overline{y}/\overline{T}}.$$

When $T_1 = \ldots = T_t$, the X^2 statistic reduces to

$$\sum_{i=1}^{t} \frac{[y_i - S/t]^2}{S/t}$$

where $S := \sum_{i=1}^{t} y_i$. Also in this case, G^2 and X^2 are the *same* as the likelihood ratio statistic and score statistics, respectively, for testing $H_0^* : \mathbf{p} = \mathbf{c}$ versus $H_A^* : \mathbf{p} \neq \mathbf{c}$ based on $\mathbf{Y} \sim M_t(S, \mathbf{p})$. As will be shown in Section 3.3, this phenomenon holds more generally. The reason is that the null conditional distribution of $\mathbf{Y}$ given S is $M_t(S, \mathbf{c})$. Thus both the likelihood ratio and score statistics are unchanged if the analysis is conducted conditionally on S. Furthermore the asymptotic null critical points for both statistics are identical as the following proposition states.

Proposition 2.3.1. *Under H_0, both G^2 and $X^2 \xrightarrow{\mathcal{L}} \chi^2_{t-1}$ as* $\min\{T_i\} \to \infty$.

Therefore

$$G^2 \geq \chi^2_{\alpha,t-1} \quad \text{and} \quad X^2 \geq \chi^2_{\alpha,t-1}$$

are both asymptotic size α tests of H_0. Potthoff and Whittinghill (1966) discuss optimality properties of the score test and several competitors which they suggest. Gbur (1979) extends the one-way layout analysis to Poisson data which follow a two-way layout.

When homogeneity is rejected, it is often of interest to determine whether only one (or a few) of the Poisson processes exhibit heterogeneity. Again consider the case when $T_1 = \ldots = T_t = T$. A graphical assessment of the fit to the homogeneous rate model is obtained by plotting the residuals based on the fitted model in various ways. The *raw residuals* $e_i = Y_i - \overline{Y}$ are not standardized and can be used only for determining observations which exhibit large deviations relative to the other e_i. The *Pearson residuals,* defined by $e_i^P = e_i/(\overline{Y})^{1/2}$, are the components of the dispersion test of Poissonness. The quantity $(\overline{Y})^{1/2}$ is the MLE of the null standard error of Y_i and hence the normalization in e_i^P ignores the variability of $\overline{Y}$. The *adjusted residuals* $e_i^a = e_i/(((t-1)/t)\overline{Y})^{1/2}$ account for variability in both terms of $e_i = Y_i - \overline{Y}$; they normalize e_i by the MLE of the null standard error of e_i. This follows from

$$\begin{aligned}\mathrm{Var}_{H_0}(Y_i - \overline{Y}) &= \mathrm{Var}_{H_0}\left(\frac{(t-1)}{t}Y_i - \frac{1}{t}\sum_{j\neq i} Y_j\right)\\ &= \frac{(t-1)^2}{t^2}\lambda T + \frac{1}{t^2}(t-1)\lambda T\\ &= \frac{(t-1)}{t}\lambda T.\end{aligned}$$

Thus e_i^a is no harder to calculate than e_i^P and is to be preferred. Two useful plots of the residuals are index plots and probability plots. The index plot is $\{(i, e_i^a) : 1 \le i \le t\}$ which can be examined for trends or patterns. If $e_{(1)}^a \le \ldots \le e_{(t)}^a$ denotes the ordered adjusted residuals then lack of linearity in the probability plot $\{(\Phi^{-1}[(i-.5)/t], e_{(i)}^a) : 1 \le i \le t\}$ can be used to check for gross deviations from H_0; other abscissa arguments are also used in the literature. As with the Poissonness plot, different nonlinearities signal different sorts of deviations. For example, one or two points unusually far from the line might be outliers while curvature of the plot as a whole might suggest the rates vary cyclically or sinusoidally.

Example 2.3.3 (continued). Returning to the data on the incidence of acute lymphatic leukemia, a fit of the homogeneity model produces an expected count of $\overline{Y} = 42.17$ in each month. The score test of homogeneity is $X^2 = 21.34$ on 11 degrees of freedom with a P-value of $.03 = P[\chi_{11}^2 \ge 21.34]$ providing evidence of different monthly rates. The squares of the vector of Pearson residuals,

$$\mathbf{e}^P = (-.33, -1.26, -1.87, .28, -.49, 2.44, 1.36, 1.98, -.95, .90, -1.41, -.64)',$$

give the components of the score test statistic. Those corresponding to the summer months June, July, and August contribute 11.74 to the X^2 value. The probability plot of the ordered adjusted residuals in Figure 2.3.1 shows a pronounced lack of linearity as the freehand straight line through the left hand portion of the plot emphasizes. The three large positive residuals for the summer months are prominent in the upper right hand corner of the plot. There is also a trend for the other residuals to lie below the line; i.e., to be smaller than expected. This curvature occurs because the large observed values during the summer pull the fitted value $\overline{Y}$ up and produce negative residuals for most of the other months.

Suppose $Y_i \sim P(T_i\lambda_i)$, $1 \le i \le t$. If $K \subset \mathbb{R}^t$ is a cone and $\mathcal{O} := K \cap X_{i=1}^t(0,\infty)$, then the likelihood ratio test of H_0: $\lambda_1 = \ldots = \lambda_t$ versus H_A: $\boldsymbol{\lambda} \in \mathcal{O}$ is the analog of the isotonic alternative test considered in Section 2.2 for multinomial data. For example, H_0 versus H_A: $\lambda_1 \le \ldots \le \lambda_t$ tests for trend. Robertson, Wright, and Dykstra (1988) describe the calculation of the LRT for isotonic alternatives and derive the null asymptotic distribution of the tests as $\min\{T_i\} \to \infty$. The limiting distribution is the same as for the corresponding multinomial problem defined by K when the number of trials goes to infinity.

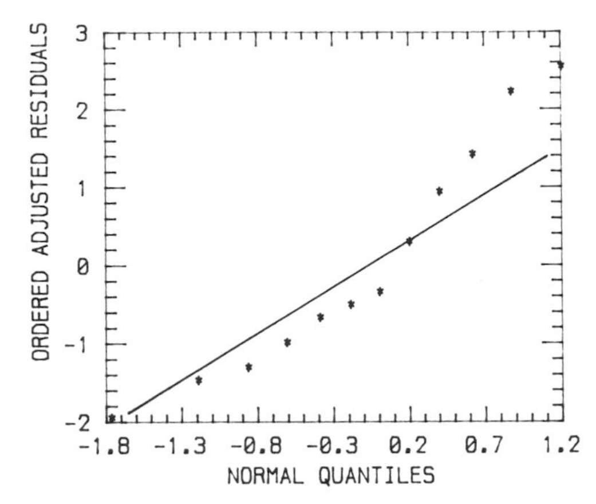

Figure 2.3.1. Probability plot of ordered adjusted residuals for British Cancer Registry data in Table 2.3.3 fitted by the homogeneity model (2.3.22).

C. Interval Estimation

Forming simultaneous confidence intervals for $\lambda_1, \dots \lambda_t$ based on independent observations $Y_1, \dots, Y_t$ with $Y_i \sim P(T_i\lambda_i)$ is fundamentally easier than forming simultaneous intervals for the vector of multinomial success probabilities. In the Poisson case it suffices to form individual confidence intervals for λ_i based on $Y_i \sim P(\lambda_i)$. This is immediate from

$$P_{\boldsymbol{\lambda}}[\underline{\lambda}_i \leq \lambda_i \leq \overline{\lambda}_i, i = 1(1)t] = \prod_{i=1}^{t} P_{\lambda_i}[\underline{\lambda}_i \leq \lambda_i \leq \overline{\lambda}_i] \tag{2.3.25}$$

and the fact that intervals for $T_i\lambda_i$ yield those for λ_i. If the individual intervals satisfy $P_{\lambda_i}[\underline{\lambda}_i \leq \lambda_i \leq \overline{\lambda}_i] \geq (1-\alpha)^{1/t}$ for all $\lambda_i > 0$, then (2.3.25) will be at least $1-\alpha$.

Suppose now that $Y \sim P(\lambda)$ where the subscript has been suppressed. Garwood (1936) derived the two-sided $100(1-\alpha)\%$ tail limits

$$(\underline{\lambda}, \overline{\lambda}) = (\underline{\lambda}(y), \overline{\lambda}(y)) = (\chi^2_{1-\alpha/2,2y}, .5\chi^2_{\alpha/2,2(y+1)}) \tag{2.3.26}$$

for λ where $\chi^2_{\alpha,\nu}$ is the upper α percentile of the chi-squared distribution with ν degrees of freedom for $\nu \geq 1$ and $\chi^2_{\alpha,0} = 0$. Formula (2.3.26) is useful for forming upper or lower confidence bounds for λ. However (2.3.26) yields excessively conservative two-sided confidence limits. Crow and Gardner (1959) proposed acceptance region based analogs of the Sterne–Crow binomial p intervals (Section 2.1). Casella (1987) generalizes his algorithm of Section 2.1 for constructing improved invariant p intervals to the Poisson case. Given a Poisson system of intervals, his algorithm generates a revised system of intervals which are at least as short as the input intervals and still achieve the target nominal level. In particular, his algorithm yields a strict improvement of the Crow and Gardner intervals.

For large λ, the continuity corrected values

$$(\underline{\lambda}, \overline{\lambda}) = \Big(\min\{\lambda : (y - \lambda - .5)^2/\lambda \leq \chi^2_{1,\alpha/2}\},$$

$$\max\{\lambda : (y - \lambda + .5)^2/\lambda \leq \chi^2_{1,\alpha/2}\}\Big)$$

are satisfactory approximate limits.

D. Selection and Ranking

Both the indifference zone and subset selection formulations have been applied to Poisson ranking and selection problems. For specificity assume the data come from t Poisson processes and that $Y_i \sim P(T\lambda_i)$ is the count from the ith process observed over $(0, T)$. Let

$$\lambda_{[1]} \leq \ldots \leq \lambda_{[t]}$$

denote the ordered hazard rates. Suppose it is desired to select the process with the largest hazard rate $\lambda_{[t]}$.

Consider the goal of selecting the *single* process associated with $\lambda_{[t]}$ subject to the probability requirement

$$P_{\boldsymbol{\lambda}}[CS] \geq 1 - \alpha \qquad (2.3.27)$$

whenever the true vector $\boldsymbol{\lambda}$ of hazard rates satisfies $\lambda_{[t]}/\lambda_{[t-1]} \geq \delta^*$. Here the constants $\delta^* > 1$ and $0 < \alpha < 1 - t^{-1}$ are specified prior to experimentation. The event CS means the process with hazard rate $\lambda_{[t]}$ is selected (a correct selection).

Alam (1971) proposed several selection procedures with inverse sampling rules that guarantee (2.3.27). One rule stops as soon as any process attains a given number N_0 of counts while another waits a specified length of time T_0. Bechhofer, Kiefer, and Sobel (1967) propose an alternative sequential procedure. For all the above stopping rules, the terminal decision rule is to select the process with the largest observed count as the one associated with $\lambda_{[t]}$.

An alternative goal is subset selection. The object is to determine a *subset* of the t processes so that the process with hazard rate $\lambda_{[t]}$ is contained in the selected subset. The probability requirement for subset selection procedures is

$$P_{\boldsymbol{\lambda}}[CS] \geq 1 - \alpha \qquad (2.3.28)$$

for all $\boldsymbol{\lambda}$ where α, $0 < \alpha < 1 - t^{-1}$ is specified. Note that the meaning of the event CS is different for (2.3.27) and (2.3.28) as is the set of $\boldsymbol{\lambda}$ configurations for which the requirement is to hold.

Gupta and Panchapakesan (1979) summarize a variety of sequential procedures which attain (2.3.28). The procedures differ with respect to their stopping rules, and terminal decision rule for choosing the processes included in the selected subset. For example, some of the procedures observe all the processes for a fixed time T_0 while others stop after a fixed number of events N_0 have occurred for at least one of the processes.

Problems

2.1. Suppose $Y \sim B(n,p)$ where n is known and $0 < p < 1$ is unknown. Verify that $\hat{p}^B = (Y + \sqrt{n}/2)/(n + \sqrt{n})$ has constant risk under SEL and is therefore minimax, since it is Bayes with respect to a positive prior.

2.2. Consider empirical Bayes estimation based on independent data $Y_i \sim B(1, p_i)$, $1 \le i \le m$, where the $\{p_i\}$ are iid Be(α, β). Derive the likelihood equations for the MLEs of α and β based on the marginal distribution of $\mathbf{Y} = (Y_1, \ldots, Y_m)'$ given (α, β).

2.3. Suppose $Y_1, Y_2, \ldots$ are iid $B(1,p)$. Observations Y_i are taken until a fixed number r of successes are obtained; let N denote the random number of trials. (This is called inverse sampling.) If p has the Be(α, β) distribution, show that given N and $(Y_1, \ldots, Y_N)$, the posterior distribution of p is Be$(\alpha + r,\ \beta + N - r)$. In particular, the mean of the posterior is $(\alpha + r)/(N + \alpha + \beta)$ which is the Bayes estimate of p under SEL. (The same Be$(\alpha + r,\ \beta + N - r)$ posterior arises when N is *fixed* and the total number r of successes is *random*. This example illustrates the general phenomenon that stopping rules are irrelevant for Bayesian analysis.)

2.4. Suppose $Y \sim B(n,p)$ where n is known and $0 < p < 1$ is unknown. Show that there do not exist unbiased estimators of $\omega = p/(1-p)$ or $\lambda = \ln(\omega)$.

2.5. Construct the Wald, score, and likelihood ratio tests of $H_0{:}\, p = p_0$ versus $H_A{:}\, p \neq p_0$ based on $Y \sim B(n,p)$ where n is known and $0 < p < 1$ is unknown.

2.6. Show that the tail intervals given by (2.1.13) are invariant under relabeling of the outcomes "success" and "failure" in the sense that

$$(\underline{p}(y), \overline{p}(y)) = (1 - \overline{p}(n-y), 1 - \underline{p}(n-y))$$

for $y \in \{0, \ldots, n\}$.

2.7. Construct 95% tail and Sterne confidence intervals for data $Y \sim B(2,p)$. Contrast the widths and coverages of the two systems.

2.8. Consider the 32 patients from Example 1.2.4 who received vitamin C therapy; 24 of these patients experienced clinical improvement. Let

$$p = P[\text{clinical improvement} \mid \text{vitamin C}]$$

and $\lambda = \ln[p/(1-p)]$ be the log odds of improvement. Construct two 95% confidence intervals for λ: one based on (2.1.11) and one based on (2.1.16).

2.9. Consider the problem of estimating both binomial parameters n *and* p based on iid observations $Y_1, \ldots, Y_m$ with $Y_i \sim B(n,p)$. This model is motivated by the following example drawn from the field of wildlife ecology. Using a light aircraft on five successive cloudless days, five highly trained wildlife officials desired to count the number of impala herds of size at least 25 in a certain wilderness preserve. The observed number of herds of size at least 25 are 15, 20, 21, 23, and 26.

(a) Show that the method of moments estimators (MME) of n and p are $\hat{n} = \hat{\mu}^2/(\hat{\mu} - \hat{\sigma}^2)$ and $\hat{p} = \hat{\mu}/\hat{n}$ where $\hat{\mu} = \sum_{i=1}^m y_i/m$ and $\hat{\sigma}^2 = \sum_{i=1}^m (y_i - \hat{\mu})^2/m$. Prove the following shortcoming of $(\hat{n}, \hat{p})$: if $\hat{\sigma}^2 > \hat{\mu}$, then $\hat{n} < 0$. Use the impala data to illustrate a second deficiency of $\hat{n}$: if $\hat{\sigma}^2 < \hat{\mu}$ but close, then $\hat{n}$ can be unstable. Show this by calculating the MME of n based on the original data and the modified data in which the largest count (only) is changed from 26 to 27. [The answers are 57 and 77, respectively.]

(b) Show the likelihood function at (n,p) given $y_1, \ldots, y_m$ is

$$L(n,p) = \prod_{i=1}^{m} \binom{n}{y_i} p^{y_i}(1-p)^{n-y_i}$$

for $0 \le p \le 1$ and $\max_{1 \le i \le m} y_i \le n < \infty$. Prove that if $\hat{\mu} \le \hat{\sigma}^2$, where these quantities are defined in (a), then the MLE of (n,p) does not exist. [Hint: Consider the limit of $L(n,p)$ as $n \to \infty$ and $np = \hat{\mu}$.]

(c) Show the instability of the MLE by computing it for the original and modified impala data in (a). [The answers are 53 and 74, respectively.]

[Olkin, Petkau, and Zidek (1981), Carroll and Lombard (1985), and Casella (1986b) study estimators of n and graphical methods for assessing the stability of the MLE.]

2.10. This problem considers Bayesian estimation of (n, p) based on $Y_1, \ldots, Y_m$ as described in Problem 2.9.

(a) Suppose the fact that n is a positive integer is ignored, and it is assumed that $n \sim P(\mu)$. Show that the components of $\mathbf{Y} := (Y_1, \ldots, Y_m)$ are jointly exchangeable with $Y_i \sim P(\lambda := \mu p)$.

(b) Consider a specification of a (second-stage) prior for (λ, p), which is presumably easier to formulate than a prior for (μ, p) because λ is the unconditional mean of $Y_1, \ldots, Y_m$. Prove the posterior probability $p(k \,|\, \mathbf{y})$ that $n = k$ given $\mathbf{Y} = \mathbf{y}$ is proportional to

$$\frac{1}{k!}\left\{\prod_{i=1}^{m}\binom{k}{y_i}\int_0^1\int_0^\infty p^{-k+S}(1-p)^{mk-S}\lambda^k e^{-\lambda/p}h(\lambda, p)d\lambda dp\right\} \tag{1}$$

for $k \geq \max\{y_i : 1 \leq i \leq m\}$ where $S = \sum_{i=1}^m y_i$ and $h(\cdot,\cdot)$ is the second stage prior for (λ, p).

(c) Consider the vague prior for (λ, p) for which $h(\lambda, p) \propto \lambda^{-1}$ (Jaynes, 1968). Prove the posterior in (1) specializes to an expression proportional to

$$\left\{\frac{(mk-S)!}{(mk+1)!k}\right\}\prod_{i=1}^{m}\binom{k}{y_i}. \tag{2}$$

Prove that when $m = 1$, the posterior mean based on (2) is $2y_1$.

(d) Consider Bayesian estimation of n based on (2) under relative mean squared error (i.e., the risk is $E[(\hat{n}/n-1)^2]$ for an estimator $\hat{n}$). Prove the Bayes estimator of n corresponding to this loss is

$$\hat{n}_{RE} = \frac{\sum_{k=y_{\max}}^{\infty} k^{-1}p(k \,|\, \mathbf{y})}{\sum_{k=y_{\max}}^{\infty} k^{-2}p(k \,|\, \mathbf{y})}$$

where $y_{\max} = \max\{y_i : 1 \leq i \leq m\}$.

(e) Calculate the mode based on (2) and $\hat{n}_{RE}$ for the original impala data and the data perturbed by replacing 26 by 27.

(f) Compute the highest posterior density credible region for (n, p) using (2) based on both the original and perturbed data. Comment on its stability.

[Raftery, 1988]

2.11. Prove that if $\mathbf{Y} \sim M_t(n, \mathbf{p})$ and $\mathbf{p}$ has prior $\mathcal{D}_t(K\boldsymbol{\mu})$ with $\boldsymbol{\mu}$ known and K unknown, then the pseudo Bayes risk minimizing choice of K under SEL is

$$K = (1 - ||\mathbf{p}||^2)/||\mathbf{p} - \boldsymbol{\mu}||^2.$$

2.12. Albert (1981a) reports the data in Table 2.P.12 on a sample of 50 students from a class of 571 at Findlay College classified with respect to their home background into one of five categories. Consider estimating the vector of population proportions $\mathbf{p} = (p_1, \ldots, p_5)'$. In this case the true vector of population proportions (based on all 571 students) is known to be $\mathbf{p} = (.200, .335, .257, .121, .088)'$. Compute and compare the actual SEL for the MLE, the minimax Bayes estimator $\hat{\mathbf{p}}^B$ $((\boldsymbol{\mu}, K) = (\mathbf{c}, 2.5))$, $\hat{\mathbf{p}}^O$, $\hat{\mathbf{p}}^T$, $\hat{\mathbf{p}}^U$, $\hat{\mathbf{p}}^G$, and $\hat{\mathbf{p}}^S$.

Table 2.P.12. Home Backgrounds for a Sample of Findlay College Students (Reprinted with permission from J. Albert: "Pseudo-Bayes Estimation of Multinomial Populations," *Communications in Statistics: Theory and Methods,* vol. 10, pg. 1608, Marcel Dekker, New York, 1981.)

Home Background	Observed Count
Farm	6
Small Town	17
Moderate-Sized Town	19
Suburb	5
City	3
Total	50

2.13. Suppose the point null hypothesis $H_0\colon \hat{\mathbf{p}} = \mathbf{p}^0$ (given) is tested against the global alternative $H_{\neq}\colon \mathbf{p} \neq \mathbf{p}^0$ based on multinomial data. Consider the sequence of contiguous alternatives $\mathbf{Y}^n \sim M_t(n, \mathbf{p}^n)$, $n \geq 1$, with $\mathbf{p}^n = \mathbf{p}^0 + \mathbf{u}/\sqrt{n}$ where $\sum_{i=1}^t u_i = 0$. Show that for any $\lambda \in \mathbb{R}$, the Cressie–Read statistic I^λ satisfies

$$I^\lambda = X^2 + o_p(1).$$

2.14. Suppose, in Example 2.2.2, that Mendel wanted to choose the sample size to guarantee power .90 at the alternative $\mathbf{p}^* = (5/16, 4/16, 4/16, 3/16)'$ based on the Pearson test X^2 of H_0. How many crossings should be used? Compare this with the sample size required by the likelihood ratio statistic G^2 using noncentrality parameter

$$2n \sum_{i=1}^t p_i^* \ln(p_i^*/p_i^0).$$

2.15. Consider the data of Example 2.2.1 on the frequencies of nine occupational groups according to a 1972 University of Michigan survey. A similar survey was conducted in 1956 and also reported in Knoke

(1976). The observed proportions in each category for the 594 individuals in the 1956 survey are

$$\hat{\mathbf{p}}^{56} = (.128, .167, .039, .059, .247, .146, .059, .086, .069)'.$$

One interesting question is whether there has been a shift between 1956 and 1972 towards more employment in white collar occupations and less in blue collar occupations. A simple categorization of the occupational groups would consider groups 1–4 as white collar and groups 5–9 as blue collar. Let $\mathbf{p}^{72}$ be the vector of cell probabilities for the 1972 survey *population* and let

$$\mathbf{u}' = (.25, .25, .25, .25, -.20, -.20, -.20, -.20, -.20).$$

Consider formulating the above question as a test of the composite null hypothesis $H_0: \mathbf{p}^{72} \in \{\mathbf{p}(\eta) : \eta \in \mathbb{R} \text{ and } \mathbf{p}(\eta) \in \mathcal{S}\}$ against the global alternative H_A: (not H_0) where $\mathbf{p}(\eta) = \hat{\mathbf{p}}^{56} + \eta\mathbf{u}$. Since $\sum_{i=1}^{9} u_i = 0$, the null hypothesis includes all vectors $\mathbf{p}$ which differ from $\hat{\mathbf{p}}^{56}$ by uniform increases in white collar occupations and uniform decreases in blue collar occupations. (Alternative formulations are obviously possible including an "isotonic" null requiring only that $p_i^{72} \geq \hat{p}_i^{56}$ for $1 \leq i \leq 4$ and $p_i^{72} \leq \hat{p}_i^{56}$ for $5 \leq i \leq 9$.) What is the MLE of η under the null hypothesis H_0 formulated above? Compute the score and LR statistics for this hypothesis and interpret your results.

2.16. The data of Example 1.2.8 concern the severity of nausea for cancer patients receiving chemotherapy either with (Cis) or without (~Cis) Cisplatinum. Consider the group of patients receiving Cisplatinum.

(a) Apply the Bonferroni method to form 90% simultaneous confidence intervals for $\{p_0 - p_i\}_{i=1}^{5}$ where $p_i = P[Y = i \mid \text{Cis}]$.

(b) Derive 90% simultaneous confidence intervals for these same linear combinations based on Gold's covariance and compare them to the intervals in (a).

2.17. Suppose $Y_1, \ldots, Y_n$ are iid with unknown continuous distribution $F(\cdot)$. Consider testing fit to the normal family

$$H_0: F(\cdot) \in \left\{\Phi\left(\frac{y-\mu}{\sigma}\right) : \mu \in \mathbb{R}, \sigma > 0\right\} \text{ versus } H_A: (\text{not } H_0)$$

using the Pearson X^2 test. Suppose boundaries $-\infty =: a_0 < a_1 \ldots < a_{t-1} < a_t := +\infty$ are selected and count data $X_1, \ldots, X_t$ are defined by $X_j := \sum_{i=1}^{n} I[a_{j-1} < Y_i \leq a_j]$ for $1 \leq j \leq t$. Derive the likelihood equations for the MLEs of μ and σ^2 based on the multinomial data $(X_1, \ldots, X_t)$ and comment on their solvability.

2.18. Consider estimating the multinomial probability vector $\mathbf{p}$ based on data $\mathbf{Y} \sim M_t(n, \mathbf{p})$ under entropy loss (EL) given by (2.2.5).

(a) Show that the risk of the MLE with respect to EL is infinite except at vertices where it is zero.

(b) Use (a) to show that the MLE is *admissible* if the parameter space is $\mathcal{S} \setminus \{\text{vertices}\}$.

(c) Suppose $\pi(\cdot)$ is an arbitrary prior on $\mathcal{S}$. Show that the Bayes estimator for $\mathbf{p}$ with respect to $\pi(\cdot)$ and EL is the mean of the posterior of the distribution. (This proves that the Bayes estimator under EL is identical to that under SEL.)

[Ighodaro, Santner, and Brown, 1982]

2.19. Given data $\mathbf{Y} \sim M_t(n, \mathbf{p})$, consider estimators for $\mathbf{p}$ of the form

$$\alpha \mathbf{c} + (1 - \alpha)\hat{\mathbf{p}},$$

where $0 \leq \alpha \leq 1$.

(a) Show that the choice of α which minimizes Stone's cross-validation prediction criterion under SEL is (2.2.21).

(b) If the modulus loss $L_M(\mathbf{p}, \mathbf{a}) = \sum_{i=1}^{t} |p_i - a_i|$ is used instead of SEL, then show that the minimizing α is

$$\alpha = \begin{cases} 0, & X^2 \geq 1 \\ 1, & X^2 \leq 1 \end{cases}$$

where

$$X^2 = \sum_{i=1}^{t} \{(Y_i - nt^{-1})^2 / nt^{-1}\}.$$

(c) If entropy loss is used, show that the minimizing α is identical to Aitchison and Aitken's (1976) choice which maximizes (2.2.18).

[Stone, 1974]

2.20. The larger the parameter α in the estimator $\alpha\mathbf{c} + (1 - \alpha)\hat{\mathbf{p}}$, the more the estimator pulls toward the homogeneity state. (The quantity α is sometimes referred to as the "squashing" power of the estimator.)

(a) Prove that for $\boldsymbol{\mu} = \mathbf{c}$, the Fienberg and Holland (1973) one-step estimator can be written in the form $\hat{\mathbf{p}}^O = \hat{\alpha}^{FH}\mathbf{c} + (1 - \hat{\alpha}^{FH})\hat{\mathbf{p}}$ where

$$\hat{\alpha}^{FH} = (n - Z)/(n + (n - 1)Z)$$

and Z is as in (2.2.19) with $\boldsymbol{\mu} = \mathbf{c}$.

(b) Prove that for any t and any data $\mathbf{y}$,

$$\hat{\alpha}^G \geq \hat{\alpha}^S \geq \hat{\alpha}^{FH}$$

so that Good's estimator squashes $\hat{\mathbf{p}}$ the most and Fienberg and Holland's squashes $\hat{\mathbf{p}}$ the least.

[Stone, 1974]

2.21. Consider pseudo Bayes estimation of $\boldsymbol{\lambda} = (\lambda_1, \ldots, \lambda_t)'$ based on independent data $Y_i \sim P(T_i\lambda_i)$, $1 \leq i \leq t$. Show that the value of c given in (2.3.13) minimizes $R_S(\hat{\boldsymbol{\lambda}}, \hat{\boldsymbol{\lambda}}^c)$.

[Albert, 1981b]

2.22. Consider the vehicle repair data of Example 1.2.1 (Table 1.2.1). Assume that

$$p_j := P[\text{a truck has } j \text{ repairs during the year}]$$

for $j = 0, 1, \ldots, 5$ and

$$p_6 := P[\text{a truck has 6 or more repairs}]$$

is the same for all trucks in the fleet. Test goodness-of-fit to the Poisson model; i.e.,

$$H_0\colon p_j = e^{-\lambda}\lambda^j/j! \text{ for some } \lambda > 0 \text{ and } j = 0, \ldots, 5$$

$$\text{and} \quad p_6 = \sum_{j=6}^{\infty} e^{-\lambda}\lambda^j/!$$

versus H_A: (not H_0) by applying the Pearson X^2 goodness-of-fit test.

2.23. Rutherford and Geiger (1910) present the data in Table 2.P.23 on the number of scintillations due to radioactive decay of polonium in each of $t = 2608$, 1/8-minute time intervals. Let F_k denote the number of time intervals in which k scintillations were observed.

(a) Construct the Hoaglin Poissonness plot (2.3.20) for the Rutherford and Geiger data. Which points fail to fall on the line?

(b) Show that when t is large, $\ln(F_k)$ is approximately normally distributed with mean $\ln(tp_k)$ and variance $(1 - p_k)/tp_k$ where $p_k = e^{-\lambda}\lambda^k/k!$.

(c) Add the line with slope $\ln(\hat{\lambda})$ and intercept $\ln(t) - \hat{\lambda}$ to the plot in (a) where $\hat{\lambda}$ is the MLE of λ assuming the data are iid Poisson. Assess the degree of lack of fit of the points (k, F_k) for $k = 0$, 13, and 14.

[Hoaglin, 1980]

Table 2.P.23. Number of Scintillations from Radioactive Decay of Polonium

k	F_k
0	57
1	203
2	383
3	525
4	532
5	408
6	273
7	139
8	45
9	27
10	10
11	4
12	0
13	1
14	1

2.24. Potthoff and Whittinghill (1966) report an examination of articles in a Saturday issue of a large North Carolina daily newspaper to determine the number Y_i of certain types of printing errors in each article. The length T_i of the ith article was measured by counting the number of lines in the article. Heterogeneous conditions of typesetting or proofreading could result in there not being a uniform error rate λ throughout the newspaper. Table 2.P.24 lists not only the Y_i and T_i, but also the section and page of the newspaper on which the article started. There are a total of 65 errors and 5429 lines in 112 articles.

(a) Apply the G^2 and X^2 tests of homogeneity of rates to these data.

(b) Consider grouping the data into those articles beginning in the A section of the paper and those beginning in the B section. Is there heterogeneity between the rates for Sections A and B?

2.25. The data in Table 2.P.25 from Bulick, Montgomery, Fetterman and Kent (1976) list the number of times during 1969–1974 that each of the 38,400 volumes acquired in 1969 by the Hillman Library of the University of Pittsburgh was circulated.

Table 2.P.24. Number of Printing Errors (Y_i) and Number of Lines (T_i) in Each of 112 Newspaper Articles (Reprinted with permission from R. Potthoff and M. Whittinghill: "Testing for Homogeneity II: The Poisson Distribution," *Biometrika*, vol. 53, 1966, Biometrika Trust.)

Part	Errors Y_i	Length T_i	Part	Errors Y_i	Length T_i	Part	Errors Y_i	Length T_i
A1	0	48	A3	0	95	B1	0	15
A1	0	70	A3	2	37	B1	0	20
A1	3	149	A3	0	23	B1	0	58
A1	0	48	A3	0	17	B1	0	28
A1	0	80	A5	0	53	B1	0	47
A1	3	150	A5	0	24	B1	0	63
A1	0	61	A5	0	10	B1	0	160
A1	1	32	A5	2	93	B1	0	49
A1	1	14	A5	1	83	B2	0	35
A1	1	96	A5	1	48	B2	0	96
A1	1	79	A5	1	73	B2	0	74
A1	1	42	A5	0	50	B2	0	119
A2	0	69	A5	0	18	B2	1	111
A2	0	44	A5	0	23	B3	0	14
A2	1	77	A6	1	52	B3	1	34
A2	0	49	A6	2	53	B3	1	19
A2	0	13	A6	2	124	B3	1	20
A2	4	101	A6	0	23	B3	0	38
A2	1	65	A6	1	35	B3	0	36
A2	0	42	A6	0	19	B3	1	28
A2	6	97	A6	0	16	B3	0	39
A2	0	10	A7	0	26	B3	0	20
A2	0	50	A8	2	82	B4	2	103
A2	0	56	A8	1	31	B4	0	43
A2	0	27	A8	0	18	B4	0	33
A2	2	61	A8	0	11	B4	0	35
A2	1	56	A8	0	18	B4	0	17
A2	1	33	A8	0	10	B4	0	13
A3	1	110	A8	0	20	B4	0	35
A3	0	29	A8	0	22	B4	1	21
A3	0	67	A8	0	11	B4	0	21
A3	0	56	A8	2	72	B4	3	162
A3	0	38	A8	0	41	B4	1	22
A3	1	18	B1	0	42	B5	1	69
A3	0	10	B1	0	42	B5	0	13
A3	1	16	B1	0	31	B5	1	35
A3	0	13	B1	0	87			
A3	1	34	B1	0	41			

Table 2.P.25. Book Circulation for the Hillman Library (Reprinted with permission from S. Bulick, K.L. Montgomery, J. Fetterman, and A. Kent, "Use of Library Materials in Terms of Age," *Journal of the American Society for Information Science,* vol. 27, p. 177, 1976, John Wiley & Sons, New York.)

Number of Volumes	Number of Times Circulated
17151	0
5201	1
3014	2
2123	3
1682	4
1310	5
1043	6
884	7
759	8
638	9
610	10
544	11
468	12
368	13
346	14
311	15
254	16
245	17
195	18
172	19
150	20
136	21
106	22
110	23
94	24
73	25
413	26+

(a) Fit a Poisson model to these data, assuming for simplicity that all volumes in the "26+" circulation category were taken out exactly 26 times. Compute the expected number of volumes for each circulation category. The data display greater heterogeneity and a much larger number of uncirculated volumes than such a model would predict; there is also some bias due to the treatment of books in the 26+ category.

(b) Consider the following two parameter model for the probability that Y, the number of times a book is circulated, is j:

$$P[Y=j] = \begin{cases} \alpha, & j=0 \\ \frac{1-\alpha}{(1-\exp\{-\lambda\})} \times \frac{\exp\{-\lambda\}\lambda^j}{j!}, & j=1(1)25 \\ \frac{1-\alpha}{(1-\exp\{-\lambda\})} \times \left(1-\sum_{i=0}^{25}\exp\{-\lambda\}\lambda^i/i!\right), & j=26+ \end{cases} \quad (1)$$

for $\lambda > 0$ and $0 < \alpha < 1$. Fit model (1) to the data and contrast the results with those from the Poisson fit.

Table 2.P.26. Number of Occurrences of Loss of Feedwater Flow and Number of Operating Years for 30 Nuclear Power Plants (Reprinted with permission from S. Kaplan: "On a 'Two-Stage' Bayesian Procedure for Determining Failure Rates from Experimental Data," *IEEE Transactions on Power Aparatus and Systems,* PAS-102, 1983, pgs. 195–202.)

Plant Number	Number of Occurrences	Number of Operating Years
1	4	15
2	40	12
3	0	8
4	10	8
5	14	6
6	31	5
7	2	5
8	4	4
9	13	4
10	4	3
11	27	4
12	14	4
13	10	4
14	7	2
15	4	3
16	3	3
17	11	2
18	1	2
19	0	2
20	3	1
21	5	1
22	6	1
23	35	5
24	12	3
25	1	1
26	10	3
27	5	2
28	16	4
29	14	3
30	58	11

2.26. Consider the data of Table 2.P.26, originally given in Kaplan (1983), on the number of occurrences of loss of feedwater flow and the number of operating years in 30 nuclear power plants. Let Y_i be the number of occurrences in the ith plant; suppose $Y_1, \ldots, Y_{30}$ are mutually independent with $Y_i \sim P(T_i \lambda_i)$ where T_i is the number of operating years for the ith plant.

(a) Compute the MLE of $\boldsymbol{\lambda} = (\lambda_1, \ldots, \lambda_{30})'$ and the Bayes estimator with respect to the prior for which λ_i are iid $\Gamma(1.70, 1.63)$. Interpret this prior.

(b) Compute the likelihood ratio statistic for testing the hypothesis of equality of yearly rates of occurrence; i.e., H_0: $\lambda_1 = \ldots = \lambda_{30}$ versus H_A: (not H_0). Does the data support the homogeneity hypothesis?

(c) For the 20th plant only, apply the tail method discussed in Section 2.3, to calculate a 95% confidence interval for λ_{20}.

2.27. Suppose $Y_1 \sim P(T\lambda_1)$ is independent of $Y_2 \sim P(T\lambda_2)$. Let $\rho = \lambda_1/\lambda_2$. It is desired to test

$$H_0\colon \lambda_1 = \lambda_2 \text{ (i.e., } \rho = 1) \text{ versus } H_A\colon \lambda_1 > \lambda_2 \ (\rho > 1).$$

Let $S = Y_1 + Y_2$.

(a) Prove the conditional distribution of Y_2 given $S = s$ and $\boldsymbol{\lambda} = (\lambda_1, \lambda_2)'$ is

$$P[Y_2 = j \mid S = s, \rho] = \binom{s}{j} \left(\frac{\rho}{1+\rho}\right)^j \left(\frac{1}{1+\rho}\right)^{s-j}$$

(b) Show that the uniformly most powerful size α test of H_0 versus H_A is

$$\phi(\mathbf{Y}) = \begin{cases} 1, & Y_2 > c \\ \gamma, & Y_2 = c \\ 0, & Y_2 < c \end{cases}$$

where $\gamma = \gamma(s)$ and $c = c(s)$ satisfy

$$P[Y_2 > c \mid S = s, 1] + \gamma P[Y_2 = c \mid S = s, 1] = \alpha. \qquad (1)$$

[Gail (1974) studies the nonrandomized conservative size α test of H_0 versus H_A

$$\phi^*(\mathbf{Y}) = \begin{cases} 1, & Y_2 > c \\ 0, & Y_2 \leq c \end{cases}$$

where $c = c(s)$ satisfies (1). For a given alternative ρ_0 (> 1) and power $1 - \beta$, he tables the total number of successes s required for $\phi^*(\cdot)$ to have power $1 - \beta$ when $\rho = \rho_0$.]

2.28. Let $Y_1, \ldots, Y_n$ be iid with geometric probability distribution

$$P[Y_i = y] = (1-p)p^y, \quad y \in \{0, 1, \ldots\}.$$

(a) Compute the MLE and the method of moments estimator of p.

(b) Show that the UMVUE for p is

$$\frac{n-1}{\sum_{i=1}^{n} y_i + n - 1}.$$

2.29. Suppose $Y_1, \ldots, Y_t$ are iid with negative binomial distribution (1.3.9) for $\mu > 0$ and $\sigma^2 > 0$. Show that $E[Y_i] = \mu$, $\text{Var}(Y_i) = \sigma^2$, and the probability function

$$P[Y_i = j] \rightarrow e^{-\mu}\mu^j/j!$$

as $\sigma^2 \rightarrow 0$. Show that the score test of $H_0: \sigma^2 = 0$ versus $H_>: \sigma^2 > 0$ rejects the null hypothesis for large values of Fisher's Dispersion Index $\sum_{i=1}^{t}(Y_i - \overline{Y})^2/\overline{Y}$.

3

Loglinear Models

3.1 Introduction

The first three sections of this chapter present the theory of maximum likelihood estimation of a vector of means which satisfy a loglinear model under Poisson, multinomial, and product multinomial sampling. Example 1.2.10 (considered in Problem 3.6), Problem 3.3, and Problem 3.4 illustrate loglinear modeling under Poisson sampling for data on valve failures in nuclear plants, breakdowns in electronic equipment, and absences of school children, respectively. Further applications are deferred to Chapter 4 where cross-classified (multinomial) data are studied and to Chapter 5 where binary regression (product multinomial) data are considered. Alternative methods of estimation and non-loglinear models are discussed in Section 3.4.

Consider data $\{Y_i : i \in \mathcal{I}\}$ where $\mathcal{I}$ is an index set of finite cardinality. For ease of notation it is assumed that $\mathcal{I} = \{1, \ldots, n\}$; formally this means that there is a 1–1 transformation from $\mathcal{I}$ to $\{1, \ldots, n\}$. Assume that $\mu_i := E[Y_i] > 0$ for all $i \in \mathcal{I}$, and let $\ell_i := \ln(\mu_i)$ be the natural logarithm of μ_i with $\boldsymbol{\ell} := \{\ell_i : i \in \mathcal{I}\}$.

Definition. Data $\{Y_i : i = 1(1)n\}$ follow a *loglinear model* (LLM) if $\boldsymbol{\ell} \in \mathcal{M}$ for some known subspace $\mathcal{M}$ of $\mathbb{R}^n$.

The definition of a loglinear model does not require Y_i be nonnegative or discrete. However in practice, statistical inference for Y_i's following LLMs is most thoroughly developed in three sampling settings: (i) Poisson (P) sampling, (ii) multinomial (M) sampling, and (iii) product multinomial (PM) sampling. The precise assumptions for (i)–(iii) will be stated in Sections 3.2 and 3.3.

The definition of a loglinear model also does not require that every vector in $\mathcal{M}$ correspond to a set of logmeans for the Y_i. Indeed, in M and PM sampling this will not be true as Examples 3.1.2–3.1.4 will illustrate. Several examples of loglinear models will now be given.

Example 3.1.1. Let $Y_1, \ldots, Y_n$ be iid $P(\lambda)$ where $\lambda > 0$ but is otherwise unknown. Then $\mu_i = E[Y_i] = \lambda$ and $\ell_i = \ln(\lambda)$ for $i = 1(1)n$. In this case $\boldsymbol{\ell} = (\ln(\lambda), \ldots, \ln(\lambda))' \in \mathcal{M} = \mathcal{C}(\mathbf{1}_n)$. Furthermore every $\boldsymbol{\nu} \in \mathcal{C}(\mathbf{1}_n)$

corresponds to a legitimate vector of logmeans for the $\{Y_i : 1 \le i \le n\}$.

Example 3.1.2. Let $\mathbf{Y} = \{Y_{ij},\ 1 \le i \le R,\ 1 \le j \le C\}$ satisfy $\mathbf{Y} \sim M_{RC}(m, \mathbf{p})$, and suppose $p_{ij} = p_{i+}p_{+j}$ for all i and j where $p_{i+} := \sum_{j=1}^{C} p_{ij}$ and $p_{+j} := \sum_{i=1}^{R} p_{ij}$. This is the model of independence of the row and column classifications. Formally, $\mathcal{I} = \{1, \ldots, R\} \times \{1, \ldots, C\}$, $n = RC$, $\mu_{ij} = mp_{ij} = mp_{i+}p_{+j}$, and $\ell_{ij} = \ln(m) + \ln(p_{i+}) + \ln(p_{+j})$. Define $R \times C$ incidence matrices $\boldsymbol{\rho}_1, \ldots, \boldsymbol{\rho}_R, \boldsymbol{\kappa}_1, \ldots, \boldsymbol{\kappa}_C$ which identify rows and columns by

$$(\boldsymbol{\rho}_r)_{ij} = \begin{cases} 1 & \text{if } i = r \\ 0 & \text{if } i \neq r \end{cases};$$

i.e.,

$$\boldsymbol{\rho}_r = \begin{bmatrix} \mathbf{0}_{r-1,C} \\ \mathbf{1}_C \\ \mathbf{0}_{R-r,C} \end{bmatrix} \} \ r\text{th row} \tag{3.1.1}$$

for $r = 1, \ldots, R$, and

$$(\boldsymbol{\kappa}_c)_{ij} = \begin{cases} 1 & \text{if } j = c \\ 0 & \text{if } j \neq c \end{cases};$$

i.e.,

$$\boldsymbol{\kappa}_c = [\ \mathbf{0}_{R,c-1} \quad \underbrace{\mathbf{1}_R}_{c\text{th column}} \quad \mathbf{0}_{R,C-c}\ ] \tag{3.1.2}$$

for $c = 1, \ldots, C$. Then $\boldsymbol{\ell} = \{\ell_{ij} : 1 \le i \le R,\ 1 \le j \le C\}$ satisfies

$$\boldsymbol{\ell} \in \mathcal{M} = \mathcal{C}(\boldsymbol{\rho}_1, \ldots, \boldsymbol{\rho}_R, \boldsymbol{\kappa}_1, \ldots, \boldsymbol{\kappa}_C).$$

Not every vector $\boldsymbol{\ell} \in \mathcal{M}$ corresponds to a set of logmeans since $\sum_i \sum_j p_{ij} = 1$ implies the ℓ_{ij} must satisfy $m = \sum_i \sum_j \exp\{\ell_{ij}\}$. However, $\mathcal{M}$ does characterize the model of independence in the sense that *every* $\boldsymbol{\ell} \in \mathcal{M}$ satisfying $\sum_i \sum_j \exp\{\ell_{ij}\} = m$ corresponds to a matrix of probabilities satisfying $p_{ij} = p_{i+}p_{+j}$. To see this assume $\boldsymbol{\ell} \in \mathcal{M}$ satisfies $\sum_i \sum_j \exp\{\ell_{ij}\} = m$ and define $p_{ij} := \exp\{\ell_{ij}\}/m$. Then $\ell_{ij} = a_i + b_j$ for all i, j for some constants $\{a_i\}_{i=1}^{R}$ and $\{b_j\}_{j=1}^{C}$, and $m = \sum_i \sum_j \exp\{\ell_{ij}\} = \sum_i \exp\{a_i\} \sum_j \exp\{b_j\}$. Algebra gives $p_{i+} = (\exp\{a_i\}/m)(\sum_j \exp\{b_j\})$, and $p_{+j} = (\exp\{b_j\}/m)(\sum_i \exp\{a_i\})$ yielding

$$p_{i+}p_{+j} = \frac{\exp\{a_i + b_j\}}{m} \frac{(\sum_i \exp\{a_i\})(\sum_j \exp\{b_j\})}{m} = p_{ij} \quad \text{for all } i, j.$$

This argument shows that $\tilde{\mathcal{M}} := \{\boldsymbol{\ell} \in \mathcal{M} : \sum_i \sum_j \exp\{\ell_{ij}\} = m\}$ is exactly the set of logmeans of $\{p_{ij}\}$ matrices which satisfy the independence hypothesis in the $R \times C$ table.

An equivalent characterization of independence, analogous to the specification of additivity in two-factor ANOVA models, is used frequently in the literature (Bishop, Fienberg and Holland, 1975; Fienberg, 1980). If the matrix of $\ell_{ij} = \ln(mp_{ij})$ can be written as

$$\ell_{ij} = \lambda + \lambda_i^1 + \lambda_j^2 \tag{3.1.3}$$

for all i and j where $\lambda_+^1 = 0 = \lambda_+^2$, then $\mathbf{p}$ satisfies independence. In this case

$$\boldsymbol{\ell} = \left(\begin{array}{c|c} \lambda + \lambda_i^1 + \lambda_j^2 & \lambda + \lambda_i^1 - \sum_{c=1}^{C-1} \lambda_c^2 \\ \hline \lambda - \sum_{r=1}^{R-1} \lambda_r^1 + \lambda_j^2 & \lambda - \sum_{r=1}^{R-1} \lambda_r^1 - \sum_{c=1}^{C-1} \lambda_c^2 \end{array}\right) \begin{array}{l} R-1 \text{ rows} \\ \\ 1 \text{ row} \end{array}$$

$$\qquad\quad \underbrace{\hphantom{\lambda - \sum \lambda_r^1 + \lambda_j^2}}_{C-1 \text{ columns}} \quad \underbrace{\hphantom{\lambda - \sum \lambda_r^1 - \sum \lambda_c^2}}_{1 \text{ column}}$$

$$= \lambda \underbrace{\begin{pmatrix} 1 & \cdots & 1 \\ \vdots & & \vdots \\ 1 & \cdots & 1 \end{pmatrix}}_{\boldsymbol{\nu}_1}$$

$$+ \lambda_1^1 \underbrace{\begin{pmatrix} 1 & \cdots & 1 \\ & \mathbf{0}_{R-2,C} & \\ -1 & \cdots & -1 \end{pmatrix}}_{\boldsymbol{\nu}_2} + \ldots + \lambda_{R-1}^1 \underbrace{\begin{pmatrix} & \mathbf{0}_{R-2,C} & \\ 1 & \cdots & 1 \\ -1 & \cdots & -1 \end{pmatrix}}_{\boldsymbol{\nu}_R}$$

$$+ \lambda_1^2 \underbrace{\begin{pmatrix} 1 & & -1 \\ \vdots & \mathbf{0}_{R,C-2} & \vdots \\ 1 & & -1 \end{pmatrix}}_{\boldsymbol{\nu}_{R+1}} + \ldots + \lambda_{C-1}^2 \underbrace{\begin{pmatrix} & 1 & -1 \\ \mathbf{0}_{R,C-2} & \vdots & \vdots \\ & 1 & -1 \end{pmatrix}}_{\boldsymbol{\nu}_{R+C-1}}$$

where $\mathbf{0}_{a,b}$ denotes the a by b matrix of zeroes. Therefore $\boldsymbol{\ell} \in \mathcal{C}(\boldsymbol{\nu}_1, \ldots, \boldsymbol{\nu}_{R+C-1})$ and it is straightforward to show that $\mathcal{C}(\boldsymbol{\nu}_1, \ldots, \boldsymbol{\nu}_{R+C-1}) = \mathcal{M}$. Furthermore, the $\{\boldsymbol{\nu}_j\}_{j=1}^{R+C-1}$ are linearly independent and the sets $\{\boldsymbol{\nu}_1\}$, $\{\boldsymbol{\nu}_2, \ldots, \boldsymbol{\nu}_R\}$, $\{\boldsymbol{\nu}_{R+1}, \ldots, \boldsymbol{\nu}_{R+C-1}\}$ are pairwise orthogonal. This spanning set is a basis for $\mathcal{M}$ which shows that the dimension of $\mathcal{M}$ is $R + C - 1$.

The final two examples of this section illustrate product multinomial sampling.

Example 3.1.3 (Logistic Regression). Brown(1980) lists the data in Table 3.1.1 on nodal involvement (1 := Yes/0 := No) in 53 prostatc cancer patients undergoing surgery. The objective of this study was to determine which of five preoperative variables were predictive of nodal involvement. The treatment of choice depends on whether the lymph nodes are involved. Two of the potential explanatory variables are quantitative (*age* at diagnosis, and level of serum *acid* phosphatase). The other three explanatory variables are qualitative assessments of the tumor according to (i) an *X-ray* reading, (ii) a pathology reading of the *grade* of a presurgical biopsy, and (iii) a *staging* of the extent of disease which is a rough measure of the size and location of the tumor obtained by palpitation. Problem 5.25 considers the analysis of these data.

This example is a special case of a general setting in which binary responses are recorded together with the level of one or more factors that affect the "success" probability. Let Z_i denote the binary response for the ith patient and assume the $Z_1, \ldots, Z_{53}$ are mutually independent Bernoulli random variables with Z_i having success probability $p_i = p(\mathbf{x}_i)$ for $i = 1(1)53$. The vector $\mathbf{x}_i = (x_{i1}, x_{i2}, x_{i3}, x_{i4}, x_{i5})'$ denotes the levels of the X-ray reading (x_{i1}), stage (x_{i2}), grade (x_{i3}), patient age (x_{i4}), and acid level (x_{i5}) in the ith patient. Define $\text{logit}(p_i) := \ln(\frac{p_i}{1-p_i})$, and for illustration suppose

$$\text{logit}(p_i) = \beta_0 + \beta_1 x_{i1} + \beta_2 x_{i4}; \tag{3.1.4}$$

i.e., $\ln(p_i) = \ln(1-p_i) + \beta_0 + \beta_1 x_{i1} + \beta_2 x_{i4}$. This model says that *only* the X-ray reading and patient age affect the probability of nodal involvement.

For $1 \le i \le 53$ define $Y_i = Z_i$ and $Y_{i+53} = 1 - Z_i$. The data $\mathbf{Y} = \{Y_i : 1 \le i \le 106\}$ satisfy

$$\mu_i = \begin{cases} p_i, & 1 \le i \le 53 \\ 1 - p_i, & 54 \le i \le 106, \end{cases}$$

and

$$\ell_i = \begin{cases} \ln(p_i) = \ln(1-p_i) + \beta_0 + \beta_1 x_{i1} + \beta_2 x_{i4}, & 1 \le i \le 53 \\ \ln(1-p_i), & 54 \le i \le 106. \end{cases}$$

Thus $n = 106$ and the vector of logmeans belongs to the linear subspace $\mathcal{M} = \mathcal{C}(\mathbf{X})$ of $\mathbb{R}^{106}$ where

$$\mathbf{X} := \left(\begin{array}{c|c|c|c} \mathbf{1}_{53} & \boldsymbol{\xi}_1 & \boldsymbol{\xi}_4 & \mathbf{I}_{53} \\ & & & \\ \mathbf{0}_{53} & \mathbf{0}_{53} & \mathbf{0}_{53} & \mathbf{I}_{53} \end{array} \right)$$

where $\boldsymbol{\xi}_j := (x_{1j}, \ldots, x_{53,j})'$ for $j = 1, 4$. As in Example 3.1.2, not every $\boldsymbol{\ell} \in \mathcal{M}$ corresponds to a legitimate logmean vector for $\mathbf{Y}$. However, $\tilde{\mathcal{M}} := \{\boldsymbol{\ell} \in \mathcal{M} : \exp\{\ell_i\} + \exp\{\ell_{i+53}\} = 1, \ i = 1(1)53\}$ characterizes the logit model in the sense that for every $\boldsymbol{\ell} \in \tilde{\mathcal{M}}$, $p_i := \exp\{\ell_i\}$, $1 \le i \le 53$ satisfies (3.1.4).

Table 3.1.1. Occurrence of Nodal Involvement in Patients with Prostate Cancer (Reprinted with permission from B.W. Brown, Jr.: "Prediction Analyses for Binary Data." *Biostatistics Casebook,* edited by R. Miles, Jr. et al., John Wiley & Sons, New York, 1980.)

X-ray	Stage	Grade	Age	Acid	Nodal Involvement
0	0	0	66	48	0
0	0	0	68	56	0
0	0	0	66	50	0
0	0	0	56	52	0
0	0	0	58	50	0
0	0	0	60	49	0
1	0	0	65	46	0
1	0	0	60	62	0
0	0	1	50	56	1
1	0	0	49	55	0
0	0	0	61	62	0
0	0	0	58	71	0
0	0	0	51	65	0
1	0	1	67	67	1
0	0	1	67	47	0
0	0	0	51	49	0
0	0	1	56	50	0
0	0	0	60	78	0
0	0	0	52	83	0
0	0	0	56	98	0
0	0	0	67	52	0
0	0	0	63	75	0
0	0	1	59	99	1
0	0	0	64	187	0
1	0	0	61	136	1
0	0	0	56	82	1
0	1	1	64	40	0
0	1	0	61	50	0
0	1	1	64	50	0
0	1	0	63	40	0
0	1	1	52	55	0
0	1	1	66	59	0
1	1	0	58	48	1
1	1	1	57	51	1
0	1	0	65	49	1
0	1	1	65	48	0
1	1	1	59	63	0

Table 3.1.1. (cont.)

X-ray	Stage	Grade	Age	Acid	Nodal Involvement
0	1	0	61	102	0
0	1	0	53	76	0
0	1	0	67	95	0
0	1	1	53	66	0
1	1	1	65	84	1
1	1	1	50	81	1
1	1	1	60	76	1
0	1	1	45	70	1
1	1	1	56	78	1
0	1	0	46	70	1
0	1	0	67	67	1
0	1	0	63	82	1
0	1	1	57	67	1
1	1	1	51	72	1
1	1	0	64	89	1
1	1	1	68	126	1

Example 3.1.4 ($2 \times 2 \times S$ Table). Consider mutually independent observations $Z_{sj} \sim B(m_{sj}, p_{sj})$ for $1 \leq s \leq S$ and $j = 1, 2$. One way to display these data is as a series of S 2×2 tables, the sth of which is Table 3.1.2.

Table 3.1.2. Generic 2 by 2 Table

Z_{s1}	Z_{s2}
$m_{s1} - Z_{s1}$	$m_{s2} - Z_{s2}$
m_{s1}	m_{s2}

A specific example of data of this form is Example 1.2.6 in which Z_{sj} is the number of mice with tumor at the end of the study period for $S = 4$ sex-by-strain combinations of mice and $j = 1(2)$ denotes control (Avadex). Consider the constant odds ratio model which assumes $\ln(\frac{p_{s1}}{(1-p_{s1})} \frac{(1-p_{s2})}{p_{s2}})$ is constant, say γ, for $1 \leq s \leq S$. This model implies that

$$\ln(p_{s1}) = \ln(1 - p_{s1}) - \ln(1 - p_{s2}) + \ln(p_{s2}) + \gamma \tag{3.1.5}$$

for $s = 1(1)S$. Let $Y_{sj1} = Z_{sj}$ and $Y_{sj2} = m_{sj} - Z_{sj}$ for $s = 1(1)S$ and $j = 1, 2$; then $n = 4 \times S$ and

$$\mu_{sjk} = \begin{cases} m_{sj} p_{sj}, & k = 1 \\ m_{sj}(1 - p_{sj}), & k = 2. \end{cases}$$

Hence for $s = 1(1)S$

$$\begin{aligned}
\ell_{s11} &= \ln(m_{s1}) + \ln(1-p_{s1}) - \ln(1-p_{s2}) + \ln(p_{s2}) + \gamma, \\
\ell_{s12} &= \ln(m_{s1}) + \ln(1-p_{s1}), \\
\ell_{s21} &= \ln(m_{s2}) + \ln(p_{s2}), \text{ and} \\
\ell_{s22} &= \ln(m_{s2}) + \ln(1-p_{s2}).
\end{aligned}$$

Thus $\boldsymbol{\ell} \in \mathcal{M} \subset \mathbb{R}^{4S}$ for a subspace defined analogous to that in Example 3.1.3; the dimension of $\mathcal{M}$ is $3S+1$. As in Examples 3.1.2 and 3.1.3, not every vector $\boldsymbol{\ell} \in \mathcal{M}$ corresponds to a legitimate set of logmeans; however, $\tilde{\mathcal{M}} := \{\boldsymbol{\ell} \in \mathcal{M} : \exp\{\ell_{sj1}\} + \exp\{\ell_{sj2}\} = m_{sj},\ 1 \le s \le S,\ 1 \le j \le 2\}$ characterizes the constant odds ratio model in that $p_{sj} := \exp\{\ell_{sj1}\}/m_{sj}$, $1 \le s \le S$, $j = 1, 2$ satisfies (3.1.5) for every $\boldsymbol{\ell} \in \tilde{\mathcal{M}}$. (See Problem 3.1.)

The following two sections detail the theory of maximum likelihood estimation for LLMs first under Poisson sampling and then under multinomial and product multinomial sampling. The final section outlines alternative approaches to maximum likelihood estimation in LLMs.

3.2 Maximum Likelihood Estimation for Loglinear Models Under Poisson Sampling

Consider mutually independent Poisson distributed observations $\{Y_i : i = 1(1)n\}$ and let μ_i denote the mean of Y_i. Suppose the logmean $\boldsymbol{\ell}$ of $\boldsymbol{\mu}$ satisfies $\boldsymbol{\ell} \in \mathcal{M}$ for a given subspace $\mathcal{M}$ of $\mathbb{R}^n$ having dimension p. Choose an $n \times p$ basis matrix

$$\mathbf{X} = \begin{bmatrix} \mathbf{x}_1' \\ \vdots \\ \mathbf{x}_n' \end{bmatrix}$$

for $\mathcal{M}$; i.e., $\mathcal{M} = \mathcal{C}(\mathbf{X})$ where $\mathbf{x}_j' = (x_{j1}, \ldots, x_{jp})$; thus $\boldsymbol{\ell} = \mathbf{X}\boldsymbol{\beta}$ for some $\boldsymbol{\beta} \in \mathbb{R}^p$. Let $\mathbf{e}^{\mathbf{X}\boldsymbol{\beta}} := (\ldots, e^{\mathbf{x}_i'\boldsymbol{\beta}}, \ldots)' \in \mathbb{R}^n$ denote pointwise exponentiation of $\mathbf{X}\boldsymbol{\beta}$. The likelihood can be equivalently regarded as a function of $\boldsymbol{\mu}$, $\boldsymbol{\ell}$ ($\in \mathcal{M}$) or $\boldsymbol{\beta}$ ($\in \mathbb{R}^p$). In terms of $\boldsymbol{\mu}$, the likelihood is

$$\prod_{i=1}^{n} \frac{e^{-\mu_i}(\mu_i)^{y_i}}{y_i!}.$$

Expressing the mean in terms of $\boldsymbol{\beta}$, the loglikelihood is, apart from constants,

$$\ln(L(\boldsymbol{\beta})) := \sum_{i=1}^{n} y_i \ln \mu_i - \sum_{i=1}^{n} \mu_i$$

$$= \sum_{i=1}^{n} y_i \mathbf{x}_i' \boldsymbol{\beta} - \sum_{i=1}^{n} e^{\mathbf{x}_i' \boldsymbol{\beta}}$$

$$= \mathbf{Y}' \mathbf{X} \boldsymbol{\beta} - \sum_{i=1}^{n} e^{\mathbf{x}_i' \boldsymbol{\beta}}.$$

The question studied in this section is the existence and uniqueness of a vector $\hat{\boldsymbol{\beta}} \in \mathbb{R}^p$ such that $\ln(L(\hat{\boldsymbol{\beta}})) = \sup_{\boldsymbol{\beta} \in \mathbb{R}^p} \ln(L(\boldsymbol{\beta}))$. It is more convenient to study this question in terms of $\boldsymbol{\beta}$ than either $\boldsymbol{\mu}$ or $\boldsymbol{\ell}$ because, as the following proposition shows, $\ln(L(\boldsymbol{\beta}))$ is strictly concave in $\boldsymbol{\beta}$.

Proposition 3.2.1. (i) *If* $\hat{\boldsymbol{\beta}}$ *solves*

$$\mathbf{X}' \mathbf{Y} = \mathbf{X}' \mathrm{e}^{\mathbf{X} \boldsymbol{\beta}} \tag{3.2.1}$$

then $\hat{\boldsymbol{\beta}}$ ($\hat{\boldsymbol{\ell}} := \mathbf{X}\hat{\boldsymbol{\beta}}$, $\hat{\boldsymbol{\mu}} := \mathrm{e}^{\mathbf{X}\hat{\boldsymbol{\beta}}}$) *is the unique* MLE *of* $\boldsymbol{\beta}$ ($\boldsymbol{\ell}, \boldsymbol{\mu}$) *under* $(\mathcal{M}, P)$ *where* $(\mathcal{M}, P)$ *denotes the* LLM *with* $\boldsymbol{\ell} \in \mathcal{M}$ *and Poisson sampling.*

(ii) *If* (3.2.1) *has no solution then the* MLE *of* $\boldsymbol{\beta}$ ($\boldsymbol{\ell}, \boldsymbol{\mu}$) *does not exist.*

Proof. The proof proceeds by showing $\ln(L(\boldsymbol{\beta}))$ is strictly concave in $\boldsymbol{\beta}$ and that the gradient $\nabla \ln(L(\boldsymbol{\beta})) = \mathbf{0}_p$ if and only if (3.2.1) holds. Calculation gives

$$\nabla \ln(L(\boldsymbol{\beta})) = \mathbf{X}' \mathbf{Y} - \sum_{i=1}^{n} \mathbf{x}_i e^{\mathbf{x}_i' \boldsymbol{\beta}}$$

$$= \mathbf{X}' \mathbf{Y} - (\mathbf{x}_1, \ldots, \mathbf{x}_n) \begin{pmatrix} \vdots \\ e^{\mathbf{x}_i' \boldsymbol{\beta}} \\ \vdots \end{pmatrix}$$

$$= \mathbf{X}' \mathbf{Y} - \mathbf{X}' \mathrm{e}^{\mathbf{X} \boldsymbol{\beta}}.$$

Thus $\nabla \ln(L(\boldsymbol{\beta})) = \mathbf{0}_p$ if and only if (3.2.1) holds. The strict concavity of $\ln(L(\boldsymbol{\beta}))$ is established by applying Proposition A.2.2 (of Appendix A.2). Calculation yields $\nabla^2 \ln(L(\boldsymbol{\beta})) = -\mathbf{X}' \mathbf{D}(\boldsymbol{\beta}) \mathbf{X}$ where $\mathbf{D} := \mathrm{Diag}(\ldots, [e^{\mathbf{x}_i' \boldsymbol{\beta}}], \ldots)$. If $\mathbf{D}^{1/2} := \mathrm{Diag}(\ldots, [e^{\mathbf{x}_i' \boldsymbol{\beta}}]^{1/2}, \ldots)$ then $\nabla^2 \ln(L(\boldsymbol{\beta})) = -\mathbf{X}'(\mathbf{D}^{1/2})' \mathbf{D}^{1/2} \mathbf{X}$. Thus $\nabla^2 \ln(L(\boldsymbol{\beta}))$ is negative definite if and only if

$$\begin{aligned} \|\mathbf{D}^{1/2} \mathbf{X} \mathbf{w}\| &> 0 \;\; \forall \, \mathbf{w} \in \mathbb{R}^p, \;\; \mathbf{w} \neq \mathbf{0}_p \\ &\Longleftrightarrow \mathbf{D}^{1/2} \mathbf{X} \mathbf{w} \neq \mathbf{0}_p \;\; \forall \, \mathbf{w} \in \mathbb{R}^p, \;\; \mathbf{w} \neq \mathbf{0}_p \\ &\Longleftrightarrow \mathrm{rank}(\mathbf{D}^{1/2} \mathbf{X}) = p \;\; (\mathbf{D}^{1/2} \text{ is nonsingular}) \\ &\Longleftrightarrow \mathrm{rank}(\mathbf{X}) = p. \end{aligned}$$

The last statement is true by assumption which completes the proof. □

Two characterizations of the MLE are particularly useful. Let $\mathbf{P}$ be the projection matrix from $\mathbb{R}^n$ onto $\mathcal{M}$. Then (3.2.1) implies $\mathbf{X}'(\mathbf{Y} - e^{\mathbf{X}\boldsymbol{\beta}}) = \mathbf{0}_p$, which holds if and only if $\mathbf{Y} - e^{\mathbf{X}\boldsymbol{\beta}}$ is orthogonal to the subspace $\mathcal{M}$. The latter is equivalent to $\mathbf{P}(\mathbf{Y} - e^{\mathbf{X}\boldsymbol{\beta}}) = \mathbf{0}_p$. So $\hat{\boldsymbol{\beta}}$ ($\hat{\boldsymbol{\mu}}$) is the MLE of $\boldsymbol{\beta}$ ($\boldsymbol{\mu}$) if and only if

$$\mathbf{PY} = \mathbf{P}e^{\mathbf{X}\hat{\boldsymbol{\beta}}} = \mathbf{P}\hat{\boldsymbol{\mu}}. \tag{3.2.2}$$

Statement (3.2.2), that the projection of the data onto an appropriate linear space be equal to the projection of the estimated mean $\hat{\boldsymbol{\mu}}$, is precisely the characterization of the MLE of $E[\mathbf{Y}]$ in the more familiar linear model case.

A second characterization of the MLE of $\boldsymbol{\mu}$ can be obtained by considering an arbitrary spanning set (not necessarily a basis) for $\mathcal{M}$. The paragraph above shows that if $\hat{\boldsymbol{\mu}} > 0$ satisfies (i) $\ln(\hat{\boldsymbol{\mu}}) \in \mathcal{M}$ and (ii) $\mathbf{PY} = \mathbf{P}\hat{\boldsymbol{\mu}}$, then $\hat{\boldsymbol{\mu}}$ is the MLE of $\boldsymbol{\mu}$. Condition (ii) is often verified in the following manner. Suppose that $\mathcal{M} = \mathcal{C}(\boldsymbol{\nu}_1, \ldots, \boldsymbol{\nu}_k)$ with $k \geq p$ then (ii) holds if and only if $\mathbf{Y}'\boldsymbol{\nu}_j = \hat{\boldsymbol{\mu}}'\boldsymbol{\nu}_j$ for $1 \leq j \leq k$ by Lemma A.1.1. Observe $\boldsymbol{\nu}_1, \ldots, \boldsymbol{\nu}_k$ need only span $\mathcal{M}$ but need not be a basis. The examples below rely on this second characterization.

Example 3.1.1 (continued). Let $Y_1, \ldots, Y_n$ be iid Poisson observations with positive mean λ; then $\boldsymbol{\mu}' = (\lambda, \ldots, \lambda)$ and $\boldsymbol{\ell} \in \mathcal{M} := \mathcal{C}(\mathbf{1}_n)$. Using (3.2.2), if $\hat{\boldsymbol{\mu}} > 0$ satisfies $\ln(\hat{\boldsymbol{\mu}}) \in \mathcal{C}(\mathbf{1}_n)$ and $\mathbf{1}_n'\mathbf{Y} = \mathbf{1}_n'\hat{\boldsymbol{\mu}}$, then $\hat{\boldsymbol{\mu}}$ is the MLE of $\boldsymbol{\mu}$. But $\hat{\boldsymbol{\mu}} > 0$ and $\ln(\hat{\boldsymbol{\mu}}) \in \mathcal{C}(\mathbf{1}_n)$ if and only if $\hat{\boldsymbol{\mu}} = \mathbf{1}_n\beta_0$ for some $\beta_0 > 0$. Further, $\mathbf{1}_n'\mathbf{Y} = \mathbf{1}_n'\mathbf{1}_n\beta_0$ if and only if $\sum_{i=1}^n Y_i = n\beta_0$, or $\beta_0 = \overline{Y}$. So if $\overline{Y} > 0$, then $\hat{\boldsymbol{\mu}} = \mathbf{1}_n\overline{Y}$ is the MLE of $\boldsymbol{\mu}$ while if $\overline{Y} = 0$ then the MLE of $\boldsymbol{\mu}$ does not exist. The latter can easily be seen by direct examination of the loglikelihood: $\overline{Y} = 0$ if and only if $Y_i = 0$ for $1 \leq i \leq n$; thus the loglikelihood is $-n\lambda$, and $\sup\{-n\lambda : \lambda > 0\}$ is not achieved. Furthermore the loglikelihood is not strictly concave in λ. However in terms of β, the loglikelihood $\ln L(\beta) = -ne^{\beta}$ *is* strictly concave as a function of β. This example shows the importance of using the β representation of the mean in the proof of Proposition 3.2.1.

Example 3.2.1. Suppose $\{Y_{ij} : 1 \leq i \leq R, 1 \leq j \leq C\}$ are mutually independent Poisson observations where Y_{ij} has mean $\mu_{ij} > 0$. Let $\boldsymbol{\mu}$ be the R by C matrix $\{\mu_{ij}\}$, $\boldsymbol{\ell} = \{\ell_{ij} = \ln(\mu_{ij})\}$, and assume that $\boldsymbol{\ell} \in \mathcal{M} = \mathcal{C}(\boldsymbol{\rho}_1, \ldots, \boldsymbol{\rho}_R, \boldsymbol{\kappa}_1, \ldots, \boldsymbol{\kappa}_C)$ where the $\boldsymbol{\rho}_i$ and the $\boldsymbol{\kappa}_j$ are defined in (3.1.1) and (3.1.2). The problem is to determine the MLE of $\boldsymbol{\mu}$. Following the technique of Example 3.1.1 to establish Proposition 3.2.1, if $\hat{\boldsymbol{\mu}} > 0$ satisfies $\ln(\hat{\boldsymbol{\mu}}) \in \mathcal{M}$, $\boldsymbol{\rho}_i'\mathbf{Y} = \boldsymbol{\rho}_i'\hat{\boldsymbol{\mu}}$ for $i = 1(1)R$, and $\boldsymbol{\kappa}_j'\mathbf{Y} = \boldsymbol{\kappa}_j'\hat{\boldsymbol{\mu}}$ for $j = 1(1)C$, then $\hat{\boldsymbol{\mu}}$ is the MLE of $\boldsymbol{\mu}$. Now $\boldsymbol{\rho}_i'\mathbf{Y} = \boldsymbol{\rho}_i'\hat{\boldsymbol{\mu}}$ if and only if $Y_{i+} = \hat{\mu}_{i+}$ and $\boldsymbol{\kappa}_j'\mathbf{Y} = \boldsymbol{\kappa}_j'\hat{\boldsymbol{\mu}}$ if and only if $Y_{+j} = \hat{\mu}_{+j}$. Thus if $Y_{i+} > 0$ and $Y_{+j} > 0$ for all i and j, then

$$\hat{\mu}_{ij} := \frac{Y_{i+}Y_{+j}}{Y_{++}} > 0$$

satisfies $\ln(\hat{\boldsymbol{\mu}}) \in \mathcal{M}$, $\hat{\mu}_{i+} = Y_{i+}$ for $i = 1(1)R$ and $\hat{\mu}_{+j} = Y_{+j}$ for $j = 1(1)C$. Hence $\hat{\boldsymbol{\mu}}$ is the unique MLE of $\boldsymbol{\mu}$. Conversely, if $Y_{i+} = 0$ for some $i = 1(1)R$ or $Y_{+j} = 0$ for some $j = 1(1)C$, then it is impossible to find $\hat{\boldsymbol{\mu}} > 0$ satisfying $\hat{\mu}_{i+} = 0$ or $\hat{\mu}_{+j} = 0$ for that i or j. Thus the MLE of $\boldsymbol{\mu}$ exists if and only if $Y_{i+} > 0$ for $i = 1(1)R$ *and* $Y_{+j} > 0$ for $j = 1(1)C$.

The previous two examples derive simple necessary and sufficient conditions for the existence of the MLE. The following result gives necessary and sufficient conditions for the existence of the MLE in a general LLM under P sampling. These conditions can be verified in specific cases by solving an associated linear programming problem.

Proposition 3.2.2. *The* MLE *of* $\boldsymbol{\mu}$ *under the* $(\mathcal{M}, P)$ *model exists if and only if there exists a vector* $\boldsymbol{\delta} \in \mathbb{R}^n$ *orthogonal to* $\mathcal{M}$ *such that* $Y_i + \delta_i > 0$ *for all* $1 \leq i \leq n$.

Proof. ($\Rightarrow$) Let $\hat{\boldsymbol{\mu}} > 0$ be the MLE of $\boldsymbol{\mu}$ and define $\boldsymbol{\delta} := \hat{\boldsymbol{\mu}} - \mathbf{Y}$. Clearly $\boldsymbol{\delta}$ is orthogonal to $\mathcal{M}$ by (3.2.2) and $\mathbf{0}_n < \hat{\boldsymbol{\mu}} = \mathbf{Y} + (\hat{\boldsymbol{\mu}} - \mathbf{Y}) = \mathbf{Y} + \boldsymbol{\delta}$.

($\Leftarrow$) Suppose there exists a $\boldsymbol{\delta}$ orthogonal to $\mathcal{M}$ with $\mathbf{Y} + \boldsymbol{\delta} > \mathbf{0}_n$. It must be shown that there exists $\hat{\boldsymbol{\beta}}$ satisfying

$$\ln(L(\hat{\boldsymbol{\beta}})) = \sup_{\boldsymbol{\beta} \in \mathbb{R}^p} \ln(L(\boldsymbol{\beta})).$$

For $\mathcal{M} = \mathcal{C}(\mathbf{X})$ the proof of Proposition 3.2.1 showed that

$$\begin{aligned} \ln(L(\boldsymbol{\beta})) &= \sum_{i=1}^{n} Y_i \mathbf{x}_i' \boldsymbol{\beta} - \sum_{i=1}^{n} \exp\{\mathbf{x}_i' \boldsymbol{\beta}\} \\ &= \sum_{i=1}^{n} (Y_i + \delta_i) \mathbf{x}_i' \boldsymbol{\beta} - \sum_{i=1}^{n} \exp\{\mathbf{x}_i' \boldsymbol{\beta}\}, \end{aligned}$$

where the second equality follows from the fact that $\boldsymbol{\delta}' \mathbf{X} = \mathbf{0}_p$ since $\boldsymbol{\delta}$ is orthogonal to $\mathcal{M}$. Fix any i between 1 and n and let $z := \mathbf{x}_i' \boldsymbol{\beta}$. Then $Q(z) := (Y_i + \delta_i) z - e^z \to -\infty$ as $|z| \to +\infty$; the convergence of $Q(z)$ to $-\infty$ as $z \to -\infty$ requires $(Y_i + \delta_i) > 0$. Obviously $Q(z)$ is bounded above. Set $\mathcal{D} := \{\boldsymbol{\beta} \in \mathbb{R}^p \colon L(\boldsymbol{\beta}) \geq L(\mathbf{0}_p)\}$. It suffices to show that $\mathcal{D}$ is closed and bounded since $L(\boldsymbol{\beta})$ will then attain its supremum over $\mathcal{D}$ and $\sup_{\mathcal{D}} L(\boldsymbol{\beta}) = \sup_{\mathbb{R}^p} L(\boldsymbol{\beta})$. The set $\mathcal{D}$ is closed as it is the inverse image of a closed set under a continuous function. To see that $\mathcal{D}$ is bounded argue by contradiction. If $\mathcal{D}$ is not bounded then there exists a sequence $\boldsymbol{\beta}^k$ in $\mathbb{R}^p$ such that $\lim_{k \to \infty} |\beta_j^k| = +\infty$ for some component $j = 1(1)p$. Choose i, $1 \leq i \leq n$, such that the jth component of $\mathbf{x}_i$ is nonzero; such an i must exist for if not then $\mathbf{X}$ contains a column of zeroes contradicting the assumption it is of full rank. Setting $z_k := \mathbf{x}_i' \boldsymbol{\beta}^k$, it follows that $|z_k| \to +\infty$ as $k \to +\infty$ which implies $Q(z_k) \to -\infty$. But $Q(z_k)$ is one of the (bounded!)

summands in $\ln L(\beta^k)$ and hence $\ln L(\beta^k) \to -\infty$ which contradicts the assumption $\{\beta^k\} \subset \mathcal{D}$. □

One immediate application of Proposition 3.2.2 is that if $Y_i > 0$ for all $i \in \mathcal{I}$ then the MLE of $\boldsymbol{\mu}$ exists for any subspace $\mathcal{M}$ because $\boldsymbol{\delta}$ can be chosen to be the zero vector.

Proposition 3.2.2 also gives a systematic method for determining whether $\hat{\boldsymbol{\mu}}$ exists. Choose a matrix $\mathbf{X}$ whose columns span $\mathcal{M}$ (but are not necessarily a basis). Then $\hat{\boldsymbol{\mu}}$ exists if and only if the objective function of the following problem is positive at its optimum:

$$\max_{\boldsymbol{\delta}} \left\{ \min_i (Y_i + \delta_i) \right\} \quad \text{subject to} \quad \boldsymbol{\delta}'\mathbf{X} = \mathbf{0}'_p.$$

This problem can be formulated as the following linear program:

$$\text{maximize } s \text{ subject to } \boldsymbol{\delta}'\mathbf{X} = \mathbf{0}_p, \quad \mathbf{Y} + \boldsymbol{\delta} \geq s\mathbf{1}_n, \tag{3.2.3}$$

and hence standard linear programming techniques can be used to solve it (Chvatal, 1980). The variable s is the largest that the minimal component of $\mathbf{Y}+\boldsymbol{\delta}$ can be made among all choices of $\boldsymbol{\delta}$ orthogonal to $\mathcal{M}$. The following example illustrates the method.

Example 3.2.2. Consider mutually independent Poisson observations $\{Y_{ijk} : i = 1,2,\ j = 1,2,\ k = 1,2\}$ where Y_{ijk} has mean μ_{ijk}. Using the standard ANOVA decomposition of means for a 3-factor experiment, consider the problem of finding the maximum likelihood estimators of the $\{\mu_{ijk}\}$ when the logmeans satisfy the "no-three-factor-interaction" model

$$\ell_{ijk} = \lambda + \lambda_i^1 + \lambda_j^2 + \lambda_k^3 + \lambda_{ij}^{12} + \lambda_{ik}^{13} + \lambda_{jk}^{23}, \tag{3.2.4}$$

where

$$\lambda_+^1 = \lambda_+^2 = \lambda_+^3 = 0,$$

and

$$\lambda_{+j}^{12} = \lambda_{+k}^{13} = \lambda_{i+}^{12} = \lambda_{+k}^{23} = \lambda_{i+}^{13} = \lambda_{j+}^{23} = 0$$

for $i, j, k = 1, 2$. The interpretation of this model will be discussed in Section 4.3.

The linear space $\mathcal{M}$ of logmeans satisfying (3.2.4) is spanned by the columns of the matrix

$$\mathbf{X} = \begin{pmatrix} 1&0&0&0&1&0&0&0&1&0&0&0 \\ 1&0&0&0&0&1&0&0&0&1&0&0 \\ 0&1&0&0&0&0&1&0&1&0&0&0 \\ 0&1&0&0&0&0&0&1&0&1&0&0 \\ 0&0&1&0&1&0&0&0&0&0&1&0 \\ 0&0&1&0&0&1&0&0&0&0&0&1 \\ 0&0&0&1&0&0&1&0&0&0&1&0 \\ 0&0&0&1&0&0&0&1&0&0&0&1 \end{pmatrix}$$

where the rows of $\mathbf{X}$ correspond to the (i, j, k)th element of $\mathbf{Y}$ in the order (111), (121), (112), (122), (211), (221), (212), (222). The linear programming problem in (3.2.3) for determining the existence of the MLE will be solved for two data sets $\mathbf{Y}$ presented in Haberman (1973b). First suppose that $\mathbf{Y}' = (0, 8, 4, 8, 9, 13, 6, 0)$; observe that $\mathbf{Y}$ is not strictly positive and so the existence of the MLE does not follow from the comment after Proposition 3.2.2. However it is easy to check that all 2-way marginal totals Y_{i+k}, Y_{+jk} and Y_{ij+} are positive which might lead one to conjecture that the MLE of $\boldsymbol{\mu}$ exists for the 2-way interaction model (3.2.4). From (3.2.3) the MLE exists if and only if the solution to

$$\max\ s \quad \text{subject to} \quad \mathbf{Y} + \boldsymbol{\delta} \geq s\mathbf{1}_8, \quad \boldsymbol{\delta}'\mathbf{X} = \mathbf{0}_{12}$$

is positive. To reformulate this as a standard linear program in nonnegative variables, let $s := s_1 - s_2$, $s_1 \geq 0$, $s_2 \geq 0$, and $\boldsymbol{\delta} := \boldsymbol{\delta}_1 - \boldsymbol{\delta}_2$ with $\boldsymbol{\delta}_i \in \mathbb{R}^8$ with $\boldsymbol{\delta}_i \geq 0$ for $i = 1, 2$. Then

$$\max\{s_1 - s_2\} \quad \text{subject to}$$

$$\begin{bmatrix} -\mathbf{I}_8 & \mathbf{I}_8 & \mathbf{1}_8 & -\mathbf{1}_8 \\ \mathbf{X}' & -\mathbf{X}' & \mathbf{0}_{12} & \mathbf{0}_{12} \\ -\mathbf{X}' & \mathbf{X}' & \mathbf{0}_{12} & \mathbf{0}_{12} \end{bmatrix} \begin{bmatrix} \boldsymbol{\delta}_1 \\ \boldsymbol{\delta}_2 \\ s_1 \\ s_2 \end{bmatrix} \leq \begin{bmatrix} \mathbf{Y} \\ \mathbf{0}_{24} \end{bmatrix} \tag{3.2.5}$$

must be solved. Standard simplex code yields an optimal solution of $(\boldsymbol{\delta}_1', \boldsymbol{\delta}_2', s_1, s_2) = \mathbf{0}_{18}$ with corresponding optimal objective function value of $s = s_1 - s_2 = 0$. Hence, the MLE does not exist.

Now consider $\mathbf{Y}' = (0, 8, 4, 8, 9, 0, 6, 10)$; again $\mathbf{Y}$ is not strictly positive but all two dimensional marginal subtables are strictly positive. Solving (3.2.5) yields optimal values $\boldsymbol{\delta}_1' = (2, 0, 0, 2, 0, 2, 2, 0)$, $\boldsymbol{\delta}_2' = (0, 2, 2, 0, 2, 0, 0, 0)$, and $(s_1, s_2) = (2, 0)$. The optimal objective function value of $s = s_1 - s_2 = 2$ is positive hence the MLE exists. Note that 2 equals the minimum element of $\mathbf{Y}' + \boldsymbol{\delta}' = (2, 6, 2, 10, 7, 2, 8, 10)$.

Example 3.2.2 shows that positivity of all two-dimensional marginals is not sufficient to guarantee the existence of the MLE under the no-three-factor interaction model (3.2.4) even though the model involves only two-dimensional λ-terms. This is in contrast to Example 3.2.1 where the positivity of the one-dimensional marginal totals Y_{i+} and Y_{+j} for two-dimensional data Y_{ij} does guarantee the existence of the MLE for the independence model (3.1.3) which involves only one-dimensional λ-terms.

3.3 Maximum Likelihood Estimation for Loglinear Models Under (Product) Multinomial Sampling

Suppose data $\{Y_i : i \in \mathcal{I}\}$ with $\mathcal{I} = \{1, \ldots, n\}$ satisfy the following sampling model. There exists $S \geq 1$ and a partition $\mathcal{I}_1, \ldots, \mathcal{I}_S$ of $\mathcal{I}$ (i.e., $\mathcal{I} = \cup_{j=1}^{S} \mathcal{I}_j$ with $\mathcal{I}_j \cap \mathcal{I}_k = \emptyset$ for all $j \neq k$) so that the sets $\{Y_i : i \in \mathcal{I}_j\}$, $j = 1(1)S$, have mutually independent multinomial distributions with $\{Y_i : i \in \mathcal{I}_j\}$ based on t_j cells and m_j trials. Note that the cardinality of $\mathcal{I}_j$ is t_j and that $\sum_{j=1}^{S} t_j = n$. Let $\mathbf{p}^j = \{p_i : i \in \mathcal{I}_j\}$ be the vector of cell probabilities for the jth multinomial with $\sum_{i \in \mathcal{I}_j} p_i = 1$. When $S = 1$, the data arise from multinomial sampling; when $S > 1$, they arise from product multinomial sampling.

Given a subspace $\mathcal{M} \subset \mathbb{R}^n$, let

$$\tilde{\mathcal{M}} := \left\{ \ell \in \mathcal{M} : \sum_{i \in \mathcal{I}_j} \exp\{\ell_i\} = m_j,\ j = 1, \ldots, S \right\}.$$

Then $\tilde{\mathcal{M}}$ consists of exactly those logmean vectors in $\mathcal{M}$ that satisfy the multinomial sampling constraints of the data. The notation $(\tilde{\mathcal{M}}, PM)$ will be used to denote the product multinomial loglinear model for $\mathbf{Y}$. At times when $S = 1$ it will be convenient to use the notation $(\tilde{\mathcal{M}}, M)$ to emphasize that the data satisfy multinomial sampling.

To derive the likelihood for $\mathbf{Y}$, proceed as in the previous section by choosing a basis matrix $\mathbf{X}$ for $\mathcal{M}$. Then the likelihood is

$$\prod_{j=1}^{S} m_j! \prod_{i \in \mathcal{I}_j} \left(\frac{p_i^{y_i}}{y_i!} \right) \frac{m_j^{y_i}}{m_j^{y_i} m_j}.$$

Ignoring constants and expressing the loglikelihood in terms of $\boldsymbol{\beta}$,

$$\begin{aligned} \ln L(\boldsymbol{\beta}) &= \sum_{j=1}^{S} \sum_{i \in \mathcal{I}_j} y_i \ln(m_j p_i) \\ &= \sum_{j=1}^{S} \sum_{i \in \mathcal{I}_j} y_i \mathbf{x}_i' \boldsymbol{\beta}, \end{aligned}$$

where for some $\boldsymbol{\beta}$, $\ln(m_j p_i) = \ell_i = \mathbf{x}_i' \boldsymbol{\beta}$ for all $j = 1(1)S$ and $i \in \mathcal{I}_j$.

The issue now addressed is the existence and uniqueness of $\hat{\boldsymbol{\beta}} \in \mathbb{R}^p$ such that $\mathbf{X}\hat{\boldsymbol{\beta}} \in \tilde{\mathcal{M}}$ and

$$\ln L(\hat{\boldsymbol{\beta}}) = \max_{\boldsymbol{\beta} : \mathbf{X}\boldsymbol{\beta} \in \tilde{\mathcal{M}}} \ln L(\boldsymbol{\beta}).$$

The following notation is required to analyze this problem. Let $\boldsymbol{\nu}^j$ be a 0/1 incidence vector defined by

$$\nu_i^j = \begin{cases} 1, & i \in \mathcal{I}_j \\ 0, & \text{otherwise.} \end{cases}$$

For example, suppose C mutually independent multinomial vectors have distributions $(Y_{1j}, \ldots, Y_{Rj}) \sim M_R(m, \mathbf{p}^j)$ for $j = 1, \ldots, C$, respectively. Then, $n = RC$, $S = C$, and $m_j = m$, and $t_j = R$ for $1 \leq j \leq C$. Further, $\mathcal{I} = \{(i,j) : 1 \leq i \leq R,\ 1 \leq j \leq C\}$, and for $1 \leq j \leq C$, $\mathcal{I}_j = \{(i,j) : 1 \leq i \leq R\}$ and $\boldsymbol{\nu}^j = \boldsymbol{\kappa}_j$ (given in (3.1.2)). The final pieces of notation required to study the existence of the MLE under PM sampling are $\hat{\boldsymbol{\mu}}^P$ $(\hat{\boldsymbol{\beta}}^P, \hat{\boldsymbol{\ell}}^P)$ which denote the (unique) MLE of $\boldsymbol{\mu}$ $(\boldsymbol{\beta}, \boldsymbol{\ell})$ under the $(\mathcal{M}, P)$ model when it exists, and $\hat{\boldsymbol{\mu}}^M$ $(\hat{\boldsymbol{\beta}}^M, \hat{\boldsymbol{\ell}}^M)$ which denote any MLE of $\boldsymbol{\mu}$ $(\boldsymbol{\beta}, \boldsymbol{\ell})$ under the $(\tilde{\mathcal{M}}, PM)$ model when one exists. The main result follows.

Proposition 3.3.1. *If* $\boldsymbol{\nu}^j \in \mathcal{M}$ *for* $j = 1(1)S$ *then the following hold:*

(i) *If* $\hat{\boldsymbol{\ell}}^P$ *exists then* $\hat{\boldsymbol{\ell}}^M$ *exists, is unique, and* $\hat{\boldsymbol{\ell}}^M = \hat{\boldsymbol{\ell}}^P$.

(ii) *If* $\hat{\boldsymbol{\ell}}^M$ *is an* $(\tilde{\mathcal{M}}, PM)$*-MLE of* $\boldsymbol{\ell}$, *then* $\hat{\boldsymbol{\ell}}^P := \hat{\boldsymbol{\ell}}^M$ *is the unique* $(\mathcal{M}, P)$*-MLE of* $\boldsymbol{\ell}$ *(and thus* $\hat{\boldsymbol{\ell}}^M$ *is the unique* $(\tilde{\mathcal{M}}, PM)$*-MLE of* $\boldsymbol{\ell}$ *by* (i)).

To restate the result in words, under the hypothesis of Proposition 3.3.1, the two MLEs exist together and when they exist are unique and equal. From a practical point of view, (i) is the most important half of the result as it shows that if one has available (computer software for evaluating) $\hat{\boldsymbol{\ell}}^P$ and knows $\boldsymbol{\nu}^j \in \mathcal{M}$ for $j = 1(1)S$, then $\hat{\boldsymbol{\ell}}^M$ is the same as $\boldsymbol{\ell}^P$.

Proof of Proposition 3.3.1(i). Suppose $\hat{\boldsymbol{\ell}}^P$ is the unique MLE of $\boldsymbol{\ell}$ under $(\mathcal{M}, P)$. Then $\mathbf{P}\hat{\boldsymbol{\mu}}^P = \mathbf{PY}$ where $\mathbf{P}$is the projection matrix from $\mathbb{R}^n$ onto $\mathcal{M}$ and $\hat{\mu}^P{}_i = \exp(\hat{\ell}_i^P)$. This implies that $\boldsymbol{\nu}_j'\hat{\boldsymbol{\mu}}^P = \boldsymbol{\nu}_j'\mathbf{Y}$ since $\boldsymbol{\nu}_j \in \mathcal{M}$ for $j = 1(1)S$. Thus

$$\sum_{i \in \mathcal{I}_j} \exp\{\hat{\ell}_i^P\} = \sum_{i \in \mathcal{I}_j} Y_i = m_j$$

for $j = 1(1)S$ and $\hat{\boldsymbol{\ell}}^P \in \tilde{\mathcal{M}}$.

Since $\hat{\boldsymbol{\ell}}^P$ is the unique MLE under the $(\mathcal{M}, P)$ model,

$$\sum_{i \in \mathcal{I}} Y_i \hat{\ell}_i^P - \sum_{i \in \mathcal{I}} \exp\{\hat{\ell}_i^P\} > \sum_{i \in \mathcal{I}} Y_i \ell_i - \sum_{i \in \mathcal{I}} \exp(\ell_i)$$

for all $\boldsymbol{\ell} \in \mathcal{M} \setminus \{\hat{\boldsymbol{\ell}}^P\}$. Equivalently, for all $\boldsymbol{\ell} \in \mathcal{M} \setminus \{\hat{\boldsymbol{\ell}}^P\}$,

$$\sum_{i \in \mathcal{I}} Y_i \hat{\ell}_i^P - \sum_{j=1}^{S} \sum_{i \in \mathcal{I}_j} \exp\{\hat{\ell}_i^P\} > \sum_{i \in \mathcal{I}} Y_i \ell_i - \sum_{j=1}^{S} \sum_{i \in \mathcal{I}_j} \exp\{\ell_i\}. \tag{3.3.1}$$

Since $\sum_{i\in\mathcal{I}_j} \exp\{\ell_i\} = m_j$ for all $\boldsymbol{\ell} \in \tilde{\mathcal{M}} \setminus \{\hat{\boldsymbol{\ell}}^P\}$ and all $j = 1(1)S$, (3.3.1) implies that

$$\sum_{i\in\mathcal{I}} Y_i\hat{\ell}_i^P > \sum_{i\in\mathcal{I}} Y_i\ell_i$$

which proves that $\hat{\boldsymbol{\ell}}^P$ is the unique $(\tilde{\mathcal{M}}, PM)$-MLE of $\boldsymbol{\ell}$. □

The proof of (ii) is more complicated and it is given in Appendix 3. The proof of Proposition 3.3.1(i) shows that the assumption $\boldsymbol{\nu}^j \in \mathcal{M}, j = 1(1)S$, is used only to insure $\hat{\boldsymbol{\ell}}^P \in \tilde{\mathcal{M}}$ by proving it has the required marginal totals over $\mathcal{I}_1, \ldots, \mathcal{I}_S$. This suggests the following method for calculating the MLE of $\boldsymbol{\ell}$ under the $(\mathcal{M}, PM)$ model.

(a) Find the MLE of $\boldsymbol{\ell}$ under the $(\mathcal{M}, P)$ model.

(b) If $\sum_{i\in\mathcal{I}_j} \exp\{\hat{\ell}_i^P\} = m_j$, $j = 1, \ldots, S$, then assert that $\hat{\boldsymbol{\ell}}^P$ is the MLE under the $(\tilde{\mathcal{M}}, PM)$ model.

Example 3.1.2 (continued). Consider data $\{Y_{ij} : 1 \le i \le R, 1 \le j \le C\} \sim M_{RC}(m, \mathbf{p})$. Suppose $\boldsymbol{\ell} \in \mathcal{M} = \mathcal{C}(\boldsymbol{\rho}_1, \ldots, \boldsymbol{\rho}_R, \boldsymbol{\kappa}_1, \ldots, \boldsymbol{\kappa}_C)$ with $\boldsymbol{\rho}_i$ and $\boldsymbol{\kappa}_j$ as in (3.1.1) and (3.1.2), respectively; then $\tilde{\mathcal{M}} = \{\boldsymbol{\ell} \in \mathcal{M}\colon \sum_i \sum_j \exp\{\ell_{ij}\} = m\}$. Example 3.2.1 showed that if $Y_{i+} > 0$ for $i = 1(1)R$ and $Y_{+j} > 0$ for $j = 1(1)C$ then

$$\hat{\mu}_{ij}^P := \frac{Y_{i+}Y_{+j}}{Y_{++}} = \frac{Y_{i+}Y_{+j}}{m}$$

is the MLE of $\boldsymbol{\mu}$ under the $(\mathcal{M}, P)$ model. Further,

$$\sum_i \sum_j \hat{\mu}_{ij}^P = \frac{1}{m}(\sum_i Y_{i+})(\sum_j Y_{+j}) = m,$$

so that $\hat{\boldsymbol{\mu}}^P$ is also the MLE of $\boldsymbol{\mu}$ under the $(\tilde{\mathcal{M}}, PM)$ model.

Example 3.3.1. Consider C mutually independent multinomial vectors $\{Y_{ij} : 1 \le i \le R\} \sim M_R(m_j, \mathbf{p}^j)$ with $\mathbf{p}^j = (p_{1j}, \ldots, p_{Rj})'$ and $\sum_{i=1}^R p_{ij} = 1$ for $j = 1(1)C$. Then $\mathcal{I} = \{(i,j) : 1 \le i \le R,\ 1 \le j \le C\}$, $S = C$ and $\mathcal{I}_j = \{(i,j) : 1 \le i \le R\}$. Let $\mathcal{M}$ be as in Example 3.1.2. The goal is to derive the MLE of $\{\mu_{ij}\}$ ($\{p_{ij}\}$) under this sampling model. The logmean space is $\tilde{\mathcal{M}} = \{\boldsymbol{\ell} \in \mathcal{M}\colon \sum_{i=1}^R \exp\{\ell_{ij}\} = m_j,\ j = 1(1)C\}$. The following argument shows that the space $\tilde{\mathcal{M}}$ characterizes homogeneity of the C multinomial probability vectors. If $\mathbf{p}^1 = \ldots = \mathbf{p}^C$ then $p_{i1} = \ldots = p_{iC}$ for all $1 \le i \le R$ so $\ell_{ij} = \ln(m_j p_{ij}) = \ln(p_{i1}) + \ln(m_j)$, which implies $\boldsymbol{\ell} \in \mathcal{M}$. Further, $\sum_{i=1}^R \exp\{\ell_{ij}\} = \sum_{i=1}^R m_j p_{ij} = m_j$ which shows $\boldsymbol{\ell} \in \tilde{\mathcal{M}}$. Conversely, suppose $\boldsymbol{\ell} \in \tilde{\mathcal{M}}$ and $\mathbf{p}$ is defined by $p_{ij} := \exp\{\ell_{ij}\}/m_j$. Then there exist constants $\{a_i\}_1^R$, $\{b_j\}_1^C$, such that $\ell_{ij} = a_i + b_j$ and

$\sum_{i=1}^{R} \exp\{a_i + b_j\} = \exp\{b_j\} \sum_{i=1}^{R} \exp\{a_i\} = m_j$ for all $j = 1(1)C$. Equivalently, $\exp\{b_j\}/m_j = (\sum_{i=1}^{R} \exp\{a_i\})^{-1} =: D$, say, is independent of $j = 1(1)C$. Hence $p_{ij} = \exp\{a_i + b_j\}/m_j = D \times \exp\{a_i\}$ which is independent of j, and therefore $\mathbf{p}^1 = \ldots = \mathbf{p}^C$.

To determine the MLE of $\{\mu_{ij}\}$ under $(\tilde{\mathcal{M}}, PM)$ sampling, observe that the following indicator matrices are all in $\mathcal{M}$:

$$\boldsymbol{\nu}^1 = (\mathbf{1}_R \,|\, \mathbf{0}_{R,C-1}),\ \boldsymbol{\nu}^2 = (\mathbf{0}_R \,|\, \mathbf{1}_R \,|\, \mathbf{0}_{R,C-2}), \ldots, \boldsymbol{\nu}^C = (\mathbf{0}_{R,C-1} \,|\, \mathbf{1}_R).$$

Hence the hypothesis of Proposition 3.3.1 holds and the MLE of $\{\mu_{ij}\}$ exists under the homogeneity model if and only if the MLE exists under the $(\mathcal{M}, P)$ model. As noted in Example 3.1.2, the latter occurs if and only if $Y_{i+} > 0$ for $i = 1(1)R$ and $Y_{+j} > 0$ for $j = 1(1)C$. Since $Y_{+j} = m_j > 0$ for $j = 1(1)C$, the MLE will exist if and only if $Y_{i+} > 0$ for $i = 1(1)R$. In this case

$$\hat{\mu}_{ij}^M = \frac{(Y_{i+})(Y_{+j})}{Y_{++}} = m_j \times \frac{Y_{i+}}{\sum_{k=1}^{C} m_k}$$

and, by invariance,

$$\hat{p}_{ij}^M = \frac{Y_{i+}}{\sum_{k=1}^{C} m_k}$$

are the MLEs of μ_{ij} and p_{ij}. The latter can intuitively be interpreted as a pooled estimate of the common probability of falling in row i obtained by collapsing the table across columns.

This section concludes by considering one consequence of Proposition 3.3.1, namely, the equivalence between likelihood ratio tests under the $(\mathcal{M}, P)$ model and the $(\tilde{\mathcal{M}}, PM)$ model. The precise setups are described in Testing Problems 3.1 and 3.2.

Testing Problem 3.1. Suppose $\{Y_i : i \in \mathcal{I}\}$ are mutually independent Poisson random variables with Y_i having positive mean μ_i. Assume the vector $\boldsymbol{\ell} = \{\ell_i = \ln(\mu_i) : i \in \mathcal{I}\}$ is in $\mathcal{M}$, a given subspace of $\mathbb{R}^n$. Given a proper subspace $V \subset \mathcal{M}$, consider testing $H_0 : \boldsymbol{\ell} \in V$ versus $H_A : \boldsymbol{\ell} \in \mathcal{M} \setminus V$.

Testing Problem 3.2. Suppose $\{Y_i : i \in \mathcal{I}\}$ satisfies $(\tilde{\mathcal{M}}, PM)$ sampling for a given $S \geq 1$ and partition $\{\mathcal{I}_j\}_{j=1}^S$ of $\mathcal{I}$ where $\tilde{\mathcal{M}} = \{\boldsymbol{\ell} \in \mathcal{M}: \sum_{i \in \mathcal{I}_j} \exp\{\ell_i\} = m_j,\ j = 1, \ldots, S\}$ and $\mathcal{M}$ is given in Problem 3.1. Let $\tilde{V} = \{\boldsymbol{\ell} \in V: \sum_{i \in \mathcal{I}_j} \exp\{\ell_i\} = m_j, j = 1(1)S\}$ where V is given in Problem 3.1. Test $\tilde{H}_0 : \boldsymbol{\ell} \in \tilde{V}$ versus $\tilde{H}_A : \boldsymbol{\ell} \in \tilde{\mathcal{M}} \setminus \tilde{V}$.

The following result stating the equality of the likelihood ratio tests for the two problems is a simple consequence of Proposition 3.2.1.

Proposition 3.3.2. *If $\boldsymbol{\nu}^j \in V$ for all $j = 1(1)S$, then the likelihood ratio statistic for testing H_0 versus H_A in Problem 3.1 exists if and only if the* *likelihood ratio statistic for $\tilde{H}_0$ versus $\tilde{H}_A$ in Problem 3.2 exists. Furthermore they are equal when they exist.*

3.4 Other Approaches

Sections 3.1–3.3 consider one class of models, loglinear models (LLMs), and one method of inference, maximum likelihood estimation. The present section will describe briefly alternative classes of models and alternative methods of inference for LLMs. First, analogs of the alternative estimators described in Chapter 2 are considered for Poisson and product multinomial data following a LLM. Second, alternative structural models to the LLM and alternative stochastic models to the Poisson and product multinomial distribution are discussed for data consisting of responses Y_i having means μ_i modeled as functions of continuous or discrete covariates.

A. Alternatives to Maximum Likelihood Estimation for Loglinear Models

The majority of research proposing alternatives to likelihood inference for LLMs is specialized to data from 2-way and 3-way tables following multinomial (or product multinomial) sampling. A Bayesian approach which has been explored more generally is described first. Then a few of the specialized methods are considered.

Bayesian methods for analyzing data following a LLM have been proposed based on both full rank and non-full rank representations of the model. If $\mathbf{X}$ is an $n \times p$ *basis* matrix for a linear space $\mathcal{M}$ describing the structural model, then the prior information can equivalently be specified in terms of $E[\mathbf{Y}] = \boldsymbol{\mu}$, $\boldsymbol{\ell}$, or $\boldsymbol{\beta}$ where

$$\ln(\boldsymbol{\mu}) = \boldsymbol{\ell} = \mathbf{X}\boldsymbol{\beta}. \tag{3.4.1}$$

Many of the LLMs for cross-classified data use the familiar ANOVA type parametrizations which lead to non-full-rank design matrices $\mathbf{Q}$ (with $\mathcal{M} = \mathcal{C}(\mathbf{Q})$), in which case Bayesian prior information can be specified for $\boldsymbol{\mu}$, $\boldsymbol{\ell}$, or $\boldsymbol{\lambda}$ where

$$\ln(\boldsymbol{\mu}) = \boldsymbol{\ell} = \mathbf{Q}\boldsymbol{\lambda}. \tag{3.4.2}$$

However identifiability constraints must be enforced in the non-full rank case to insure well-defined estimators because $\boldsymbol{\lambda}_1 = \boldsymbol{\lambda}_2$ does not imply $\mathbf{Q}\boldsymbol{\lambda}_1 = \mathbf{Q}\boldsymbol{\lambda}_2$.

To illustrate the difference between the formulations (3.4.1) and (3.4.2), consider the independence model (3.1.3) of Example 3.1.2 for an $R \times C$ matrix of cross-classified data. The model can be written as $\boldsymbol{\ell} = \mathbf{X}\boldsymbol{\beta}$ where $\mathbf{X} = [\boldsymbol{\nu}_1, \ldots, \boldsymbol{\nu}_{R+C-1}]$ denotes the $RC \times (R + C - 1)$ (full column rank) matrix with $\boldsymbol{\nu}_j$ a vector of length RC obtained by writing the elements of the matrix $\boldsymbol{\nu}_j$ of Example 3.1.2 in lexicographic order and where $\boldsymbol{\beta} \in \mathbb{R}^{R+C-1}$. Alternatively, the representation $\ell_{ij} = \lambda + \lambda_i^1 + \lambda_j^2$ leads to $\boldsymbol{\ell} = \mathbf{Q}\boldsymbol{\lambda}$ with the familiar $RC \times (R+C+1)$ (non-full column rank) incidence matrix $\mathbf{Q}$

$$\underbrace{\begin{bmatrix} \mathbf{1}_{RC} \\ \vdots \\ \end{bmatrix}}_{\text{1 col}} \underbrace{\begin{matrix} \mathbf{I}_R \\ \mathbf{I}_R \\ \vdots \\ \mathbf{I}_R \end{matrix}}_{\text{R cols}} \underbrace{\begin{matrix} \mathbf{1}_R & \mathbf{0}_{R,C-1} & & \\ \mathbf{0}_R & \mathbf{1}_R & \mathbf{0}_{R,C-2} & \\ \vdots & & & \\ \mathbf{0}_R & \mathbf{0}_{R,C-2} & \dots & \mathbf{1}_R \end{matrix}}_{C \text{ cols}}$$

Of course, the parameters in the latter representation must be subject to identifiability constraints such as $\lambda_+^1 = 0 = \lambda_+^2$.

In practice, it is usually easiest to model prior beliefs in terms of $\boldsymbol{\beta}$ or $\boldsymbol{\lambda}$. The former is more useful in the case of full rank representations used with continuous covariates since they describe the effects of covariates on $\boldsymbol{\mu}$. The latter are more useful in the case of the ANOVA models used for cross-classified data where the λ-terms are interpreted as main effects and higher order interactions. In general there is no natural conjugate prior for $\boldsymbol{\beta}$ or $\boldsymbol{\lambda}$ when data $\mathbf{Y}$ follow any of the models $(\mathcal{M}, P)$, $(\tilde{\mathcal{M}}, M)$, or $(\tilde{\mathcal{M}}, PM)$. Instead, the literature has focused on flat and normal priors. In particular, many of the Bayesian *estimators* described in Section 2.2 have been generalized to multinomial data for 2-way and 3-way tables. The examples below are selected to illustrate (i) both testing and estimation problems, (ii) full and non-full rank models, and (iii) normal and flat priors.

El-Sayyad (1973) considers the (full rank) LLM

$$\ell_i = \beta_1 + \beta_2 x_i, \quad 1 \le i \le n$$

for Poisson data with $\mathbf{x}' = (x_1, \dots, x_n)$ given. In our general notation, $p = 2$ and $X = [\mathbf{1}_n \,|\, \mathbf{x}]$. El-Sayyad's interest is in testing $\beta_2 = 0$; i.e., testing for a linear trend in the sequence of logmeans. He puts (improper) flat prior distributions on β_1 and β_2, and compares inferences based on the exact posterior distribution (obtained numerically) with those based on a computationally-simpler normal approximation to the posterior. His examples illustrate good agreement between the two approaches. Another full rank example in the same spirit is considered in detail in Section 5.4. It concerns Bayesian and related point estimators of regression coefficients based on normal priors in the logistic regression model for binary regression data (see Example 3.1.3).

Leonard (1975), Laird (1978a), and Nazaret (1987) study Bayesian and empirical Bayesian estimation based on non-full rank models for cross-classified data. Leonard (1975) starts with the fully saturated model

$$\ell_{ij} = \lambda + \lambda_i^1 + \lambda_j^2 + \lambda_{ij}^{12}$$

for the 2-way data of Example 3.1.2. He sets the normalizing constant $\lambda = 0$ and proposes the following hierarchical 2-stage model for prior belief based on the exchangeability of row and column classifications.

Stage 1: The $\{\lambda_i^1\}$, $\{\lambda_j^2\}$, and $\{\lambda_{ij}^{12}\}$ are mutually independent with $\{\lambda_i^1\}$ iid $N(\mu_1, \sigma_1^2)$, $\{\lambda_j^2\}$ iid $N(\mu_2, \sigma_2^2)$, and $\{\lambda_{ij}^{12}\}$ iid $N(\mu_{12}, \sigma_{12}^2)$.

Stage 2: The parameters μ_1, σ_1^2, μ_2, σ_2^2, μ_{12}, and σ_{12}^2 are mutually independent with μ_1, μ_2, and μ_{12} having flat priors and $\sigma_1^2/\nu_1\tau_1$, $\sigma_2^2/\nu_2\tau_2$, and $\sigma_{12}^2/\nu_{12}\tau_{12}$ having the reciprocals of χ^2 distributions with ν_1, ν_2, and ν_{12} degrees of freedom, respectively.

The quantities ν_1, ν_2, ν_{12}, τ_1, τ_2, and τ_{12} are *given* hyperparameters where τ_i is interpreted as the prior estimate for σ_i^2 because

$$E[\sigma_i^2] = E[\nu_i\tau_i/\chi^2_{\nu_i}] = \nu_i\tau_i/(\nu_i - 2) \simeq \tau_i$$

for $\nu_i \geq 3$, and ν_i is interpreted as the prior measure of conviction in τ_i because of a similar calculation of $\text{Var}(\sigma_i^2)$. Leonard separately considers the situations where the λ_{ij}^{12} are set to zero, and where they are unknown; in both cases he advocates calculation of posterior mode estimators for λ-parameters and, in the second stage, for σ_1^2, σ_2^2, and σ_{12}^2 as well. He gives algorithms for calculating the particular set of modes which satisfy the identifiability constraint $\sum_{i=1}^R \exp\{\lambda_i^1\} = n$. For example, when $\lambda_{ij}^{12} = 0$ for all i and j, the posterior mode estimates of λ_i^1 are pulled towards the *average* of the MLE's of λ_i^1 under the constraint $\sum_{i=1}^R \exp\{\lambda_i^1\} = n$; the same is true for the posterior mode estimates of λ_j^2 by symmetry. When the λ_{ij}^{12} are unknown, the effect is more complex as the estimates for λ_i^1 depend on τ_{12} and ν_{12} as well as τ_1 and ν_1.

Nazaret (1987) extends Leonard's Bayesian estimators to 3-way tables under both M and PM sampling. He also derives equations for calculating the posterior mode estimators and discusses the rate of convergence of iterative schemes to solve these estimating equations.

Laird (1978a) proposes an *empirical Bayes* estimator for a model very similar to that of Leonard (1975). In her formulation, λ is a normalizing constant, the λ_i^1 and λ_j^2 have flat prior distributions subject to the constraints $\lambda_+^1 = 0 = \lambda_+^2$, while the λ_{ij}^{12} are iid $N(0, \sigma^2)$. However instead of postulating a second-stage distribution for the single (first stage) hyperparameter σ, Laird considers estimating σ by maximum likelihood based on the marginal distribution of $\mathbf{Y}$ given σ. Although the EM algorithm can be applied to estimate σ, computational considerations lead Laird to suggest two approximations to the full EM calculations. Her empirical Bayes estimates of the means $\{\mu_i\}$ have the intuitive property that they approach those of the independence model (3.1.3) as σ^2 approaches zero, and they approach the observed cell proportions as σ^2 approaches infinity.

Good (1976) and Crook and Good (1980) propose two stage hierarchical Bayesian methods for *testing* the independence assumption in 2-way tables. The first stage prior is Dirichlet with (*known*) mean $\boldsymbol{\mu} = \mathbf{c}$ and *unknown* precision parameter K; the second stage prior for K is log-Cauchy.

Albert and Gupta (1982) derive Bayes *estimators* for 2-way tables based on a two stage prior similar to that of Good (1976) and Crook and Good (1980). In their formulation the first stage prior is Dirichlet with *unknown* K and *unknown* mean $\boldsymbol{\mu}$ satisfying either (i) the symmetry model ($p_{ij} = p_{ji}$ for all i and j) or (ii) the independence model ($p_{ij} = p_{i+}p_{+j}$ for all i and j). In the symmetry case, the second stage prior is uniform over symmetric $\{p_{ij}\}$; in the independence case, the vectors $\mathbf{p}_R := (p_{1+}, \ldots, p_{R+})'$ and $\mathbf{p}_C := (p_{+1}, \ldots, p_{+C})'$ have uniform priors over the $(R-1)$ and $(C-1)$ dimensional simplices, respectively. Albert and Gupta (1983a and 1983b) also consider Bayesian estimators based on two stage priors. The first stage employs a Dirichlet distribution as above but the second stage incorporates prior information about the cross product ratios or the correlation coefficient. Albert (1987) develops Bayesian and empirical Bayesian estimators for the case of prior belief in independence when the second stage prior on $\mathbf{p}_R$ and $\mathbf{p}_C$ is Jeffreys's noninformative prior rather than the uniform noninformative prior. Sutherland, Fienberg, and Holland (1974) derive pseudo-Bayes estimators which shrink towards the independence model; these and other pseudo-Bayes estimators are summarized in Möhner (1986).

Smoothing techniques are the second class of methods introduced in Section 2.2 for estimation of multinomial cell probabilities. This brief discussion of alternatives to MLE for arbitrary loglinear models concludes with a description of several generalizations of these methods.

In 2-way tables with cross-classified data $\{Y_{ij}\}$ and $m = \sum_{i,j} Y_{ij}$, one of the earliest proposals is Brown and Muenz (1976) who take into account only the independence model and the model of arbitrary cell probabilities $\{p_{ij}\}$. They adopt a testimator approach which estimates the set of p_{ij} either by their observed proportions Y_{ij}/m or the independence model estimates $Y_{i+}Y_{+j}/m^2$; the choice of estimator is based on the Pearson chi-squared test of the independence null hypothesis $H_0 : p_{ij} = p_{i+}p_{+j}$. There are also a variety of more sophisticated approaches. Simonoff (1983) generalizes his MPLE for unconstrained multinomial data by suggesting the following penalized likelihood function

$$\sum_{i=1}^{R}\sum_{j=1}^{C} Y_{ij}\ln(p_{ij}) - \beta \sum_{i=1}^{R-1}\sum_{j=1}^{C-1} \ln\left(\frac{p_{ij}p_{i+1,j+1}}{p_{i+1,j}p_{i,j+1}}\right). \tag{3.4.3}$$

Here, as in the one dimensional multinomial case (2.2.15), β is the smoothing parameter. The penalty in (3.4.3) smooths the estimates towards the *independence* model since the log cross-product ratios of each 2×2 subtable of adjacent rows and columns is pulled to zero. Aitchison and Aitken (1976), Titterington (1980), and Brown and Rundell (1985) discuss kernel methods for general d-dimensional tables.

Bunke (1985) proposes a complicated ad-hoc, although very effective, smoothing estimate of $\mathbf{p}$ based on the "Q-ratios," $\{Q_{ij} := p_{ij}/p_{i+}p_{+j} : 1 \leq i \leq R, 1 \leq j \leq C\}$, which measure deviation in the cells from the independence model. His estimator is based on averaging the estimated Q-ratios

$$\hat{Q}_{ij} = \frac{Y_{ij}/m}{Y_{i+}Y_{+j}/m^2}$$

in a data-dependent neighborhood $N(i,j)$ of the (i,j)-cell consisting of cells (i^*, j^*) such that $d((i,j), (i^*, j^*)) < s$ and $|\hat{Q}_{ij} - \hat{Q}_{i^*j^*}| < t$ where s and t are positive constants and d is a distance function. Two distance functions considered by Bunke are:

$$d((i,j),(i^*,j^*)) = |i - i^*| + |j - j^*|$$

and

$$d((i,j),(i^*,j^*)) = \begin{cases} 0 & \text{if } i = i^* \text{ and } j = j^* \\ 2 & \text{if } i \neq i^* \text{ and } j \neq j^* \\ 1 & \text{otherwise.} \end{cases}$$

If $\overline{Q}_{ij}$ is the average of the $\hat{Q}_{ij}$ in $N(i,j) \setminus \{(i,j)\}$, then smoothed Q-ratios are defined by $Q_{ij}^{SM} = \alpha\hat{Q}_{ij} + (1-\alpha)\overline{Q}_{ij}$ where $0 < \alpha < 1$ is a constant. The smoothed Q-ratios lead to smoothed estimates of the p_{ij} defined by

$$p_{ij}^{SM} = Q_{ij}^{SM}\left(\frac{Y_{i+}Y_{+j}}{m^2}\right).$$

Bunke's estimator is defined as the vector $\mathbf{p}^{BU}$ which fits the observed margins (i.e., $p_{i+}^{BU} = Y_{i+}/m$ and $p_{+j}^{BU} = Y_{+j}/m$) and which minimizes the Kullback–Liebler distance to $\mathbf{p}^{SM}$ (i.e., $\sum_{i=1}^{R}\sum_{j=1}^{C} p_{ij}^{BU}\ln(p_{ij}^{BU}/p_{ij}^{SM})$ is minimized). Bunke proposes choosing α and the constants s, t defining $d(\cdot,\cdot)$ and $N(\cdot,\cdot)$ by cross-validation.

Mōhner (1986) compares a number of different Bayesian and smoothing estimators in 2×2 and 2×3 tables. Included are the MLE, pseudo-Bayes estimators, cross-validation estimators, and Bunke's (1985) estimator. Bunke's estimator performs particularly well in Mōhner's study.

B. Alternatives to Loglinear Models

While loglinear models have been predominant in the literature on the analysis of multiple binomial, multinomial, and Poisson random variables, other structural models have been considered. The paragraphs below give a brief introduction to some of these approaches.

Marhoul (1984) studies a nonparametric "ACE-like" algorithm (Alternating Conditional Expectation, Brieman and Friedman (1985)) for estimation which is useful for analyzing large sparse tables as well as tables with large strata. The method involves estimating score vectors S_r and S_c for the rows and columns, and a smooth function $\varphi(\cdot)$ such that

$$\sum_{i=1}^{R}\sum_{j=1}^{C}(Y_{ij} - \varphi(S_{ri} + S_{cj}))^2 \tag{3.4.4}$$

is minimized. Marhoul considers higher dimensional tables and extensions of (3.4.4).

Another class of alternative models is based on applying a variance-stabilizing transformation $T(\cdot)$ to each element of the response vector $\mathbf{Y}$ and assuming $E[\mathbf{T}]$ follows a linear model $\mathbf{X}\boldsymbol{\beta}$ where $\mathbf{T}' = (T_1, \ldots, T_n)$ and $T_i = T(Y_i)$, $1 \leq i \leq n$. Then a regression estimator such as least squares, or an alternative such as a ridge, James–Stein, or empirical Bayes estimator, can be applied to the "data" $\mathbf{T}$ yielding results for the original problem after transformation back via $T^{-1}(\cdot)$. Two proposals of this nature are described.

Hudson (1985) considers transformation-based estimators for mutually independent Poisson responses. His approach is motivated by the results of Peng (1975) and Hudson and Tsui (1981) who consider independent Poisson data with unrelated means (see Section 2.3). Define the (approximate ln) transformation $T(\cdot)$ by $T(y) := \sum_{i=1}^{y} 1/i$ if y is positive and $T(y) := 0$ otherwise. Assume that the transformed responses $T_i = T(Y_i)$, $1 \leq i \leq n$, follow a linear model with $n \times p$ design matrix $\mathbf{X}$ having rank$(\mathbf{X}) = p$. Denote the least squares estimate of $E[\mathbf{T}]$ by $\hat{\mathbf{T}} := \mathbf{X}(\mathbf{X}'\mathbf{X})^{-1}\mathbf{X}'\mathbf{T}$ and the corresponding estimates of Y_i by

$$\hat{Y}_i := \begin{cases} .56(\exp\{\hat{T}_i\} - 1) & \text{if } \hat{T}_i \geq 0 \\ 0 & \text{otherwise} \end{cases} \tag{3.4.5}$$

where the inverse transformation (3.4.5) is based on the fact that $\ln(\frac{y+.56}{.56})$ closely approximates $T(\cdot)$. Hudson proposes the following Stein-type estimator which smooths the raw counts Y_i toward the transformation-based estimates $\hat{Y}_i$. Let $S^2 = \sum_{i=1}^{n}(T_i - \hat{T}_i)^2$ and $r = \max\{(n - p - 2 - n_0), 0\}$ where n_0 is the number of observed zeros among the Y_i. Hudson's Stein-type estimates defined by

$$Y_i^H = \begin{cases} Y_i - \frac{r}{S^2}(T_i - \hat{T}_i) & \text{if } Y_i + .56 > r/S^2 \\ \hat{Y}_i & \text{if } Y_i + .56 \leq r/S^2 \end{cases}$$

have greatest smoothing toward $\hat{Y}_i$ for small Y_i; in fact if $Y_i + .56 \leq r/S^2$, then $Y_i^H = \hat{Y}_i$. Otherwise, the adjustment to Y_i is proportional to the discrepancy $T_i - \hat{T}_i$ and has the property that there is more smoothing when the linear model fits the transformed data better (S^2 decreases). Hudson derives approximations to the MSE of $\mathbf{Y}^H$ and illustrates this estimator with two examples.

Brier, Zacks and Marlow (1986) and Zacks, Brier and Marlow (1988) also use a transformation approach as they study both Stein and empirical Bayes estimation for mutually independent binomial observations. Specifically, they consider data of the form

$$Y_{ij} \sim B(m_{ij}, p_{ij}), \quad 1 \le i \le R, \; 1 \le j \le C, \tag{3.4.6}$$

where the $\{p_{ij}\}$ are unrelated. Applying the transformation

$$T(y,n) = 2 \text{ Arcsine} \left(\frac{y + 3/8}{n + 3/4} \right)^{1/2}$$

yields $T_{ij} := T(Y_{ij}, m_{ij})$ which is approximately distributed $N(\mu_{ij}, 1/m_{ij})$ when m_{ij} is large where $\mu_{ij} := 2 \text{ Arcsine}(\sqrt{p_{ij}})$. Brier, Zacks and Marlow (1986) consider estimating $\boldsymbol{\mu}$ by both the positive-part James–Stein estimator and an empirical Bayes estimator. The p_{ij} can then be estimated by inversion yielding $\hat{p}_{ij} = \sin^2(\hat{\mu}_{ij}/2)$. They compare the MLE $\hat{p}_{ij} = Y_{ij}/m_{ij}$ to their two proposals by cross-validation and simulation and conclude that the empirical Bayes procedure has the best (frequentist) MSE properties. It should be noted that following the development in Example 3.1.3, data of the form (3.4.6) can alternatively be considered as an $R \times C \times 2$ table of counts Y_{ij} and $m_{ij} - Y_{ij}$ which could be modeled by a LLM.

A third alternative to the loglinear model, which has received increasing attention in the literature, consists of proposals for modeling "overdispersion." Overdispersion occurs when count data have larger observed variance than the binomial (or multinomial or Poisson) distribution would predict. Recall from Section 1.3 that the above distributions have variances which are functions *only* of the means; thus they do not have the flexibility to model extra variation.

Example 1.3.1 illustrates data displaying overdispersion compared to that predicted by the Poisson model. Section 2.3 considers tests for the Poisson assumption based on data $(Y_1, \ldots, Y_n)$ with unstructured mean $(\lambda_1, \ldots, \lambda_n)$ using overdispersed models as alternatives.

Hinde (1982), Breslow (1984), and Lawless (1987a) present three approaches to testing and estimation based on over-dispersed Poisson regression models which are linear in the means. (Further examples of overdispersed Poisson responses are Problem 2.25 concerning book borrowing from the Hillman library at the University of Pittsburgh, and Problem 3.3 concerning the school absences of Australian children.) Most of the other work on overdispersion is focused on extra-binomial variation in logistic regression models. Section 5.1 mentions one class of such models for the case where there are dependencies among some of the Bernoulli trials. Problem 5.32 explores this further in the context of an experiment in plant biology.

Problems

3.1. Show that $\boldsymbol{\ell} \in \tilde{\mathcal{M}}$ for the loglinear model of Example 3.1.4 is characterized by constant odds ratios across strata.

3.2. Suppose the mean $\boldsymbol{\mu} := n\mathbf{p}$ of $\mathbf{Y} \sim M_{RCK}(n, \mathbf{p})$ satisfies the loglinear model

$$\ln(\mu_{ijk}) = \lambda + \lambda_i^1 + \lambda_j^2 + \lambda_k^2 + \lambda_{ij}^{12} + \lambda_{ik}^{13}$$

for $1 \le i \le R$, $1 \le j \le C$, $1 \le k \le K$ where the usual identifiability constraints $\lambda_+^1 = \lambda_+^2 = \lambda_+^3 = \lambda_{+j}^{12} = \lambda_{i+}^{12} = \lambda_{+k}^{13} = \lambda_{i+}^{13} = 0$ hold for all i, j, k. Show that necessary and sufficient conditions for the existence of the MLE $\hat{\boldsymbol{\mu}}$ are:

(a) $Y_{i+k} > 0$ for all i, k and
(b) $Y_{ij+} > 0$ for all i, j.

(Hint: See the discussion of Examples 3.1.1 and 3.2.1 in Section 3.2.)

3.3. The data in Table 3.P.3 from Quine (1975) record the number of days absent from school for 113 Australian children. Four factors which may be relevant to the number of absences are age (coded as 1 for primary, 2 for 1st form = 7th grade, and 3 for 2nd form = 8th grade), sex (M = male and F = female), cultural background (A = aboriginal and W = white), and learning ability (S = slow and A = average). (As collected, the data includes 33 additional children in the 3rd form = 9th grade. Only one of these 33 children is classified as having slow learning ability. Thus the inclusion of the 9th grade creates three observed empty cells which must be accounted for separately. For this reason these children have been omitted from Table 3.P.3.)

(a) Assuming the number of days absent is Poisson distributed, fit a loglinear model to these data. Can you find a good fitting model based on the four available factors?
(b) Hinde (1982) finds evidence of extra-Poisson variation in these data. Do you agree? (Plot the sample mean versus sample variance for each group of observations having the same explanatory variables.)

3.4. Jorgenson (1961) presents the data in Table 3.P.4 concerning the number of breakdowns of a complex piece of electronic equipment during a 9-week period. The machine is operated in one of two modes. Consider the model

$$\ell_i = \beta_1 T_{1i} + \beta_2 T_{2i}$$

for the natural logarithm of the mean number of breakdowns during the ith week where T_{ji} is the number of hours spent in mode j during the ith week, $j = 1, 2$ and $1 \le i \le 9$. Does this model fit well? Is there any evidence of a "week" trend?

Table 3.P.3. Number of Absences of Australian School Children (Reprinted with permission from "Achievement Orientation of Aboriginal and White Australians." Ph.D. thesis by S. Quine, 1975.)

Case Number	Days Absent	Age	Sex	Cultural Background	Learning Ability
1	2	1	F	A	S
2	11	1	F	A	S
3	14	1	F	A	S
4	5	1	F	A	A
5	5	1	F	A	A
6	13	1	F	A	A
7	20	1	F	A	A
8	22	1	F	A	A
9	6	2	F	A	S
10	6	2	F	A	S
11	15	2	F	A	S
12	7	2	F	A	A
13	14	2	F	A	A
14	6	3	F	A	S
15	32	3	F	A	S
16	53	3	F	A	S
17	57	3	F	A	S
18	14	3	F	A	A
19	16	3	F	A	A
20	16	3	F	A	A
21	17	3	F	A	A
22	40	3	F	A	A
23	43	3	F	A	A
24	46	3	F	A	A
25	3	1	M	A	A
26	5	1	M	A	A
27	11	1	M	A	A
28	24	1	F	A	A
29	45	1	F	A	S
30	5	2	F	A	A
31	6	2	M	A	S
32	6	2	M	A	S
33	9	2	M	A	S
34	13	2	M	A	S
35	23	2	M	A	S
36	25	2	M	A	S
37	32	2	M	A	S
38	53	2	M	A	S

Table 3.P.3. (cont.)

Case Number	Days Absent	Age	Sex	Cultural Background	Learning Ability
39	54	2	M	A	S
40	5	2	M	A	A
41	5	2	M	A	A
42	11	2	M	A	A
43	17	2	M	A	A
44	19	2	M	A	A
45	2	3	M	A	A
46	8	3	M	A	S
47	13	3	M	A	S
48	14	3	M	A	S
49	20	3	M	A	S
50	47	3	M	A	S
51	48	3	M	A	S
52	60	3	M	A	S
53	81	3	M	A	S
54	6	1	F	W	S
55	17	1	F	W	S
56	67	1	F	W	S
57	0	1	F	W	A
58	0	1	F	W	A
59	2	1	F	W	A
60	7	1	F	W	A
61	11	1	F	W	A
62	12	1	F	W	A
63	0	2	F	W	S
64	0	2	F	W	S
65	5	2	F	W	S
66	5	2	F	W	S
67	5	2	F	W	S
68	11	2	F	W	S
69	17	2	F	W	S
70	3	2	F	W	A
71	4	2	F	W	A
72	22	3	F	W	S
73	30	3	F	W	S
74	36	3	F	W	S
75	0	3	F	W	A

Table 3.P.3. (cont.)

Case Number	Days Absent	Age	Sex	Cultural Background	Learning Ability
76	1	3	F	W	A
77	5	3	F	W	A
78	7	3	F	W	A
79	8	3	F	W	A
80	16	3	F	W	A
81	27	3	F	W	A
82	25	1	M	W	S
83	10	1	M	W	A
84	11	1	M	W	A
85	20	1	M	W	A
86	33	1	M	W	A
87	0	2	M	W	S
88	1	2	M	W	S
89	5	2	M	W	S
90	5	2	M	W	S
91	5	2	M	W	S
92	5	2	M	W	S
93	5	2	M	W	S
94	7	2	M	W	S
95	7	2	M	W	S
96	11	2	M	W	S
97	15	2	M	W	S
98	5	2	M	W	A
99	6	2	M	W	A
100	6	2	M	W	A
101	7	2	M	W	A
102	14	2	M	W	A
103	28	2	M	W	A
104	1	2	M	W	A
105	0	3	M	W	S
106	2	3	M	W	S
107	2	3	M	W	S
108	3	3	M	W	S
109	5	3	M	W	S
110	8	3	M	W	S
111	10	3	M	W	S
112	12	3	M	W	S
113	14	3	M	W	S

Table 3.P.4. Number of Breakdowns of a Piece of Electronic Equipment (Reprinted with permission from D.W. Jorgenson: "Multiple Regression Analysis of a Poisson Process," *Journal of the American Statistical Association,* 1961, vol. 56, p. 242. American Statistical Association.)

Week	Number of Breakdowns	T_1	T_2
1	15	33.3	25.3
2	9	52.2	14.4
3	14	64.7	32.5
4	24	137.0	20.5
5	27	125.9	97.6
6	27	116.3	53.6
7	23	131.7	56.6
8	18	85.0	87.3
9	22	91.9	47.8

3.5. The data in Table 3.P.5 from Piegorsch, Weinberg, and Margolin (1988) are the counts of the number of cells undergoing differentiation after exposure to doses of tumor necrosis factor (TNF), Interferen-γ (IFN), or both. Following Piegorsch, Weinberg, and Margolin (1988),

Table 3.P.5. Cellular Differentiation after Exposure to TNF and/or IFN (Reproduced from W.W. Piegorsch, C.R. Weinberg, and B.H. Margolin, "Exploring Simple Independent Action in Multifactor Tables," *Biometrics* **44**, pp. 595–603, 1988. With permission from the Biometric Society.)

Number of Cells Differentiating	Dose of TNF (U/ml)	Dose of IFN (U/ml)
11	0	0
18	0	4
20	0	20
39	0	100
22	1	0
38	1	4
52	1	20
69	1	100
31	10	0
68	10	4
69	10	20
128	10	100
102	100	0
171	100	4
180	100	20
193	100	100

all counts are assumed to be based on 200 cells. An important scientific question is whether, and if so to what extent, a synergistic effect (an enhanced effect relative to that predicted by an additive model) occurs between TNF and IFN.

(a) Apply the transformation method of Brier, Zacks, and Marlow, (1986) to fit the model containing an overall mean and single parameters for TNF and for IFN.

(b) Use the same transformation technique to fit the model which adds the TNF by IFN interaction to the model in (a). Compare the fits to determine whether the two factors interact.

3.6. Consider the data of Example 1.2.10 on the numbers of failures for 90 valves from one pressurized nuclear reactor. Assume the number of failures for the ith valve is Poisson distributed with mean $\lambda_i T_i$ where T_i is the operating time (in 100 hours) and λ_i depends on the five factors: System, Operator Type, Valve Type, Head Size, Operation Mode.

Fit the main-effects-only model

$$\ell_{ghijk} = \lambda + \lambda_g^1 + \lambda_h^2 + \lambda_i^3 + \lambda_j^4 + \lambda_k^5 \tag{1}$$

where

λ_g^1 = the main effect of the gth system,

λ_h^2 = the main effect of the hth operator type,

λ_i^3 = the main effect of the ith valve type,

λ_j^4 = the main effect of the jth head size, and

λ_k^5 = the main effect of the kth operation mode

to these data. Moore and Beckman (1988) use (1) to develop one-sided tolerance bounds on the number of failures. Suppose instead, that an understanding of possible interactions between the factors is desired. Since the model with all 2-way interactions is overparametrized for these data, consider adding interactions to (1) according to their decrease in likelihood ratio fit statistic G^2 (equation 2.2.4). In other words, add that interaction which gives the smallest P-value (greatest reduction in G^2 relative to the decrease in degrees of freedom). What reasonably fitting models does this procedure suggest?

[Moore and Beckman, 1988]

4

Cross-Classified Data

4.1 Introduction

This chapter describes the use of classical likelihood methods and loglinear models to analyze cross-classified data. Cross-classified data arise when a random sample $\mathbf{W}^1, \mathbf{W}^2, \ldots, \mathbf{W}^m$, say, is drawn from a discrete d-variate distribution where each trial $\mathbf{W}^k = (W_1^k, \ldots, W_d^k)'$ has common joint probability mass function:

$$p_{\mathbf{i}} := P[W_1^k = i_1, \ldots, W_d^k = i_d], \tag{4.1.1}$$

for $\mathbf{i} := (i_1, \ldots, i_d)' \in \mathcal{I} := \times_{j=1}^{d}\{1, \ldots L_j\}$. Here the support of W_j^k is taken to be $\{1, \ldots, L_j\}$ without loss of generality. The symbol $\mathbf{W}$, without a superscript, will be used to denote a generic classification variable with probability mass function (4.1.1). By sufficiency, the data can be summarized as the counts $\{Y_{\mathbf{i}} : \mathbf{i} \in \mathcal{I}\}$ in a d-dimensional contingency table where $Y_{\mathbf{i}}$ is the number of vectors $\mathbf{W}$ which equal $\mathbf{i}$. Thus the counts $\{Y_{\mathbf{i}} : \mathbf{i} \in \mathcal{I}\}$ have the $M_t(m, \mathbf{p})$ multinomial distribution where $\mathbf{p} = \{p_{\mathbf{i}} : \mathbf{i} \in \mathcal{I}\}$, $\sum_{\mathbf{i} \in \mathcal{I}} p_{\mathbf{i}} = 1$, $t = \prod_{j=1}^{d} L_j$, and $m = \sum_{\mathbf{i} \in \mathcal{I}} Y_{\mathbf{i}}$.

Several examples of cross-classified data are described next.

Example 4.1.1. Davies (1961) reports the data in Table 4.1.1 on piston ring failures in an Imperial Chemical Industries plant. The failures are cross-classified by compressor number (1, 2, 3, and 4) and leg position (North, Center, and South). The table has $d = 2$ dimensions with dimension 1 being compressor number with $L_1 = 4$ values and dimension 2 being leg position with $L_2 = 3$ values (North/Central/South).

Table 4.1.1. Piston Ring Failures Classified by Compressor Number and Leg (Reprinted with permission from *Statistical Methods in Research and Production* by O.L. Davies, Hafner Publishers, New York, 1961.)

Compressor	Leg			
Number	North	Center	South	Total
1	17	17	12	46
2	11	9	13	33
3	11	8	19	38
4	14	7	28	49
Total	53	41	72	166

Example 4.1.2. The data of Table 4.1.2 is an oft-used example reported by Korff, Taback, and Beard (1952). After an outbreak of food poisoning following a company picnic, the participants were questioned as to whether or not they (i) ate crabmeat (yes/no), (ii) ate potato salad (yes/no), and (iii) became ill after the picnic (yes/no). The epidemiological problem is to determine what caused the illness. This data is cross-classified with $d = 3$ dimensions and $L_1 = L_2 = L_3 = 2$.

Table 4.1.2. Cross-Classification of 304 Picnickers by Illness and Eating Pattern

		Potato Salad	Sick Y	Sick N
Crabmeat	Y	Y	120	80
		N	4	31
	N	Y	22	24
		N	0	23

Example 4.1.3. The survey data of rental property residents in Example 1.2.3 is cross-classified. Here $d = 4$, dimension 1 is housing type with $L_1 = 4$ values (tower blocks/apartments/atrium houses/terraced houses), dimension 2 is influence on apartment management with $L_2 = 3$ values (low/medium/high), dimension 3 is contact with other residents with $L_3 = 2$ values (low/high), and dimension 4 is satisfaction with $L_4 = 3$ values (low/medium/high).

The goal of this chapter is to study loglinear models for the means of cross-classified data. This class of models corresponds to the ANOVA-type decompositions of normal means which are used in experimental design. As an example, consider a $d = 2$ dimensional table with $L_1 = R$ rows and $L_2 = C$ columns. There are unique values of λ, $\{\lambda_i^1\}$, $\{\lambda_j^2\}$, and $\{\lambda_{ij}^{12}\}$ for which

$$\ell_{ij} := \ln(mp_{ij}) = \lambda + \lambda_i^1 + \lambda_j^2 + \lambda_{ij}^{12}$$

subject to the identifiability constraints $\lambda_+^1 = 0 = \lambda_+^2$, $\lambda_{i+}^{12} = 0$ for $i = 1(1)R$, and $\lambda_{+j}^{12} = 0$ for $j = 1(1)C$. The notation in which a subscript is replaced by $+$ means the sum has been computed over that subscript. The exact formulae for the λ-terms are given in Section 4.2.

In general d-dimensional tables, if $\ell_{i_1,\ldots,i_d} := \ln(mp_{i_1,\ldots,i_d})$ then there are unique terms for which

$$\ell_{i_1,\ldots,i_d} = \lambda + \sum_{j=1}^{d} \lambda_{i_j}^{j} + \sum_{1 \le j < k \le d} \lambda_{i_j,i_k}^{jk} + \cdots + \lambda_{i_1,\ldots,i_d}^{1,\ldots,d} \qquad (4.1.2)$$

subject to the identifiability conditions that the λ-terms sum to zero over every subscript. These λ-terms can be constructed recursively as will be illustrated in Section 4.3. Interesting models arise when various λ-terms are set to zero. Such models can be thought of as Taylor series approximations to the function

$$f(i_1, i_2, \ldots, i_d) = \ln(mp_{i_1,\ldots,i_d})$$

over the discrete variable domain $(i_1, i_2, \ldots, i_d) \in \mathcal{I}$. For instance, in the two-dimensional case Example 3.1.2 showed that if $\lambda_{ij}^{12} = 0$ for all i and j, then the ℓ_{ij} possesses a simple additive structure which is the analog of the linear structure $f(x_1, x_2) = a_1x_2 + a_2x_2$ for a function of two continuous variables.

An important class of loglinear models for $\boldsymbol{\ell}$, called hierarchical loglinear models, are those that correspond to certain restrictions on the types of (non-zero) interactions present in the model. To define these models it is necessary to define the notion of a higher-order relative.

Definition. Fix $1 \leq k_1 < \ldots < k_s \leq d$ and $1 \leq j_1 < \ldots < j_t \leq d$. The set of λ-terms

$$\left\{\lambda_{\mathbf{i}}^{j_1,\ldots,j_t} : \mathbf{i} \in \times_{q=1}^{t}\{1, \ldots, L_{j_q}\}\right\}$$

are a *higher order relative* of the set of λ-terms

$$\left\{\lambda_{\mathbf{i}}^{k_1,\ldots,k_s} : \mathbf{i} \in \times_{q=1}^{s}\{1, \ldots, L_{k_q}\}\right\}$$

means $\{k_1, \ldots, k_s\} \subset \{j_1, \ldots, j_t\}$.

For example, $\boldsymbol{\lambda}^{123} := \{\lambda_{ijk}^{123} : i = 1(1)L_1,\ j = 1(1)L_2,\ k = 1(1)L_3\}$ is a higher order relative of $\boldsymbol{\lambda}^1 := \{\lambda_i^1 : i = 1(1)L_1\}$ for any d (≥ 3) dimensional table. More informally $\boldsymbol{\lambda}^{123}$ is a higher order relative of $\boldsymbol{\lambda}^1$.

Definition. Data $\mathbf{Y}$ follow a *hierarchical loglinear model* (HLLM) provided the vector of logarithms of cell means, $\boldsymbol{\lambda}$, has the property that *all* higher order relatives of any zero λ-term in the decomposition (4.1.2) are also zero.

For example, when $d = 3$ the loglinear model $\ell_{ijk} = \lambda + \lambda_i^1 + \lambda_j^2 + \lambda_k^3$ is hierarchical but the model $\ell_{ijk} = \lambda + \lambda_i^1 + \lambda_j^2 + \lambda_{ik}^{13}$ is not since $\boldsymbol{\lambda}^3 = 0$ while its higher order relative $\boldsymbol{\lambda}^{13}$ is not. The convention is used throughout that the absence of a term from a model expression signifies that all its components are identically zero.

It is convenient to develop an alternate notation to denote HLLMs. For a given HLLM, let $C_1, \ldots, C_r$ denote the maximal sets of pairwise noncomparable (with respect to set containment) interactions; this collection of sets is unique. It is called the *generating set* of the model. The corresponding HLLM will be denoted by $[C_1, \ldots, C_r]$. For example, if $d = 3$ and

$$\ell_{ijk} = \lambda + \lambda_i^1 + \lambda_j^2 + \lambda_k^3 + \lambda_{ij}^{12} + \lambda_{jk}^{23} \tag{4.1.3}$$

then $C_1 = \{1, 2\}$, and $C_2 = \{2, 3\}$ are the maximal interaction sets. Neither $C_1 \subset C_2$ nor the converse. The shortened notation [12,23] is used to designate the generating set of model (4.1.3). Equivalently the model can be specified by listing the minimal sets of pairwise noncomparable interactions which are zero. This quantity is called the *dual of the generator set* and the model corresponding to zero (and higher order) interactions $D_1, \ldots, D_s$ is denoted by $[D_1, \ldots, D_s]^d$. For example, the model (4.1.3) is denoted by $[13]^d$ in the dual notation. This statement says $\boldsymbol{\lambda}^{13} = \mathbf{0}$ and hence $\boldsymbol{\lambda}^{123} = \mathbf{0}$ since the model is hierarchical. Additional examples of these two notations will be presented in Section 4.3. The only model which the generator set notation is unable to represent is the constant model $\ell_{ijk} = \lambda$ which is best denoted as $[1, 2, 3]^d$ in the dual notation; the only model which the dual notation is unable to represent is the fully saturated model (4.1.2) which is best denoted as $[12 \ldots d]$ in generator set notation.

While this chapter will restrict attention to cross-classified multinomial data, several remarks are in order regarding data obtained under other sampling schemes. First, the exact interpretation of a given HLLM depends on the sampling scheme. The same HLLM will have a different physical meaning for independent multivariate Poisson data than for multinomial data or for product multinomial data. For example, the interpretation of several HLLMs for $d = 3$ dimensional tables when one margin is fixed will be given in Section 5.5 for 2 by 2 by S tables and should be contrasted with the trivariate discrete case illustrated in Section 4.3. Second, only certain HLLMs make sense when product multinomial sampling is used. To illustrate, consider the Danish housing data of Example 4.1.3. These data were actually collected by surveying a fixed number of residents of each housing type. Thus the margin of the data table corresponding to the variable "housing type" was fixed by the sampling design, and the data comes from 4 trivariate discrete distributions. Hierarchical loglinear models of the form (4.1.2) can still be used, but it will be shown in Section 4.4 that only HLLMs which include the main effect $\boldsymbol{\lambda}^1$ have maximum likelihood estimators with the same housing-type margin as the sampling model. Thus these are the only HLLMs applicable for these data. Similar comments apply to data with more than one fixed margin. In the extreme situation where all margins except one, say the first, are fixed, the first dimension can be considered as the response and the remaining dimensions as explanatory. In this case, any HLLM including the term $\boldsymbol{\lambda}^{23\ldots d}$ (and hence all possible interaction terms involving $2, \ldots, d$), will yield estimates having the same marginal tools as the sampling design. Asmussen and Edwards (1983) discuss response models in greater generality. Data with exactly one response dimension or *un*fixed margin can alternatively be modeled by logistic regression methods; such methods will be considered in Sections 5.3 and 5.4.

4.2 Two-Dimensional Tables

This section will study two-dimensional tables. First, the various possible HLLMs will be developed for this case. Then estimation and testing goodness-of-fit to the independence model will be examined. The latter is assessed using residuals and other data analytic techniques. HLLMs for higher dimensional tables are considered in Section 4.3 and residuals appropriate for these cases are constructed in Section 4.4. Alternative loglinear and nonloglinear models for two-dimensional tables with one or both classification variables *ordinal* are discussed in Chapter 5 of Agresti (1984) and Goodman (1984). Alternative models for *square* tables are considered in Problem 4.10 (b) while an alternative for general tables based on "canonical analysis" is described in Problem 4.12.

Models and Interpretation

Suppose data $\mathbf{Y} = \{Y_{ij} : i = 1(1)R, j = 1(1)C\}$ are observed which follow the $M_{RC}(m, \mathbf{p})$ distribution. Let $\ell_{ij} = \ln(mp_{ij})$ denote the natural logarithm of the mean of Y_{ij} and write

$$\ell_{ij} = \lambda + \lambda_i^1 + \lambda_j^2 + \lambda_{ij}^{12}$$

where $\lambda_+^1 = 0 = \lambda_+^2$, $\lambda_{i+}^{12} = 0$ for $i = 1(1)R$, and $\lambda_{+j}^{12} = 0$ for $j = 1(1)C$. It is well-known (Problem 4.1) that the component terms in this case are

$$\lambda := \ell_{\bullet\bullet} := \frac{1}{RC}\sum_{r=1}^{R}\sum_{c=1}^{C}\ell_{rc} = \ln(m) + \ln\left[\left(\prod_{r,c} p_{rc}\right)^{1/RC}\right],$$

$$\lambda_i^1 := \ell_{i\bullet} - \lambda = \ln\left[\left(\prod_{c} p_{ic}\right)^{1/C} \Big/ \left(\prod_{r,c} p_{rc}\right)^{1/RC}\right],$$

$$\lambda_j^2 := \ell_{\bullet j} - \lambda = \ln\left[\left(\prod_{r} p_{rj}\right)^{1/R} \Big/ \left(\prod_{r,c} p_{rc}\right)^{1/RC}\right], \text{ and}$$

$$\lambda_{ij}^{12} := \ell_{ij} - \ell_{i\bullet} - \ell_{\bullet j} + \ell_{\bullet\bullet} = \ln\left[\frac{p_{ij}\{\prod_{r,c} p_{rc}\}^{1/RC}}{\{\prod_c p_{ic}\}^{1/C}\{\prod_r p_{rj}\}^{1/R}}\right]. \tag{4.2.1}$$

Replacing a subscript by a dot means an *average* has been computed over that subscript. Thus the λ-terms are geometric means of the logarithms of the cell means analogous to the ANOVA decomposition of the matrix of mean responses for all treatment combinations of two qualitative factors.

The four nonsaturated HLLMs in the $R \times C$ case can be interpreted as follows. The simplest possibility is that $\ell_{ij} = \lambda$; it is straightforward to see that this holds if and only if (W_1, W_2) are (independently) uniformly distributed over $\{(i,j) : i = 1(1)R \text{ and } j = 1(1)C\}$. Similarly $\ell_{ij} = \lambda + \lambda_i^1$ ($\ell_{ij} = \lambda + \lambda_j^2$) if and only if W_2 (W_1) is uniformly distributed independent of W_1 (W_2) with the distribution of W_1 (W_2) arbitrary. Finally Example 3.1.2 showed that

$$\ell_{ij} = \lambda + \lambda_i^1 + \lambda_j^2 \tag{4.2.2}$$

if and only if W_1 is independent of W_2. Chapter 3 also showed that the MLE of p_{ij} exists under (4.2.2) if and only if all marginal totals Y_{i+}, $i = 1(1)R$ and Y_{+j}, $j = 1(1)C$ are positive; when this holds $\hat{p}_{ij} := (Y_{i+}/m)(Y_{+j}/m) = \hat{p}_{i+}\hat{p}_{+j}$ is the MLE of p_{ij}.

Assessing Fit

From Section 2.2, for multinomial data $\mathbf{Y} \sim M_t(m, \mathbf{p})$ the two most commonly used omnibus tests of fit of the model $H_0 : \mathbf{p} \in \{\mathbf{p}(\boldsymbol{\eta}) : \boldsymbol{\eta} \in \mathcal{N} \subset \mathbb{R}^r\}$ where $1 \leq r < t$ versus H_A: (not H_0), are the Pearson and likelihood ratio tests. If $\hat{\mathbf{p}} = \mathbf{p}(\hat{\boldsymbol{\eta}})$ is the MLE of $\mathbf{p}$ under H_0, then the Pearson chi-squared statistic is

$$X^2 = \sum_{i=1}^{t} \frac{[Y_i - m\hat{p}_i]^2}{m\hat{p}_i},$$

and the likelihood ratio chi-squared statistic is

$$G^2 = 2\sum_{i=1}^{t} Y_i \ln(Y_i/m\hat{p}_i).$$

Both are special cases of the Cressie–Read family of test statistics

$$I^\lambda := \frac{2}{\lambda(\lambda+1)} \sum_{i=1}^{t} Y_i\{(Y_i/m\hat{p}_i)^\lambda - 1\}.$$

Further, I^λ has a null asymptotic χ^2_{t-1-r} distribution for all $\lambda \in \mathbb{R}$.

In the case of $R \times C$ tables with null independence model, $t = RC$ and $r = (R-1) + (C-1)$ yielding null degrees of freedom $t - 1 - r = RC - R - C + 1 = (R-1)(C-1)$. When independence is rejected, the experimenter will want to identify the cell(s) which display lack-of-fit. A variety of data analytic techniques have been proposed for this purpose with much recent work devoted to large sparse tables. The basic building blocks which define these procedures are residuals of various types. Four different residuals based on fitting the independence model are considered below.

Letting $\hat{p}_{ij}$ be the MLE of p_{ij} under (4.2.2), the *raw residuals* are defined as

$$e_{ij}^r := Y_{ij} - m\hat{p}_{ij}$$

and, like their counterparts in Section 2.3, they are not standardized. The *Pearson residuals*, defined by

$$e_{ij}^P := e_{ij}^r / \{m\hat{p}_{ij}\}^{1/2},$$

satisfy $X^2 = \sum_{i,j}(e_{ij}^P)^2$. Some authors call the e_{ij}^P chi-components or standardized residuals. A second calibration of e_{ij}^r is motivated by the following theorem due to Haberman (1973a).

Proposition 4.2.1. *If* $\mathbf{Y}^m \sim M_{RC}(m, \mathbf{p})$ *for* $m \geq 1$ *and* $\mathbf{p}$ *satisfies the independence model, then*

$$\sqrt{m}(Y_{ij}^m/m - (Y_{i+}Y_{+j})/m^2) = e_{ij}^r/\sqrt{m} \xrightarrow{\mathcal{L}} N(0, V_\infty^{ij}),$$

as $m \to \infty$ *where* $V_\infty^{ij} = V_\infty^{ij}(\mathbf{p}) = p_{i+}(1 - p_{i+})p_{+j}(1 - p_{+j})$.

Haberman's result suggests estimating the asymptotic variance of e_{ij}^r by

$$m\hat{V}_\infty^{ij} = m\hat{p}_{i+}(1 - \hat{p}_{i+})\hat{p}_{+j}(1 - \hat{p}_{+j}),$$

and defining *adjusted residuals* as

$$e_{ij}^a := e_{ij}^r / \{m\hat{V}_\infty^{ij}\}^{1/2} = e_{ij}^P / \{(1 - \hat{p}_{i+})(1 - \hat{p}_{+j})\}^{1/2}.$$

The adjustment in e_{ij}^a multiplies e_{ij}^P by a number greater than or equal to unity which suggests that e_{ij}^P is too small in absolute value for large-sample normal comparisons.

Analogous to the Pearson residuals, some authors suggest using the individual components of G^2 as residuals. These quantities are called *deviance residuals* and defined as

$$e_{ij}^d := \pm[2Y_{ij}\,|\ln(Y_{ij}/m\hat{p}_{ij})|]^{1/2}$$

with the sign of e_{ij}^d defined to be that of $\ln(Y_{ij}/m\hat{p}_{ij})$; i.e., the "+" sign used if and only if $\ln(Y_{ij}/m\hat{p}_{ij}) > 0$ (i.e., $Y_{ij} > m\hat{p}_{ij}$). The e_{ij}^d, like the e_{ij}^r, are uncalibrated, and hence not as informative as the e_{ij}^a.

The list of residuals identified above is not exhaustive. For example, Brown (1974) and Simonoff (1988) define deleted residuals based on the independence model for two-way tables. For each cell they compute a normalized difference between the cell count and its mean estimated from the model of independence fit to the table with that cell *deleted;* i.e., treated as a structural zero. (Independence models for tables with structural zeroes have been called models of quasi-independence. See Goodman (1984).) The calculation of deleted residuals requires fitting RC additional models.

Residuals have been incorporated into more formal procedures for determining which cells, if any, contain outliers. Simonoff (1988) suggests a backward stepping procedure for outlier detection based on deleted residuals. Another class of outlier detection methods examines the estimated logarithms of the cross product ratios of 2×2 subtables to identify unusual cells (Fienberg, 1969; Kotze and Hawkins, 1984).

Any data analysis of an $R \times C$ table should include an examination of residuals using (i) searches for systematic patterns of $\pm$ signs, (ii) plots of the residuals versus their row or column indices, and (iii) if m is large, normal probability plots of the ordered adjusted residuals. To illustrate, consider the piston ring data of Example 4.1.1.

Example 4.1.1 (continued). Fitting the independence model to the piston ring data produces the expected counts in Table 4.2.1, and residuals in Table 4.2.2. The goodness-of-fit statistics on 6 degrees of freedom are $X^2 = 11.72$ and $G^2 = 12.06$ with P-values of .07 and .06, respectively. The overall fit is neither extremely good nor extremely bad.

Comparing the three different residuals in Table 4.2.2 shows that

$$|e_{ij}^P| < |e_{ij}^a| < |e_{ij}^d|$$

holds for all cells i, j. The first inequality is always true whereas the second is specific to these data. There is a striking pattern to the signs of the residuals. The residuals associated with compressors 1 and 2 have componentwise opposite signs from those associated with compressors 3 and 4, and the residuals associated with the north and center legs having componentwise opposite signs from those associated with the south leg. The largest residual is that associated with compressor number 1 and the south leg. Looking back at the raw data in Table 4.1.1 this is not surprising because compressor 1 is the only one of the four compressors which did not have more failures in the south leg than in the other two legs.

Table 4.2.1. Expected Counts Under Independence Model Based on Table 4.1.1 Data

Compressor	Leg			
Number	North	Center	South	Total
1	14.69	11.36	19.95	46
2	10.54	8.15	14.31	33
3	12.13	9.39	16.48	38
4	15.64	12.10	21.25	49
Total	53	41	72	166

Since $m = 166$ is large relative to the number of cells and to the estimated cell probabilities, it is reasonable to plot the ordered adjusted residuals against normal quantiles as in Figure 4.2.1. The reference line in the figure is the line $y = x$; for large m and data following the independence model, the plotted points will lie near the line. The key question is what is meant by "near the line." Even when m is large and independence holds, there is naturally more variability at the extreme points rather than at the center points. Also, the $\{e_{ij}^a\}$ are not independent for any m due to the estimated variances. Thus there is some deviation from the line even when the underlying assumptions hold.

This natural deviation can be assessed by simulation, and bands can be added to the plot to aid interpretation in the following way. Generate N random tables from the fitted model; N is often taken on the order of 25. Fit the independence model to each table and calculate the associated ordered adjusted residual vector, denoted $\mathbf{O}^k$, for $1 \leq k \leq N$. Here $\mathbf{O}^k$ is a vector of length RC containing the adjusted residuals from the kth table in increasing order. Bands are formed by plotting the nth smallest and the nth largest values of the ith coordinates of $\mathbf{O}^k$, $1 \leq k \leq N$, against the ith normal score for $1 \leq i \leq RC$, and connecting these values. If the fitted model is accurate, then at any point i, $1 \leq i \leq RC$, the confidence associated with these bounds is approximately $100(1-2n/N)\%$. The usual cautions about simultaneous inference apply here and it is recommended that the bands be used as rough visual guides.

Figure 4.2.2 shows two sets of these bands for the piston data. The solid bands are based on $N = 25$ and $n = 2$ while the dotted bands use $N = 50$ and $n = 4$. The differences between the two sets of bands are negligible and both highlight four points as being unusually large. These four points correspond to compressors 1 and 4 and leg positions center and south.

Figure 4.2.2 is easier to interpret than Figure 4.2.1 because it takes into account the null variability. It does not, however, address the adequacy of the asymptotic normal assumption which leads to the use of normal scores as plotting positions. A further enchancement which considers this issue is the empirical probability plot. (For further discussion and examples see Landwehr, Pregibon, and Shoemaker (1984).) The empirical probability plot differs from the normal probability plot with simulated bands (Figure 4.2.2) in that instead of the ith normal score as the plotting position, the median of ith coordinates of the $\mathbf{O}^k$ for $1 \leq k \leq N$, is used as the abscissa. Figure 4.2.3 is an empirical probability plot for the piston data based on $N = 25$ and $n = 2$. The major difference between Figures 4.2.2 and 4.2.3 is that the bands in Figure 4.2.3 are centered around the line $y = x$, whereas those in Figure 4.2.2 are not. Since empirical probability plots are no more difficult to make than simulated bands for normal probability plots, it is recommended that the empirical plot be made.

To conclude this example, the four largest residuals corresponding to compressors 1 and 4 and legs center and south lie noticeably off the line $y = x$ and are outside the simulated bands. These facts, combined with the sign pattern noted before, suggest not only that the simple independence model does not fully explain the data, but give information about where lack-of-fit occurs.

Table 4.2.2. (a) Pearson Residuals Under Independence Model for Table 4.1.1 Data

Compressor	Leg		
Number	North	Center	South
1	.60	1.67	−1.78
2	.14	.30	−.35
3	−.33	−.45	.62
4	−.42	−1.47	1.46

(b) Adjusted Residuals Under Independence Model for Table 4.1.1 Data

Compressor	Leg		
Number	North	Center	South
1	.86	2.27	−2.78
2	.19	.38	−.52
3	−.45	−.59	.94
4	−.60	−2.01	2.32

(c) Deviance Residuals Under Independence Model for Table 4.1.1 Data

Compressor	Leg		
Number	North	Center	South
1	2.23	3.70	−3.49
2	.97	1.36	−1.58
3	−1.47	−1.60	2.32
4	−1.76	−2.77	3.93

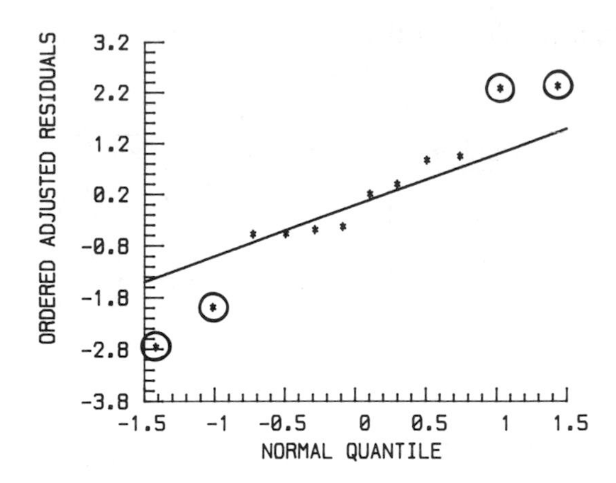

Figure 4.2.1. Normal probability plot of adjusted residuals for compressor data of Table 4.1.1 fitted by the independence model and the line $y = x$.

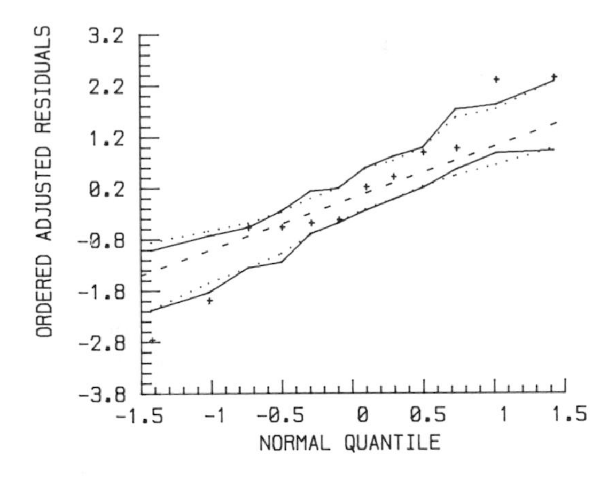

Figure 4.2.2. Normal probability plot of adjusted residuals for compressor data, the line $y = x$ (dashed), and simulated confidence intervals for $(N, n) = (25, 2)$ (solid) and $(N, n) = (50, 4)$ (dotted).

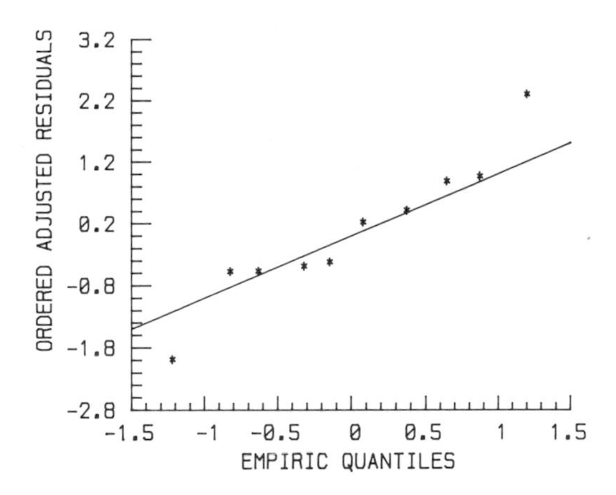

Figure 4.2.3. Empiric probability plot for compressor data based on $(N, n) = (25, 2)$ and the line $y = x$.

4.3 Hierarchical, Graphical, and Direct Loglinear Models for Arbitrary Dimensional Tables

This section considers the definition and interpretation of HLLMs for three and higher-dimensional tables. Two important subclasses of the class of hierarchical models are introduced. Algorithms for computing maximum likelihood estimators of cell means under HLLMs are described in Section 4.4.

Models and Interpretation for Three-Dimensional Tables

Consider a three-dimensional contingency table with rows, columns, and layers. Denote the cell counts by $\mathbf{Y} = \{Y_{ijk} : i = 1(1)R,\ j = 1(1)C,\ k = 1(1)L\}$ which will be assumed to be distributed according to the multinomial distribution $M_{RCL}(m, \mathbf{p})$. Alternately, let $\mathbf{W} = (W_1, W_2, W_3)'$ denote the classification variables from which the data are derived. Denote the joint probability mass function of $\mathbf{W}$ as

$$p_{ijk} = P[W_1 = i, W_2 = j, W_3 = k].$$

Let $\ell_{ijk} = \ln(mp_{ijk})$ be the natural logarithm of the mean of Y_{ijk}; decompose

$$\ell_{ijk} = \lambda + \lambda_i^1 + \lambda_j^2 + \lambda_k^3 + \lambda_{ij}^{12} + \lambda_{ik}^{13} + \lambda_{jk}^{23} + \lambda_{ijk}^{123} \tag{4.3.1}$$

where $\lambda_+^1 = \lambda_+^2 = \lambda_+^3 = 0$, and $\lambda_{i+}^{12} = \lambda_{+j}^{12} = \lambda_{i+}^{13} = \lambda_{+k}^{13} = \lambda_{j+}^{23} = \lambda_{+k}^{23} = \lambda_{ij+}^{123} = \lambda_{i+k}^{123} = \lambda_{+jk}^{123} = 0$ for $i = 1(1)R$, $j = 1(1)C$, and $k = 1(1)L$. The λ-terms can be defined in terms of quantities derived from the λ-terms of the two-dimensional layer tables in the following way. Fix a layer k, $k = 1(1)L$, and decompose

$$\ell_{ijk} = {}^k\nu + {}^k\nu_i^1 + {}^k\nu_j^2 + {}^k\nu_{ij}^{12} \tag{4.3.2}$$

where ${}^k\nu_+^1 = 0 = {}^k\nu_{i+}^{12}$ for all i, and ${}^k\nu_+^2 = 0 = {}^k\nu_{+j}^{12}$ for all j as in Section 4.2. Then the λ-terms in equation (4.3.1) are defined in terms of the ν-terms of equation (4.3.2) by,

$$\begin{aligned}
&\lambda := {}^\bullet\nu := \frac{1}{L}\sum_{k=1}^{L} {}^k\nu, \\
&\lambda_i^1 := {}^\bullet\nu_i^1,\ \lambda_j^2 :={}^\bullet \nu_j^2,\ \lambda_k^3 := {}^k\nu - {}^\bullet\nu, \qquad (4.3.3)\\
&\lambda_{ij}^{12} := {}^\bullet\nu_{ij}^{12},\ \lambda_{ik}^{13} := {}^k\nu_i^1 -{}^\bullet \nu_i^1,\ \lambda_{jk}^{23} := {}^k\nu_j^2 - {}^\bullet\nu_j^2, \text{ and} \\
&\lambda_{ijk}^{123} := {}^k\nu_{ij}^{12} - {}^\bullet\nu_{ij}^{12}
\end{aligned}$$

(Problem 4.2). Equations (4.3.3) are used below to interpret various HLLMs for the $\{p_{ijk}\}$.

Model 1: $[1,2,3] = [12,13,23]^d$. Model 1 is the main-effects-only model $\ell_{ijk} = \lambda + \lambda_i^1 + \lambda_j^2 + \lambda_k^3$ for all (i,j,k) or, equivalently,

$$mp_{ijk} = \exp\{\lambda + \lambda_i^1 + \lambda_j^2 + \lambda_k^3\}.$$

It is immediate that

$$m = m \times \sum_{i,j,k} p_{ijk} = \exp\{\lambda\}(\sum_i \exp\{\lambda_i^1\})(\sum_j \exp\{\lambda_j^2\})(\sum_k \exp\{\lambda_k^3\}). \tag{4.3.4}$$

Similarly, for all (i,j,k)

$$\begin{aligned}
mp_{i++} &= \exp\{\lambda\}\exp\{\lambda_i^1\}\left(\sum_j \exp\{\lambda_j^2\}\right)\left(\sum_k \exp\{\lambda_k^3\}\right), \\
mp_{+j+} &= \exp\{\lambda\}\exp\{\lambda_j^2\}\left(\sum_i \exp\{\lambda_i^1\}\right)\left(\sum_k \exp\{\lambda_k^3\}\right), \text{ and} \qquad (4.3.5)\\
mp_{++k} &= \exp\{\lambda\}\exp\{\lambda_k^3\}\left(\sum_i \exp\{\lambda_i^1\}\right)\left(\sum_j \exp\{\lambda_j^2\}\right).
\end{aligned}$$

Solving (4.3.5) for the marginal probabilities and applying (4.3.4) yields

$$\begin{aligned}
p_{i++}p_{+j+}p_{++k} &= \frac{\exp\{\lambda + \lambda_i^1 + \lambda_j^2 + \lambda_k^3\}}{m} \\
&\quad \times \frac{(\exp\{\lambda\})^2(\sum_i \exp\{\lambda_i^1\})^2(\sum_j \exp\{\lambda_j^2\})^2(\sum_k \exp\{\lambda_k^3\})^2}{m^2} \\
&= p_{ijk}.
\end{aligned}$$

Therefore the main-effects-only model implies mutual independence of W_1, W_2, and W_3 in the discrete multivariate model; it is easy to see the converse also holds.

Model 2: $[1,23] = [12,13]^d$. Model 2 says that

$$\ell_{ijk} = \lambda + \lambda_i^1 + \lambda_j^2 + \lambda_k^3 + \lambda_{jk}^{23}, \tag{4.3.6}$$

for all (i,j,k). After algebraic manipulations similar to those performed for Model 1, it can be shown that (4.3.6) implies $p_{ijk} = p_{i++}p_{+jk}$ for all (i,j,k). In terms of the classification variables (W_1, W_2, W_3), (4.3.6) states that

$$P[W_1 = i, W_2 = j, W_3 = k] = P[W_1 = i] \cdot P[W_2 = j, W_3 = k]$$

for all (i,j,k), so that W_1 is independent of the pair (W_2, W_3).

By symmetry $[2,13] = [12,23]^d$ says that W_2 is independent of the pair (W_1, W_3), and $[3,12] = [13,23]^d$ means that W_3 is independent of the pair (W_1, W_2).

Model 3: $[13,23] = [12]^d$. Model 3 says that

$$\ell_{ijk} = \lambda + \lambda_i^1 + \lambda_j^2 + \lambda_k^3 + \lambda_{ik}^{13} + \lambda_{jk}^{23}$$

for all (i,j,k). Observe that $\boldsymbol{\lambda}^{12} = \mathbf{0}$ if and only if ${}^{\bullet}\boldsymbol{\nu}^{12} = \mathbf{0}$ and that $\boldsymbol{\lambda}^{123} = \mathbf{0}$ if and only if ${}^{k}\boldsymbol{\nu}^{12} = {}^{\bullet}\boldsymbol{\nu}^{12}$. Thus $\boldsymbol{\lambda}^{12} = \mathbf{0} = \boldsymbol{\lambda}^{123}$ implies that ${}^{k}\boldsymbol{\nu}^{12} = \mathbf{0}$ for $k = 1(1)L$ so that additivity holds in all layers. Additional algebra shows that this implies

$$p_{ijk} = \frac{p_{i+k}p_{+jk}}{p_{++k}} \Leftrightarrow \frac{p_{ijk}}{p_{++k}} = \frac{p_{i+k}p_{+jk}}{p_{++k}^2}$$

$$\Leftrightarrow P[W_1 = i, W_2 = j \mid W_3 = k] = P[W_1 = i \mid W_3 = k]P[W_2 = j \mid W_3 = k],$$

for all (i,j,k). Hence W_1 and W_2 are conditionally independent given W_3.

Again by symmetry, $[12,23] = [13]^d$ implies W_1 and W_3 are conditionally independent given W_2, and $[13,12] = [23]^d$ implies W_2 and W_3 are conditionally independent given W_1.

Model 4: $[12,13,23] = [123]^d$. Model 4 states that

$$\ell_{ijk} = \lambda + \lambda_i^1 + \lambda_j^2 + \lambda_k^3 + \lambda_{ij}^{12} + \lambda_{ik}^{13} + \lambda_{jk}^{23}$$

for all (i,j,k). Further, $\boldsymbol{\lambda}^{123} = \mathbf{0}$ implies ${}^{k}\boldsymbol{\nu}^{12} = {}^{\bullet}\boldsymbol{\nu}^{12}$ for $k = 1(1)L$; i.e. ${}^{k}\boldsymbol{\nu}^{12}$ is independent of k. There is no interpretation of this model in terms of conditional or unconditional independence among the components of $\mathbf{W}$. However it will be shown below that Model 4 is equivalent to the constancy of odds ratios across layers. Before proving this it will be useful to digress and discuss odd ratios.

Consider a 2×2 table with (i,j)th cell probability p_{ij} for $1 \leq i,j \leq 2$. For level 1 of the first factor, the odds for the first level of the second factor is p_{11}/p_{12}. Similarly, for level 2 of the first factor, the odds for the first level of the second factor is p_{21}/p_{22}. The odds ratio is the ratio of these odds:

$$\frac{p_{11}/p_{12}}{p_{21}/p_{22}} = \frac{p_{11}p_{22}}{p_{21}p_{12}}.$$

The form of this expression shows why it is also known as the cross product ratio. The odds ratio is invariant under (i) simultaneous permutation of rows and columns and (ii) multiplication of any row or column by a positive constant. (Monotone functions of the odds ratio are important measures of association between factors in a 2×2 table. See Chapter 11 of Bishop, Fienberg and Holland (1975).)

For tables with more than two dimensions or more than two levels per factor, local odds ratios can be calculated between two levels of two factors for fixed values of the levels of the remaining factors. For example, in a d-dimensional table with L_j levels for the jth dimension, $1 \leq j \leq d$, the odds ratio between levels 1 and 2 of factor 1 and levels 1 and L_2 of factor 2 relative to level 1 of factors 3 through d is calculated from Table 4.3.1 as

$$\frac{p_{111\dots1}p_{2L_21\dots1}}{p_{1L_21\dots1}p_{2111\dots1}}.$$

The permutation invariance of the odds ratio means this definition is independent of row and column labeling.

Table 4.3.1. 2×2 Table for Calculating Odds Ratio

		Level of Factor 2	
		1	L_2
Level of Factor 1	1	$p_{111\dots1}$	$p_{1L_21\dots1}$
	2	$p_{211\dots1}$	$p_{2L_21\dots1}$

Returning to Model 4 (i.e., [12, 13, 23]) for the $R \times C \times L$ table, algebraic manipulation shows that the model holds if and only if the odds ratios between levels i and R of factor 1 and levels j and C of factor 2 are constant across level k of factor 3 for all i and j; i.e.,

$$\frac{p_{ijk}p_{RCk}}{p_{iCk}p_{Rjk}} = \frac{p_{ijL}p_{RCL}}{p_{iCL}p_{RjL}} \quad \text{for all} \quad (i,j,k).$$

Since the definition of $\boldsymbol{\lambda}^{123}$ is invariant with respect to the specification of which variable is the "layer" variable, similar odds ratio statements hold between factors 1 and 3 across levels of factor 2 and between factors 2 and 3 across levels of factor 1.

Interaction Graphs and Graphical Models

Of the four HLLMs $[1,2,3]$, $[1,23]$, $[13,23]$, and $[12,13,23]$ for the $R \times C \times L$ table, the first three can be completely interpreted in terms of independence relationships, while the last cannot. Another characteristic of these models (and HLLMs in general) is that some have closed form expressions for the MLEs of the cell means in terms of marginal table totals, and others do not.

In order to better understand these model properties, the subclasses of graphical and direct models will be defined. The former were first considered by Darroch, Lauritzen, and Speed (1980) (see also Edwards and Kreiner (1983), and Wermuth and Lauritzen (1983)). To understand the definitions, several concepts from graph theory must first be introduced. Throughout the discussion consider a fixed table of dimension d.

Definition. A *graph* is a pair (V, E) where V is a finite nonempty set ("vertices") and E is a collection of unordered pairs of distinct points of V ("edges").

Graphs can be pictured by plotting the vertices of the graph and connecting those pairs which form edges. Figure 4.3.1 uses this representation to show three graphs; for example, Figure 4.3.1(a) is formally $V = \{1,2,3\}$ and $E = \{\{1,2\}, \{2,3\}\}$.

For each given HLLM $[C_1, \ldots, C_r]$, a graph can be associated with the model called its interaction graph or two-factor interaction graph.

Definition. The *interaction graph* of $[C_1, \ldots, C_r]$ is $V = \cup_{j=1}^{r} C_j$, $E = \{\{i,j\}\colon \{i,j\} \subset C_k \text{ for some } k = 1(1)r\}$.

The interaction graph shows all the 2-factor interactions λ^{ij} present in the model as edges. Figure 4.3.1(b) is the interaction graph of the model $[12,13,23]$ for a $d = 3$ (or $d > 3$) table; it is also the interaction graph of the $d = 3$ saturated model $[123]$. This example shows that even when d is fixed, different HLLMs can have the *same* interaction graph. The vertex set of the interaction graph associated with any HLLM containing the main effects for all variables is $V = \{1, \ldots, d\}$. However when a HLLM omits a main effect as, for example,

$$\ell_{ijk} = \lambda + \lambda_i^1 + \lambda_j^2 \tag{4.3.7}$$

in a $d = 3$ dimensional table, then the interaction plot simply omits that vertex. Figure 4.3.2 displays the interaction graph for the model defined by (4.3.7) which is $[1,2]$ in generator set notation. For the extreme case of the constant model $\ell_{\mathbf{i}} = \lambda$, there is no associated interaction graph.

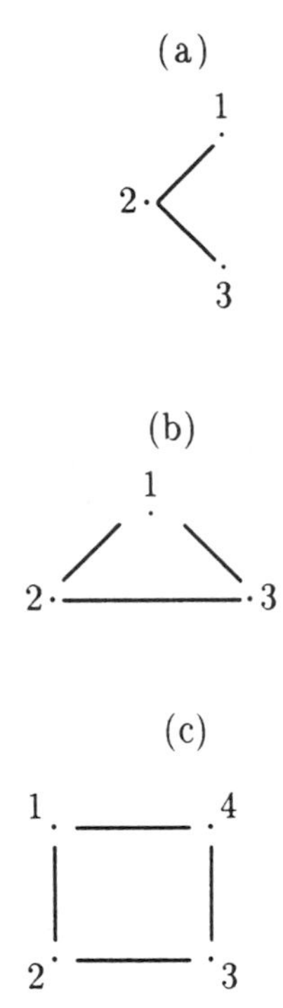

Figure 4.3.1. Three examples of graphs.

Figure 4.3.2. Interaction graph of $[1,2]$.

Now consider the converse of this process by associating a HLLM with a *given* graph (V, E). Several additional definitions must be introduced.

Definition. A pair of vertices $i, j \in V$ are *adjacent* means $\{i, j\} \in E$.

Definition. A set of vertices $C \subset V$ is *complete* means every pair of vertices in C is adjacent.

In other words, C is complete if all possible edges between vertices in C are in the graph.

Definition. A *clique* of the graph (V, E) is a maximal complete set of vertices of V; i.e., a complete set which cannot be extended by the addition of one (or more) vertices.

For example, Table 4.3.2 lists the cliques for each of the graphs in Figure 4.3.1.

Table 4.3.2. Set of All Cliques for Graphs in Figure 4.3.1

Figure 4.3.1	Cliques
(a)	$\{1,2\}$, $\{2,3\}$
(b)	$\{1,2,3\}$
(c)	$\{1,2\}$, $\{2,3\}$, $\{3,4\}$, $\{4,1\}$

The HLLM associated with a given graph (V, E) is $[C_1, \ldots, C_r]$ where $C_1, \ldots, C_r$ is the set of all cliques of (V, E); $[C_1, \ldots, C_r]$ is called the *graphical model* associated with (V, E). Table 4.3.3 lists the graphical models associated with the graphs in Figure 4.3.1.

Table 4.3.3. Graphical Models Corresponding to Graphs in Figure 4.3.1

Figure 4.3.1	Graphical Model
(a)	[12, 23]
(b)	[123]
(c)	[12, 23, 34, 41]

Since the generating sets of the model $[C_1, \ldots, C_r]$ defined by the graph (V, E) are cliques, it is easy to see that the interaction graph of this model must be (V, E). Hence this association identifies one member of the class of HLLMs having the same 2-factor interaction structure as the graphical model associated with (V, E). The HLLM associated with a given graph (V, E) has the maximal permissible higher-order interactions corresponding to the graph. The class of graphical models is the set of HLLMs $[C_1, \ldots, C_r]$ whose generator sets are the cliques corresponding to some interaction graph on $V = \cup_{j=1}^{r} C_j$. Darroch, Lauritzen, and Speed (1980) prove that graphical models are exactly those HLLMs that can be interpreted in terms of independence, conditional independence, and equiprobability of the underlying discrete classification variables. Two additional graph theoretic notions are required to state this interpretation.

The graph (V, E) is *connected* if for any pair of distinct vertices $v_1, v_2 \in V$ there is a path joining them. A *path* in (V, E) is a finite sequence $v_0, \ldots, v_n$ of vertices such that every consecutive pair (v_{i-1}, v_i), $i = 1(1)n$ in the sequence is joined by an edge.

To interpret graphical models, consider first random variables W_j whose index $j \notin V$.

Rule 1. If $V \subset \{1, \ldots, d\}$ strictly, then the random variables $\{W_j : j \in \{1, \ldots, d\} \backslash V\}$ are jointly uniformly distributed over the lattice $\times_{j \notin V} \{1, \ldots, L_j\}$.

Second, consider graphical models corresponding to graphs which are not connected (and made up of Q, say, pieces).

Rule 2. Suppose $V = \cup_{q=1}^{Q} V_q$ where for any $j \neq k$ and any pair of vertices $v \in V_k$ and $v^* \in V_j$ there is no path joining v to v^*. Then the set of random variables $\{\{W_j : j \in V_q\}\}_{q=1}^{Q}$ are mutually independent.

The notation

$$\{W_j : j \in V_1\} \perp \{W_j : j \in V_2\} \perp \cdots \perp \{W_j : j \in V_Q\}$$

will be used to indicate the sets of random variables in braces that are mutually independent. To illustrate, consider the graphical model $[1, 23, 45, 46]$ corresponding to Figure 4.3.3. The classification variables satisfy $W_1 \perp (W_2, W_3) \perp (W_4, W_5, W_6)$.

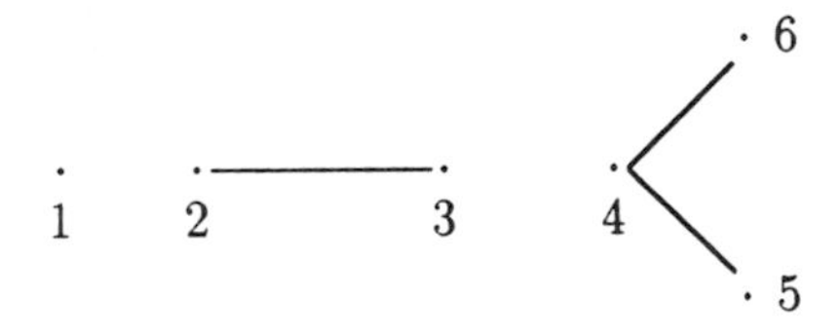

Figure 4.3.3. An interaction graph which is not connected.

Lastly, Rule 3 is applied to each connected subgraph to determine conditional independence relationships.

Rule 3. If V_1, V_2, V_3 are pairwise disjoint sets of vertices such that for any pair of vertices (v_1, v_2) with $v_1 \in V_1$ and $v_2 \in V_2$ it is the case that every path in (V, E) connecting v_1 and v_2 includes at least one vertex from V_3, then $\{W_j : j \in V_1\}$ and $\{W_j : j \in V_2\}$ are conditionally independent given $\{W_j : j \in V_3\}$.

The notation

$$\{W_j : j \in V_1\} \perp \{W_j : j \in V_2\} \,|\, \{W_j : j \in V_3\}$$

will be used to indicate conditional independence of $\{W_j : j \in V_1\}$ and $\{W_j : j \in V_2\}$ given $\{W_j : j \in V_3\}$.

The application of the three rules is illustrated with a series of examples.

Example 4.3.1. The graphical model corresponding to Figure 4.3.1(a) is [12, 23]. In a $d = 3$ dimensional table this models says $W_1 \perp W_3 \mid W_2$. In a $d = 4$ dimensional table it says that W_4 is (marginally) uniformly distributed over $1, \ldots, L_4$ and $W_1 \perp W_3 \mid W_2$.

Example 4.3.2. The graphical model corresponding to Figure 4.3.1(b) is [123]. In a $d = 3$ dimensional table this is the saturated model which imposes no restrictions on the $\{p_{ijk}\}$.

Example 4.3.3. The graphical model corresponding to Figure 4.3.1(c) is [12, 23, 34, 41]. For a $d = 4$ dimensional table this model says that $W_1 \perp W_3 \mid (W_2, W_4)$ and $W_2 \perp W_4 \mid (W_1, W_3)$.

Example 4.3.4. In a $d = 6$ dimensional table the graphical model [1, 23, 45, 46] corresponding to Figure 4.3.3 says that $W_1 \perp (W_2, W_3) \perp (W_4, W_5, W_6)$ and $W_5 \perp W_6 \mid W_4$.

Darroch, Lauritzen, and Speed provide a complete catalog of the interpretations of all graphical models of dimension $d \leq 5$.

Direct Models

The second class of HLLMs receiving attention in the literature are the direct (decomposable, multiplicative) models which are defined to be those HLLMs which have MLEs $\hat{\boldsymbol{\mu}}$ which are closed form expressions of the marginal totals in certain subtables. Goodman (1971a), Haberman (1974), and Bishop, Fienberg, and Holland (1975) provide algorithms for determining whether a given HLLM is direct. Darroch, Lauritzen, and Speed (1980) give an alternate simple characterization of the class of direct models as a subset of the class of graphical models. Two definitions are required to state their result.

Definition. A *cycle* of a graph (V, E) is a sequence of vertices $v_0, v_1, \ldots, v_n$ for which (i) $v_0 = v_n$, (ii) $v_1, \ldots, v_n$ are distinct, and (iii) each consecutive pair (v_{i-1}, v_i), $1 \leq i \leq n$, is connected by an edge. The *length* of the cycle $v_0, \ldots, v_n$ is n.

For example, (1,2,3,1) is a cycle of length $n = 3$ in Figure 4.3.1(b) and (1,2,3,4,1) is a cycle of length $n = 4$ in Figure 4.3.1(c). In general, cycles must have length $n \geq 3$ since edges must connect distinct points.

Definition. A *chord* of the cycle $v_0, \ldots, v_n$ in the graph (V, E) is an edge connecting nonconsecutive vertices of the cycle.

Cycles of length $n = 3$ cannot have chords. If the edge $\{2,4\}$ were added to the graph in Figure 4.3.1(c) then it would be a chord for the cycle (1,2,3,4,1).

Let $\mathcal{D}$, $\mathcal{G}$, and $\mathcal{H}$ denote the classes of direct, graphical, and hierarchical loglinear models, respectively. The Darroch, Lauritzen, and Speed (1980) result relating these classes is as follows.

Proposition 4.3.1. *The model classes $\mathcal{D}$, $\mathcal{G}$, and $\mathcal{H}$ satisfy*

$$\mathcal{D} \subset \mathcal{G} \subset \mathcal{H}.$$

A graphical model $[C_1, \ldots, C_r]$ is direct if and only if its interaction graph contains no cycles of length $n \geq 4$ without a chord.

Given a HLLM, one need only first check if the model is graphical and if so, whether all cycles of length 4 or more contain chords in order to assert it is direct. For example, when $d = 3$ the model $[12, 13, 23]$ is not graphical (hence not direct) since its interaction graph is Figure 4.3.1(b) which corresponds to the graphical model [123]. Every other HLLM model for 3 dimensions is both graphical and direct. One must move to $d = 4$ dimensions to find a graphical model which is not direct. The simplest example is the graphical model [12,14,23,34] corresponding to Figure 4.3.1(c). The cycle (1,2,3,4,1) has length 4 and contains no chord.

Table 4.3.4 from Darroch, Lauritzen, and Speed lists the numbers of models in the classes $\mathcal{D}$, $\mathcal{G}$, and $\mathcal{H}$ for dimensions $d = 2, 3, 4, 5$. It is clear that some sort of variable selection algorithm is needed to determine reasonable fitting models for a given data set once $d \geq 5$ or even $d \geq 4$. This topic will be discussed in Section 4.5.

Table 4.3.4. Numbers of Hierarchical, Graphical, and Direct Models for $d = 2, 3, 4, 5$ (Reprinted with permission from J.N. Darroch, S.L. Lauritzen and T.P. Speed: "Markov Fields and Log-Linear Interaction Models for Contingency," *The Annals of Statistics,* **8**, # 3, 1980. Institute of Mathematical Statistics, Hayward, CA.)

	d			
Class	2	3	4	5
$\mathcal{H}$	5	19	167	7,580
$\mathcal{G}$	5	18	133	1,450
$\mathcal{D}$	5	18	110	1,233

4.4 Numerical Evaluation of Maximum Likelihood Estimators and Residual Analysis

Two issues are considered in this section: algorithms for the numerical evaluation of MLEs based on loglinear models, and generalizations of the various residuals introduced in Section 4.2. Throughout this section and the remainder of the text the notation $\mathcal{M}[C_1, \ldots, C_q]$ is used to denote the linear space defined by the generator sets $C_1, \ldots, C_q$.

A. Calculation of Maximum Likelihood Estimators

To determine the MLEs for the three sampling models considered in Chapter 3, it suffices to consider the calculation of MLEs for independent Poisson data. This follows since Proposition 3.3.1 proves the MLEs for multinomial and product multinomial data are the same as for Poisson data provided the Poisson estimates satisfy the correct sampling constraints. Two iterative techniques are described for finding the MLE of $\boldsymbol{\mu} > \mathbf{0}$ under the Poisson model when $\boldsymbol{\ell} = \ln(\boldsymbol{\mu}) \in \mathcal{M}$.

Newton-Raphson Algorithm

Express the subspace $\mathcal{M}$ in the form $\mathcal{C}(\mathbf{X})$, where $\mathbf{X}$ is a *basis* for $\mathcal{M}$, then solve the Poisson likelihood equations $\mathbf{X}'(\mathbf{Y} - \exp\{\mathbf{X}\hat{\boldsymbol{\beta}}\}) = \mathbf{0}$ for $\hat{\boldsymbol{\beta}}$. By Proposition 3.2.1, $\hat{\boldsymbol{\mu}} := \exp\{\mathbf{X}\hat{\boldsymbol{\beta}}\}$ is the MLE of $\boldsymbol{\mu}$. The Newton–Raphson method is used to solve this nonlinear system in $\boldsymbol{\beta}$. Recall that the Newton–Raphson method for solving the system $\mathbf{f}(\mathbf{w}) = \mathbf{0}$ with current guess $\mathbf{w}^g$ has update

$$\mathbf{w}^{g+1} = \mathbf{w}^g - [\nabla \mathbf{f}(\mathbf{w})]^{-1}\mathbf{f}(\mathbf{w}).$$

In the case of the Poisson likelihood equations the update is

$$\begin{aligned}\boldsymbol{\beta}^{g+1} &= \boldsymbol{\beta}^g - (\mathbf{X}'\mathbf{D}\mathbf{X})^{-1}\mathbf{X}'(\mathbf{Y} - \exp\{\mathbf{X}\boldsymbol{\beta}^g\}) \\ &= (\mathbf{X}'\mathbf{D}\mathbf{X})^{-1}\mathbf{X}'\mathbf{D}[\mathbf{X}\boldsymbol{\beta}^g + \mathbf{D}^{-1}(\exp\{\mathbf{X}\boldsymbol{\beta}^g\} - \mathbf{Y})] \qquad (4.4.1)\end{aligned}$$

where, as in Chapter 3, $\mathbf{D} = \mathbf{D}(\boldsymbol{\beta}^g)$ is a diagonal matrix with diagonal elements $\exp\{\mathbf{X}\boldsymbol{\beta}^g\}$ and the inverse exists because $\mathbf{X}$ has full column rank. Equation 4.4.1 shows that each Newton–Raphson iteration formally corresponds to a weighted least squares (WLS) analysis with response ("working") vector

$$\mathbf{Z}^g = \mathbf{X}\boldsymbol{\beta}^g + \mathbf{D}^{-1}(\exp\{\mathbf{X}\boldsymbol{\beta}^g\} - \mathbf{Y})$$

and weight ("working") matrix $\mathbf{D}$. For this reason, the Newton–Raphson method is sometimes called the method of *iteratively reweighted least squares* (IRLS). Similar IRLS interpretations hold for Newton–Raphson maximum likelihood methods in generalized linear models (Nelder and Wedderburn (1972)).

Two criteria are commonly used to test convergence for the Newton–Raphson algorithm. The first is to stop when the change in successive (relative) estimates of all components of $\boldsymbol{\beta}$ are sufficiently small. The second criterion is to stop when the change in successive (relative) estimates of the loglikelihood is sufficiently small. If one does not first check $\mathbf{Y}$ to determine whether the MLE exists then the second criterion has at least one advantage over the first. Changes in loglikelihood become small even if the MLE does not exist because the loglikelihood converges to a limiting value (either 0 or a strictly negative quantity), while $\boldsymbol{\beta}$ can be diverging to $+\infty$ or $-\infty$ in one or more coordinates. This point will be illustrated more explicitly for the case of logistic regression in Section 5.3.

Haberman (1974, Chapter 3) proposes a modified Newton–Raphson algorithm for fitting loglinear models to Poisson data. He discusses asymptotic convergence rates, starting values, and implementation issues in detail. The main practical disadvantage of the IRLS method is that the matrix $\mathbf{X}$ must be a basis for $\mathcal{M}$. In cross-classified data problems it is relatively easy to determine a spanning set for $\mathcal{M}$ but more difficult to obtain a basis although, in principal, Gram–Schmidt could be applied to a given spanning set to generate a basis.

Iterative Proportional Fitting Algorithm

The iterative proportional fitting (IPF) algorithm is adapted from Deming and Stephan (1940). It can be applied to models specified by a nonbasis spanning set such as HLLMs for cross-classified data. From the viewpoint of numerical methods, the IPF algorithm is a Gauss–Seidel (or cyclic ascent) algorithm for determining the zeros of the nonlinear system $\mathbf{f}(\mathbf{w}) = \mathbf{0}_R$ where $\mathbf{f}(\cdot) = (f_1(\cdot), \ldots, f_R(\cdot))'$ and $\mathbf{w} = (w_1, \ldots, w_Q)'$. In the case where the number of equations equals the number of variables ($R = Q$), each cycle of the Gauss–Seidel algorithm performs R steps. The gth step updates the component w_g of $\mathbf{w}$ by solving $f_g(\mathbf{w}) = 0$ for w_g when all other components are *fixed* at their previously computed values. In the application to solving the Poisson likelihood equations, the system $\mathbf{f}(\mathbf{w}) = \mathbf{0}_R$ contains *fewer* equations than variables w_j ($R < Q$). The gth step of the algorithm solves $f_g(\mathbf{w}) = 0$ for a *subset* of $\mathbf{w}$ variables while holding the remainder fixed at their previously computed values. (See Thisted, 1988, Sections 3.11.2 and 4.3.4.)

To determine the MLE of the mean $\boldsymbol{\mu}$ for a d-dimensional table satisfying the HLLM $\boldsymbol{\ell} \in \mathcal{M} = \mathcal{C}[\boldsymbol{\nu}_1, \ldots, \boldsymbol{\nu}_R]$ where each $\boldsymbol{\nu}_r$ consists of zeros and ones, a $\hat{\boldsymbol{\mu}} > \mathbf{0}$ satisfying

$$\boldsymbol{\nu}_r'\mathbf{Y} = \boldsymbol{\nu}_r'\boldsymbol{\mu}, \quad 1 \le r \le R \tag{4.4.2}$$

and $\ln(\hat{\boldsymbol{\mu}}) \in \mathcal{M}$ must be determined. In brief, the rth step of each cycle of the IPF algorithm updates $\boldsymbol{\mu}^{g-1}$ to $\boldsymbol{\mu}^g$ so that

$$f_r(\boldsymbol{\mu}^g) := \boldsymbol{\nu}_r'\mathbf{Y} - \boldsymbol{\nu}_r'\boldsymbol{\mu}^g = 0$$

where $r = g \bmod R$ by adjusting *only* the components μ_j^{g-1} with $(\boldsymbol{\nu}_r)_j = 1$.

The IPF algorithm is defined by the following steps. First, an initial guess $\boldsymbol{\mu}^0$ with $\ln(\boldsymbol{\mu}^0) \in \mathcal{M}$ is selected. The algorithm uses update

$$\ln(\boldsymbol{\mu}^g) = \ln(\boldsymbol{\mu}^{g-1}) + \sigma\boldsymbol{\nu}_r$$

for $r = g \bmod R$ where

$$\sigma = \ln(\boldsymbol{\nu}_r'\mathbf{Y}/\boldsymbol{\nu}_r'\boldsymbol{\mu}^{g-1}). \tag{4.4.3}$$

Several aspects of the algorithm deserve comment. First, the scalar σ in (4.4.3) maximizes the log likelihood

$$\ln L(\boldsymbol{\mu}) = \mathbf{Y}'\ln(\boldsymbol{\mu}) - \mathbf{1}'\boldsymbol{\mu}$$

in the direction $\boldsymbol{\nu}_r$. This is true because

$$\ln L(\boldsymbol{\mu}^g) = \mathbf{Y}'\ln(\boldsymbol{\mu}^{g-1}) + \sigma\mathbf{Y}'\boldsymbol{\nu}_r - e^{\sigma}\boldsymbol{\nu}_r'\boldsymbol{\mu}^{g-1} - (\mathbf{1} - \boldsymbol{\nu}_r)'\boldsymbol{\mu}^{g-1} \tag{4.4.4}$$

is strictly concave in σ and

$$\frac{\partial}{\partial\sigma}\ln L(\boldsymbol{\mu}^g) = \boldsymbol{\nu}_r'\mathbf{Y} - e^{\sigma}\boldsymbol{\nu}_r'\boldsymbol{\mu}^{g-1} \stackrel{set}{=} 0$$

holds if and only if (4.4.3) holds. Second, since the rth step of an IPF cycle alters only those components μ_j^{g-1} for which $(\boldsymbol{\nu}_r)_j = 1$, if $\boldsymbol{\nu}_r, \ldots, \boldsymbol{\nu}_{r+k}$ have pairwise disjoint coordinate positions where the 1's are located, then $\boldsymbol{\mu}^{g-1}$ can be updated simultaneously in these coordinates. This observation is used in Examples 3.1.2 and 4.4.1 below. Lastly, the Poisson (log) likelihood is nondecreasing at each step. By the strict log concavity in σ of (4.4.4)

$$\ln L(\boldsymbol{\mu}^g) = \ln L(\boldsymbol{\mu}^{g-1}(\mathbf{1} - \boldsymbol{\nu}_r) + \boldsymbol{\mu}^{g-1}\boldsymbol{\nu}_r\sigma) \geq \ln L(\boldsymbol{\mu}^{g-1})$$

with strict inequality unless (4.4.3) is zero; i.e., $\boldsymbol{\nu}_r'\mathbf{Y} = \boldsymbol{\nu}_r'\boldsymbol{\mu}^{g-1}$. The latter can be interpreted as saying the previous update fits the current margins in which case $\boldsymbol{\mu}^g = \boldsymbol{\mu}^{g+1}$.

Example 3.1.2 (continued). Fit the independence model $\mathcal{M}[1,2] = \mathcal{C}(\boldsymbol{\rho}_1, \ldots, \boldsymbol{\rho}_R, \boldsymbol{\kappa}_1, \ldots, \boldsymbol{\kappa}_C)$ to independent Poisson observations $\{Y_{ij} : i = 1(1)R, j = 1(1)C\}$. A vector $\hat{\boldsymbol{\mu}} > \mathbf{0}$ is the MLE of $\boldsymbol{\mu}$ if and only if

$$\hat{\mu}_{i+} = Y_{i+}, \quad i = 1(1)R \quad \text{and} \quad \hat{\mu}_{+j} = Y_{+j}, \quad j = 1(1)C.$$

The IPF based on $\boldsymbol{\rho}_1, \ldots, \boldsymbol{\rho}_R, \boldsymbol{\kappa}_1, \ldots, \boldsymbol{\kappa}_C$ is

Initialization: Set $\boldsymbol{\mu}^0 = \mathbf{1}_{RC}$, $c = 1$, and perform cycle c.

Cycle c: For $1 \leq i \leq R$ and $1 \leq j \leq C$ set

$$\mu_{ij}^{2c-1} := \mu_{ij}^{2c-2}(Y_{i+}/\mu_{i+}^{2c-2}) \quad \text{and}$$

$$\mu_{ij}^{2c} := \mu_{ij}^{2c-1}(Y_{+j}/\mu_{+j}^{2c-1}), \quad 1 \le i \le R \text{ and } 1 \le j \le C.$$

Recursion: Check for convergence. If not, set $c = c + 1$ and perform cycle c.

It is straightforward to check that $\mu_{i+}^{2c-1} = Y_{i+}$ and $\mu_{+j}^{2c} = Y_{+j}$. Geometrically, if $\mathbf{P}_1$ is the projection matrix onto $\mathcal{M}[1] = \mathcal{C}(\boldsymbol{\rho}_1, \ldots, \boldsymbol{\rho}_R)$ and $\mathbf{P}_2$ is the projection matrix onto $\mathcal{M}[2] = \mathcal{C}(\boldsymbol{\kappa}_1, \ldots, \boldsymbol{\kappa}_C)$ then $\mathbf{P}_1\boldsymbol{\mu}^{2c-1} = \mathbf{P}_1\mathbf{Y}$ and $\mathbf{P}_2\boldsymbol{\mu}^{2c} = \mathbf{P}_2\mathbf{Y}$.

The model in this example is direct. In the first cycle, $\mu_{ij}^1 = Y_{i+}/C =: Y_{i\bullet}$ and $\mu_{ij}^2 = Y_{i\bullet}(Y_{+j}/\sum_{i=1}^R Y_{i\bullet}) = (Y_{i+})(Y_{+j})/m$; this shows that $\boldsymbol{\mu}^2$ fits the row margins as well as the column margins. Furthermore if $Y_{i+} > 0$ for $i = 1(1)R$ and $Y_{+j} > 0$ for $j = 1(1)C$ then $\ln(\boldsymbol{\mu}^2) \in \mathcal{M}[1,2]$ and thus $\boldsymbol{\mu}^2$ is the MLE of $\boldsymbol{\mu}$ under $\mathcal{M}[1,2]$. If the algorithm continues for additional cycles it is easy to see that $\boldsymbol{\mu}^g = \boldsymbol{\mu}^2$ $(g \ge 3)$. Thus the IPF algorithm converges to the MLE of $\boldsymbol{\mu}$ in one cycle. If any row or column margin is zero then the algorithm still converges to $\boldsymbol{\mu}^2$ although the corresponding rows or columns of $\boldsymbol{\mu}^2$ are zero. The adjustments the algorithm makes to $\ln(\boldsymbol{\mu}^{g-1})$ are:

$$\begin{aligned} \ln(\mu_{ij}^{2c-1}) &= \ln(\mu_{ij}^{2c-2}) + \ln[Y_{i+}/\mu_{i+}^{2c-2}] \\ &=: \ln(\mu_{ij}^{2c-2}) + a_{ij}^{2c-2} \end{aligned}$$

with $\mathbf{a}^{2c-2} \in \mathcal{M}[1]$ and

$$\begin{aligned} \ln(\mu_{ij}^{2c}) &= \ln(\mu_{ij}^{2c-1}) + \ln[Y_{+j}/\mu_{+j}^{2c-1}] \\ &=: \ln(\mu_{ij}^{2c-1}) + a_{ij}^{2c-1} \end{aligned}$$

with $\mathbf{a}^{2c-1} \in \mathcal{M}[2]$.

In the general case of calculating the MLE based on the HLLM $[C_1, \ldots, C_q]$, the jth step simultaneously solves the $\boldsymbol{\nu}_r'\mathbf{Y} = \boldsymbol{\nu}_r'\boldsymbol{\mu}$ corresponding to the pairwise disjoint $\boldsymbol{\nu}_r$ generating $\mathcal{M}[C_j]$. Equivalently, the jth step of the algorithm solves $\mathbf{P}_j\mathbf{Y} = \mathbf{P}_j\boldsymbol{\mu}$ where $\mathbf{P}_j$ is the projection matrix onto $\mathcal{M}[C_j]$. The update $\boldsymbol{\mu}^g$ generated at the jth step of the cth cycle satisfies $\boldsymbol{\mu}^g = \boldsymbol{\mu}^{g-1} + \mathbf{a}^{g-1}$ where the adjustment $\mathbf{a}^{g-1} \in \mathcal{M}[C_j]$. Hence if $\ln(\boldsymbol{\mu}^0) \in \mathcal{M}$, it automatically follows that $\ln(\boldsymbol{\mu}^g) \in \mathcal{M}$ for $g \ge 1$. These points will be illustrated in a second example.

Example 4.4.1. Find the MLE of $\boldsymbol{\mu}$ under the no three-factor interaction model $\mathcal{M}[12,13,23]$ for independent Poisson data $\{Y_{ijk} : i = 1(1)R,\ j = 1(1)C,\ k = 1(1)L\}$. The space $\mathcal{M}[12,13,23]$ of $\boldsymbol{\ell}$ for which $\boldsymbol{\lambda}^{123} = \mathbf{0}$ is spanned by

$$\{\{\boldsymbol{\rho}^{ik} : i,k\}, \{\boldsymbol{\kappa}^{jk} : j,k\}, \{\boldsymbol{\Lambda}^{ij} : i,j\}\},$$

where the ρ^{ik}, κ^{jk} and Λ^{ij} are $R \times C \times L$ 0/1 incidence matrices defined by

$$(\rho^{ik})_{q_1,q_2,q_3} = \begin{cases} 1 & \text{if } q_1 = i, q_3 = k \\ 0 & \text{otherwise,} \end{cases}$$

$$(\kappa^{jk})_{q_1,q_2,q_3} = \begin{cases} 1 & \text{if } q_2 = j, q_3 = k \\ 0 & \text{otherwise,} \end{cases}$$

$$(\Lambda^{ij})_{q_1,q_2,q_3} = \begin{cases} 1 & \text{if } q_1 = i, q_2 = j \\ 0 & \text{otherwise.} \end{cases}$$

The matrix ρ^{ik} identifies the ith row of the kth layer of an $R \times C \times L$ table and similar interpretations hold for κ^{jk} and Λ^{ij}. Another way of saying that $\mathcal{M}[12,13,23] = \mathcal{C}(\{\rho^{ik}: i,k\} \cup \{\kappa^{jk}: j,k\} \cup \{\Lambda^{ij}: i,j\})$ is that $\boldsymbol{\ell}$ satisfies $\boldsymbol{\lambda}^{123} = \mathbf{0}$ if and only if there exist constants $\{a_{ij}\}$, $\{b_{ik}\}$, $\{c_{jk}\}$ such that $\ell_{ijk} = a_{ij} + b_{ik} + c_{jk}$. To see this latter statement, note that if $\boldsymbol{\lambda}^{123} = \mathbf{0}$ then these constants can be defined as $a_{ij} = \lambda + \lambda_i^1 + \lambda_j^2 + \lambda_{ij}^{12}$, $b_{ik} = \lambda_k^3 + \lambda_{ik}^{13}$ and $c_{jk} = \lambda_{jk}^{23}$, while if $\ell_{ijk} = a_{ij} + b_{ik} + c_{jk}$ then calculation gives $\boldsymbol{\lambda}^{123} = \mathbf{0}$.

By Proposition 3.2.1, the MLE $\hat{\boldsymbol{\mu}}$ is that positive $R \times C \times L$ matrix, if it exists, such that $\ln(\hat{\boldsymbol{\mu}}) \in \mathcal{M}[12,13,23]$ and $\mathbf{P}\hat{\boldsymbol{\mu}} = \mathbf{PY}$ where $\mathbf{P}$ is the projection matrix from $\mathbb{R}^{RCL}$ onto $\mathcal{M}[12,13,23]$, or equivalently by

$$\left.\begin{aligned} \Lambda^{ij}\hat{\boldsymbol{\mu}} &= \Lambda^{ij}\mathbf{Y} \\ \rho^{ik}\hat{\boldsymbol{\mu}} &= \rho^{ik}\mathbf{Y} \\ \kappa^{jk}\hat{\boldsymbol{\mu}} &= \kappa^{ik}\mathbf{Y} \end{aligned}\right\} \tag{4.4.5}$$

for all (i,j,k) which is equal to

$$\left.\begin{aligned} \hat{\mu}_{ij+} &= Y_{ij+} \\ \hat{\mu}_{i+k} &= Y_{i+k} \\ \hat{\mu}_{+jk} &= Y_{+jk} \end{aligned}\right\} \tag{4.4.6}$$

for all (i,j,k). Thus the MLE $\hat{\boldsymbol{\mu}}$ "fits the margins" defined by the generating sets [12], [13], [23]. The steps of the IPF algorithm calculate successive guesses $\boldsymbol{\mu}^g$ solving the three subsystems of equations in (4.4.6).

Initialization: Set $\boldsymbol{\mu}^0 := \mathbf{1}_{RCL}$, $c = 1$, and perform cycle c.

Cycle c: For all i, and set

$$\mu_{ijk}^{3c-2} := \mu_{ijk}^{3c-3}(Y_{ij+}/\mu_{ij+}^{3c-3}),$$

$$\mu_{ijk}^{3c-1} := \mu_{ijk}^{3c-2}(Y_{i+k}/\mu_{i+k}^{3c-2}), \quad \text{and}$$

$$\mu_{ijk}^{3c} := \mu_{ijk}^{3c-1}(Y_{+jk}/\mu_{+jk}^{3c-1}).$$

Recursion: Check for convergence. If not set $c = c + 1$ and perform cycle c.

It is straightforward algebra to check that the results of these three steps successively adjust previous estimates of $\boldsymbol{\mu}$ to fit $\{Y_{ij+}\}$, $\{Y_{i+k}\}$, and $\{Y_{+jk}\}$, respectively. Geometrically the $\hat{\boldsymbol{\mu}}^g$ satisfy $\mathbf{P}_{12}\boldsymbol{\mu}^{3c-2} = \mathbf{P}_{12}\mathbf{Y}$, $\mathbf{P}_{13}\boldsymbol{\mu}^{3c-1} = \mathbf{P}_{13}\mathbf{Y}$, and $\mathbf{P}_{23}\boldsymbol{\mu}^{3c} = \mathbf{P}_{23}\mathbf{Y}$, where $\mathbf{P}_{ij}$ is the projection matrix onto $\mathcal{M}[ij]$. Clearly the adjustments to $\ln(\boldsymbol{\mu}^{g-1})$ at steps 1, 2, and 3 belong to $\mathcal{M}[12]$, $\mathcal{M}[13]$, and $\mathcal{M}[23]$, respectively.

The first two steps in cycle $c = 1$ yield

$$\mu^1_{ijk} = Y_{ij+}/L = Y_{ij\bullet}$$

and

$$\mu^2_{ijk} = (Y_{ij+}/L)(Y_{i+k}/(\sum_j Y_{ij\bullet}) = Y_{ij+}Y_{i+k}/Y_{i++}\ .$$

Observe that μ^2_{ijk} fits not only the [13] margin $\{Y_{i+k}\}$ as designed, but also the [12] margin $\{Y_{ij+}\}$. The third step gives,

$$\mu^3_{ijk} = (Y_{ij+}Y_{i+k}/Y_{i++})(Y_{+jk}/\sum_i \mu^2_{ijk}).$$

While $\mu^3_{+jk} = Y_{+jk}$, it need not fit either previous set of margins $\{Y_{ij+}\}$ or $\{Y_{i+k}\}$.

There are several additional points that can be made regarding this example. First the algorithm is not defined for all data sets $\mathbf{Y}$ since at some point it can involve a division by zero (see μ^2_{ijk}). Second, if $\boldsymbol{\mu}^2 > \mathbf{0}$ then $\boldsymbol{\mu}^2$ is the MLE of $\boldsymbol{\mu}$ for the (direct) model $\mathcal{M}[13, 12]$ (conditional independence of W_2 and W_3 given W_1). This follows since $\boldsymbol{\mu}^2$ has the same inner product as does $\mathbf{Y}$ with the matrices $\{\boldsymbol{\rho}^{ik}: i, k\}$ and $\{\boldsymbol{\Lambda}^{ij}: i, j\}$ which span $\mathcal{M}[13, 12] = \{\boldsymbol{\ell}: \boldsymbol{\lambda}^{23} = \mathbf{0} = \boldsymbol{\lambda}^{123}\}$.

Haberman (1974) contains a detailed discussion of the properties of the IPF; he shows that if it is applied to cross-classified data $\mathbf{Y}$ of 6 or fewer dimensions for which a direct MLE exists, then the algorithm converges to the MLE in one iteration. Csiszar (1975) develops a dual algorithm for the IPF. Darroch and Ratcliff (1972) propose a generalization of IPF called the Generalized Iterative Scaling Method. Finally, Meyer (1982) shows several classes of examples where it is possible to transform a contingency table and the associated model into a form where the existence of direct estimates can be easily recognized or the IPF easily applied.

B. Residual Analysis of Contingency Tables

This section concludes by generalizing the different types of residuals introduced in Section 4.2 to tables of arbitrary dimension. The extensions of $e^r_{\mathbf{i}}$, $e^P_{\mathbf{i}}$, and $e^d_{\mathbf{i}}$ are obvious; i.e.,

$$e^r_{\mathbf{i}} = Y_{\mathbf{i}} - m\hat{p}_{\mathbf{i}},$$

$$e_i^P = e_i^r/(m\hat{p}_i)^{1/2},$$

and

$$e_i^d = \pm[2Y_i \,|\, \ln(Y_i/m\hat{p}_i)|]^{1/2}$$

with the "−" sign used if and only if $\ln(Y_i/m\hat{p}_i) < 0$ (i.e., $Y_i < m\hat{p}_i$). Asymptotically equivalent versions of e_i^P include

$$2[Y_i^{1/2} - (m\hat{p}_i)^{1/2}]$$

and the Freeman–Tukey residuals

$$Y_i^{1/2} + (Y_i + 1)^{1/2} - (4m\hat{p}_i + 1)^{1/2}$$

(Freeman and Tukey (1950)). More general linear combinations of the e_i^P are studied by Cox and Snell (1968) and Haberman (1976).

A heuristic calculation is now given which yields the formula for the analog of the adjusted residuals introduced in Section 4.2. Start with the IRLS interpretation of the MLE introduced in the discussion of the Newton–Raphson algorithm; it shows that $\hat{\boldsymbol{\beta}}$ is formally the WLS estimate of $\boldsymbol{\beta}$ based on responses

$$\mathbf{Z} = \mathbf{X}\hat{\boldsymbol{\beta}} + \mathbf{D}^{-1}(\exp\{\mathbf{X}\hat{\boldsymbol{\beta}}\} - \mathbf{Y}) \tag{4.4.7}$$

having mean vector $E[\mathbf{Z}] = \mathbf{X}\boldsymbol{\beta}$ and variance covariance matrix $\mathbf{D}^{-1}$ where $\mathbf{D} = \mathbf{D}(\hat{\boldsymbol{\beta}})$ is diagonal with entries $\exp\{\mathbf{X}\hat{\boldsymbol{\beta}}\}$. Applying the standard linear model calculation to the residual vector

$$\mathrm{e} = \mathbf{Z} - \mathbf{X}\hat{\boldsymbol{\beta}} = [\mathbf{I} - \mathbf{X}(\mathbf{X}'\mathbf{D}\mathbf{X})^{-1}\mathbf{X}'\mathbf{D}]\mathbf{Z}$$

and treating $\mathbf{D}$ as fixed yields $E[\mathrm{e}] = \mathbf{0}$ and

$$\mathrm{Var}(\mathrm{e}) = \mathbf{D}^{-1} - \mathbf{X}(\mathbf{X}'\mathbf{D}\mathbf{X})^{-1}\mathbf{X}'. \tag{4.4.8}$$

Applying (4.4.7) gives

$$\mathbf{e} = \mathbf{D}^{-1}(\exp\{\mathbf{X}\hat{\boldsymbol{\beta}}\} - \mathbf{Y}) = -\mathbf{D}^{-1}\mathbf{e}^r$$

or componentwise $e_i^r = -m\hat{p}_i e_i$. Treating $m\hat{p}_i$ as constant and applying (4.4.8) yields the following adjusted version of e_i^r

$$\begin{aligned} e_i^a &= \frac{e_i^r}{\left(m\hat{p}_i[1 - m\hat{p}_i\{\mathbf{X}(\mathbf{X}'\mathbf{D}\mathbf{X})^{-1}\mathbf{X}'\}_{ii}]\right)^{1/2}} \\ &= \frac{e_i^P}{(1 - m\hat{p}_i\{\mathbf{X}(\mathbf{X}'\mathbf{D}\mathbf{X})^{-1}\mathbf{X}'\}_{ii})^{1/2}}. \end{aligned} \tag{4.4.9}$$

Under mild assumptions, Haberman (1973a; 1974, Chapter 4) proves that the set of $\{e_i^a\}$ is asymptotically standard normally distributed (for any of the three sampling models).

As in the $R \times C$ case, examination of the residuals to assess model fit is important. The adjusted residuals can be difficult to compute from (4.4.9) because they require a basis matrix; where a basis is not readily available the $\{e_i^P\}$ can be used. The residuals should be examined for patterns, and the types of plots illustrated in Example 4.1.1 constructed. The case study at the end of Section 4.5 illustrates some of these techniques.

4.5 Model Selection and a Case Study

This section will discuss the problem of choosing models for cross-classified data. This process amounts to choosing the order and identity of the interactions to be used in the model and is particularly difficult when the number of classifying variables is large. Algorithms for choosing models are called either variable selection procedures or model selection procedures and the two terms will be used interchangeably. Many of the algorithms are analogs of variable selection methods for linear regression.

General Notions

Five important general notions for model selection are parsimony, interpretability, significance of effects, percent decrease in goodness-of-fit statistic, and coherence (see Benedetti and Brown, 1978). *Parsimony* means a model should have as few parameters as possible while adequately explaining the data. Of course, models with more parameters fit data better; this motivated the development of principles such as the Akaike Information Criteria (AIC) which balance parsimony and goodness-of-fit (Sakamoto and Akaike, 1978).

A model's *interpretability* is determined in part, by whether its interaction structure is compatible with the experimental sampling. In addition, models with interactions but without main effects are difficult to interpret and this has led most authors to restrict attention to the family of hierarchical loglinear models when analyzing cross-classified data.

A model has all *significant effects* (at a given α level) means that based on some test of significance the removal of any term is significant and the addition of any term is not significant.

Considerations of *percentage decrease* in some goodness-of-fit statistic are relevant for very large data sets, such as census data, where it is often the case that all effects are significant even if some are very close to zero. One might then select a model with only those effects which offer a sufficiently large percentage decrease in some goodness-to-fit statistic relative to some baseline model. Since our primary concern is with the analysis of small-to-moderate size data sets, the problem of all significant effects will not be discussed further.

Lastly the *coherence principle* (Gabriel, 1969) states that if a model is rejected then all submodels should be rejected and, conversely, if a model is accepted then all models containing it should be accepted.

None of the above considerations necessarily leads to a unique "best" model, even when notions such as adequacy of fit, significance, and large percentage increase are quantified. Furthermore, specific subject-matter applications often suggest additional criteria important in variable selection. These last two considerations suggest the strategy of identifying a *subset* of models which, relative to the important criteria, explain the data equally well. When practically feasible, entertaining several models both broadens our understanding of the data and highlights the sensitivity of various conclusions to the chosen model.

With one exception, model selection procedures are characterized by four elements:

(i) the class of models under consideration,

(ii) a starting model,

(iii) a rule for stepping from one model to the next model, and

(iv) a termination criterion.

The exception to the above characterization is an analog of the all-possible regressions procedure which is proposed by Fowlkes, Freeny, and Landwehr (1988). Their fundamental premise is that in many applications it is computationally feasible to fit all possible HLLMs to a contingency table. They construct a scatterplot of df := degrees of freedom versus G^2 for all HLLMs. Well-fitting and simple models lie near the line $y = x$ and have higher df values. Suggestions are given for selecting a set of models in this region for further analysis.

The following paragraphs review some of the specific proposals in the literature for (i)–(iv).

The first component of a variable selection algorithm is the choice of model class. Sections 4.1 and 4.3 introduced three classes of loglinear models: hierarchical, graphical and direct models. The direct models form a proper subset of the graphical models which in turn are a proper subset of the hierarchical models. Furthermore the class of graphical models are exactly those HLLMs which have interpretations in terms of equiprobability and (conditional) independence. The latter is an important motivation for considering graphical models. Model selection algorithms have been proposed within all three classes of models. Goodman (1971a, 1973), Brown (1976), Benedetti and Brown (1978), and Edwards and Havranek (1985) discuss selection of hierarchical models; Edwards and Kreiner (1983), Havranek (1984), and Edwards and Havranek (1985) discuss selection of graphical models; and Wermuth (1976) and Havranek (1984) discuss selection of direct models. Our discussion will focus on model selection within the class of *hierarchical models* although comments will be made when appropriate about the selection of graphical and direct models.

Many of the starting models and all of the stepping rules that have been proposed in the literature depend on goodness-of-fit tests. As reviewed in Section 4.2 (see also Section 2.2), the two commonly used goodness-of-fit statistics are the Pearson and likelihood ratio chi-squared statistics X^2 and G^2. Both X^2 and G^2 have null asymptotic χ^2_{t-1-r} distributions where t is the number of cells and r is the number of linearly independent parameters in the model. Despite the discussion in Chapter 2 indicating that X^2 has more nearly nominal size than G^2, G^2 tends to be preferred in the variable selection literature. This is because G^2 has the following attractive additive property: if $\mathcal{M}'$ is a submodel of $\mathcal{M}$ then

$$G^2(\mathcal{M}') = G^2(\mathcal{M}' \,|\, \mathcal{M}) + G^2(\mathcal{M}) \tag{4.5.1}$$

where $G^2(\mathcal{M}')$ $[G^2(\mathcal{M})]$ is the likelihood ratio statistic for testing H_0: $\mathbf{p}$ satisfies $\mathcal{M}'$ $[\mathcal{M}]$ versus the global alternative H_A: $\mathbf{p}$ does not satisfy $\mathcal{M}'$ $[\mathcal{M}]$ and $G^2(\mathcal{M}' \,|\, \mathcal{M})$ is the likelihood ratio statistic for testing H_0: $\mathbf{p}$ satisfies $\mathcal{M}'$ versus H_A: $\mathbf{p}$ satisfies $\mathcal{M}$ but not $\mathcal{M}'$. In particular, (4.5.1) implies that $G^2(\mathcal{M}') \geq G^2(\mathcal{M})$; that is, the simpler model has at least as large a goodness-of-fit statistic. Not only does X^2 fail to satisfy property (4.5.1), but $X^2(\mathcal{M}') < X^2(\mathcal{M})$ can occur. Haberman (1974, p. 108) gives conditions under which $G^2(\mathcal{M}' \,|\, \mathcal{M})$ is asymptotically null distributed as a $\chi^2_{r-r'}$ random variable where r (r') is the rank of $\mathcal{M}(\mathcal{M}')$. Furthermore these same conditions guarantee that $X^2(\mathcal{M}') - X^2(\mathcal{M})$ has an asymptotic $\chi^2_{r-r'}$ null distribution.

The power of a sequence of tests of the form (4.5.1) has only more recently been studied. Oler (1985) and Fenech and Westfall (1988) consider the joint asymptotic distribution of the set of LR tests for successive hypotheses in the nested family $H_0 \subset H_1 \subset \ldots \subset H_K$ under a sequence of Pitman alternatives $\{\mathbf{p}_n\}_{n \geq 1}$; here H_i states that $\mathbf{p}$ satisfies model $\mathcal{M}_i$. The sequence $\{\mathbf{p}_n\}_{n \geq 1}$ for which the joint asymptotic distribution is computed, is assumed to satisfy $\ln(\mathbf{p}_n) = \ln(\mathbf{p}_0) + \boldsymbol{\eta}/\sqrt{n} + o(n^{-1/2})$ where $\ln(\mathbf{p}_0)$ satisfies H_0 and $\boldsymbol{\eta}$ is fixed. Their calculations show the power of $G^2(\mathcal{M}_i \,|\, \mathcal{M}_{i+1})$ depends not only on $\mathcal{M}_{i+1}$ but also on the models $\mathcal{M}_{i+2}, \ldots, \mathcal{M}_K$ "further out" in the sequence and can be adversely affected. Simulation studies in both papers confirm the effect occurs for moderate size samples. These results about the power of conditional tests have practical implications for choosing a method of stepping from one model to the next. One class of methods, called forward selection procedures, starts with a simple model and adopts the more complicated $\mathcal{M}_{i+1}$ over $\mathcal{M}_i$ when $G^2(\mathcal{M}_i \,|\, \mathcal{M}_{i+1})$ *rejects* $\mathcal{M}_i$. The results above show that forward selection procedures can stop early because a particular conditional test in the sequence has low power. In contrast, backward selection procedures start with a complicated model and adopt the simpler $\mathcal{M}_i$ over $\mathcal{M}_{i+1}$ when $G^2(\mathcal{M}_i \,|\, \mathcal{M}_{i+1})$ *accepts*

$\mathcal{M}_i$. Backward selection procedures are not affected by the power phenomenon described above. A final general remark concerns testing models when the table is sparse. Edwards and Kreiner (1983) briefly discuss this problem; recall from Section 2.2 that the accuracy of the asymptotic null approximation for G^2 and X^2 is questionable in this case. One alternative is to perform a permutation test. The test is carried out either by completely enumerating or randomly sampling tables which follow the null model being tested. The proportion of tables for which the G^2 (or X^2) statistic exceeds the observed value of G^2 (X^2) approximates the significance level. In the case of randomly sampled tables, the accuracy of the approximation increases when it is based on a larger sample of random tables. Mehta and Patel (1983) describe a very fast network-based algorithm for computing the exact P-value of the independence test in a 2 dimensional table by "smart" enumeration. See also Baglivo, Olivier, and Pagano (1988). Random tables can also be generated by the methods discussed in Mehta, Patel, and Senchaudhuri (1988).

Tests for Order

There are two broad categories of tests used in determining starting models and stepping rules. These are tests for order and tests for specific $\boldsymbol{\lambda}$ interactions. Tests for order are used to determine the order of the maximum interaction required to adequately describe the data. To define procedures for these two problems in a d-dimensional table, let M_0 denote the constant model and let M_k denote the hierarchical loglinear model containing *all* $\binom{d}{k}$ k-factor interactions for $k = 1(1)d$. Thus M_1 is the main-effects-only model and M_d is the fully saturated model.

One method of establishing the model order is to test

$$H_0\colon \mathbf{p} \text{ satisfies } M_k \text{ versus } H_A\colon (\mathbf{p} \text{ does not satisfy } M_k).$$

The null hypothesis is that *all* $(k+1)$-way and higher interactions are zero; the alternative is global. The minimum k for which H_0 is *accepted* gives the maximum degree interaction needed in the model. A second method of testing for model order is to test

$$H_0\colon \mathbf{p} \text{ satisfies } M_{k-1} \text{ versus } H_A\colon (\mathbf{p} \text{ satisfies } M_k \text{ but not } M_{k-1}).$$

The null hypothesis is that all k-way and higher interactions are zero; the alternative is that *all* $(k+1)$-way and higher interactions are zero but at least one k-way interaction is not zero. The maximum k for which H_0 is *rejected* gives the maximum degree interaction needed in the model.

The difference between these two procedures is that the first procedure compares M_k to M_d and is thus *unconditional* while the second procedure compares M_{k-1} to M_k and is *conditional* on all $(k+1)$-way and higher interactions being zero. It is theoretically possible for the two procedures to choose different values of k based on a given test and α level. For example, suppose for a $2 \times 2 \times 2$ table with $\alpha = .05$ that $G^2(M_2) = 3.2$ (< 3.84) and $G^2(M_1) = 10$ (> 9.49). Then $k = 2$ for the unconditional method while $G^2(M_1 \mid M_2) = G^2(M_1) - G^2(M_2) = 6.8$ (< 7.81) so that $k < 2$ for the condition method.

Tests for Interactions

A second type of hypothesis that model selection algorithms test is whether a specific interaction, say $\boldsymbol{\lambda}^{1\ldots k}$, is zero (in all components). If $\boldsymbol{\ell}$ is the vector of logmeans, a general strategy for conducting a test of $\boldsymbol{\lambda}^{1\ldots k} = \mathbf{0}$ is to choose a pair of HLLMs $\mathcal{M}'$ and $\mathcal{M}$ which differ *only* in that $\mathcal{M}$ contains the extra interaction term $\boldsymbol{\lambda}^{1\ldots k}$ and test $H_0\colon \boldsymbol{\ell} \in \mathcal{M}'$ versus $H_A\colon \boldsymbol{\ell} \in \mathcal{M}$. It is easy to see that the reduced model $\mathcal{M}'$ will be hierarchical only if $[1\ldots k]$ is one of the generators of $\mathcal{M}$. Unfortunately there are almost always multiple models $\mathcal{M}$ for which $[1\ldots k]$ is a generator set and which could be used to perform the test. For example, consider deciding whether $\boldsymbol{\lambda}^{12} = \mathbf{0}$ in a 3-dimensional table. There are 18 non-constant HLLMs, 5 of which have $[12]$ as a generator, and hence 5 possible models $\mathcal{M}$ on which to base the test. These five models are listed in Table 4.5.1. The conclusions of the test can depend on the model in which $\boldsymbol{\lambda}^{12}$ is embedded. This is an example of the familiar phenomenon that the additional explanatory ability of a term depends on the terms already in the model. This problem is chronic in any model building effort for cross-classified data and more generally. In a 3-dimensional table it is feasible to compute all 5 tests statistics; if all are significant it seems reasonable to reject the hypothesis while if some of them are significant and others are not, the situation is less clear. In higher dimensions there are many more ways to test a specific interaction in a pair of hierarchical models. Computing and interpreting all such test statistics is generally impractical. Thus several specific choices of $\mathcal{M}'$ and $\mathcal{M}$ have been emphasized in the literature.

Table 4.5.1. Five Model Pairs for Testing $H_0\colon \boldsymbol{\lambda}^{12} = \mathbf{0}$

Test	$\mathcal{M}$	$\mathcal{M}'$
1	[12]	[1, 2]
2	[12, 3]	[1, 2, 3]
3	[12, 13]	[2, 13]
4	[12, 23]	[1, 23]
5	[12, 13, 23]	[13, 23]

Brown (1976) proposed performing tests of "marginal" and "partial" association. His test of *marginal association* compares $\mathcal{M} = [1\ldots k]$ to the submodel $\mathcal{M}'$ obtained from $\mathcal{M}$ by deleting $\boldsymbol{\lambda}^{1\ldots k}$. For example when $d = 3$, Line 1 of Table 4.5.1 ($\mathcal{M}' = [1,2]$ and $\mathcal{M} = [12]$) gives the marginal test of

$\boldsymbol{\lambda}^{12} = \mathbf{0}$. These *same* two models are used to perform the marginal association test of $\boldsymbol{\lambda}^{12} = \mathbf{0}$ for higher dimensional tables. Thus the name derives from the fact that the test statistic depends only on the [1, 2] marginal data.

Brown's test of *partial association* compares the model $\mathcal{M} = M_k$ containing all $\binom{d}{k}$ kth order interactions with the model $\mathcal{M}'$ obtained from M_k by removing $\boldsymbol{\lambda}^{1\ldots k}$. For example when $d = 3$, Line 5 of Table 4.5.1 ($\mathcal{M}' = [13, 23]$ and $\mathcal{M} = [12, 13, 23]$) gives the partial association test of $\boldsymbol{\lambda}^{12} = \mathbf{0}$. When $d = 4$ the partial association test of $\boldsymbol{\lambda}^{12} = \mathbf{0}$ compares $\mathcal{M}' = [13, 14, 23, 24, 34]$ with $\mathcal{M} = [12, 13, 14, 23, 24, 34]$.

Brown's two tests are intended to give an indication of the range of significance of all the tests of $\boldsymbol{\lambda}^{1\ldots k}$ without requiring extensive computation. In any specific data set it need not be the case that all other comparisons of HLLMs for testing a specific interaction will have X^2 or G^2 values between those of the tests for marginal and partial association.

For the *special* case of testing 2-way interactions, a third choice of $\mathcal{M}'$ and $\mathcal{M}$ arises by considering tests of zero partial association (zpa) as defined by Birch (1964, 1965). In a $d = 3$ dimensional table, Birch defined a pair of classification variables W_i and W_j to have *zero partial association* if W_i and W_j are conditionally independent given the third classification variable. From the model discussion of Section 4.3 this means mathematically that $\boldsymbol{\lambda}^{ij} = \boldsymbol{\lambda}^{123} = \mathbf{0}$. In a $2 \times 2 \times 2$ table, the row and column classifications exhibit zpa if the odds ratio is unity in each layer of the table. More generally in a d-dimensional table, W_i and W_j have zpa if they are conditionally independent given the remaining classification variables $\{W_q : q \in \{1, \ldots, d\} \setminus \{i, j\}\}$. Analytically this means $\boldsymbol{\lambda}^{ij}$ and all its higher order relatives are zero.

Birch proposed testing $\boldsymbol{\lambda}^{ij} = \mathbf{0}$ by taking $\mathcal{M}'$ to be the model of zpa between W_i and W_j; thus he would take $H_0 : \boldsymbol{\ell} \in \mathcal{M}' = [ij]^d$ versus $H_A : \boldsymbol{\ell} \in \mathcal{M} = [[ij]^d, ij]$. For example, in the case of a $d = 3$ dimensional table the Birch zpa test for $\boldsymbol{\lambda}^{ij} = \mathbf{0}$ is

$$H_0 : \boldsymbol{\lambda}^{ij} = \boldsymbol{\lambda}^{123} = \mathbf{0} \text{ versus } H_A : \boldsymbol{\lambda}^{123} = \mathbf{0}\ (\neq \boldsymbol{\lambda}^{ij}). \tag{4.5.2}$$

In the $2 \times 2 \times S$ case, (4.5.2) tests the null hypothesis that the odds ratio is unity in each layer of the table versus the alternative of a constant odds ratio (not unity) across layers. Either the likelihood ratio or Pearson chi-squared statistics can be used to perform the test of (4.5.2); however the alternative hypothesis is not direct and does require iterative calculation of the means. Alternatively, the Cochran–Mantel–Haenszel test (Cochran, 1954; Mantel and Haenszel, 1959) is a *closed-form* large sample test of (4.5.2) (see Section 5.5). In $d = 3$ dimensions, the hypothesis (4.5.2) agrees with Brown's partial association test. However in higher dimensions, the tests use different models. For example, Birch's zpa test of $\boldsymbol{\lambda}^{12} = \mathbf{0}$ in $d = 4$

dimensions compares $\mathcal{M}' = [12]^d = [134, 234]$ with $\mathcal{M} = [12, 134, 234]$ while, as noted above, Brown's partial association test compares $\mathcal{M} = [13, 14, 23, 24, 34]$ with $\mathcal{M} = [12, 13, 14, 23, 24, 34]$.

For completeness's sake it should be noted that several authors discuss an alternative test of zpa between W_i and W_j (Bishop, Fienberg, and Holland, 1975; Benedetti and Brown, 1978; or Wermuth, 1976). When $d = 3$ the alternative test uses

$$H_0: \boldsymbol{\lambda}^{ij} = \boldsymbol{\lambda}^{123} = \mathbf{0} \text{ versus } H_A: (\text{not } H_0) \tag{4.5.3}$$

which has the global alternative. Since the models $\mathcal{M}'$ and $\mathcal{M}$ differ by more than the [12] generator set, (4.5.3) is not appropriate for testing the *single* interaction $\boldsymbol{\lambda}^{ij} = \mathbf{0}$ in a model selection algorithm.

In stepping from one model to another within a *subclass* of the HLLMs, the problem of testing whether a specific interaction, say $\boldsymbol{\lambda}^{i_1 \ldots i_q}$, is zero also occurs. The general strategy of performing such a test by comparing two models within the class which differ only in the interaction term $\boldsymbol{\lambda}^{i_1 \ldots i_1}$ still applies.

Starting Models and Stepping Rules

With the paragraphs above as background on goodness-of-fit tests, specific proposals for starting models and stepping rules can be discussed. Some starting models that have been proposed in the literature are:

(i) the main-effects-only model,

(ii) the saturated model,

(iii) M_k where k is chosen by testing for order,

(iv) the model including all interactions for which both Brown's marginal and partial tests of interaction are significant,

(v) the model including all 2-way interactions which have significant tests of zpa plus any 3 and higher way interactions all of whose lower order relatives are included in the model, and

(vi) a model motivated by the given application.

The third element of a variable selection scheme is a rule for stepping from one model to a new model. As noted previously, both backward stepping rules and forward stepping rules have been proposed. For example, one can step forward from starting model (iv) by trying to add terms which are significant in only one of Brown's two tests, or from model (v) by trying to add higher interactions if some (but not all) of their 2-way relatives had significant tests. Backward steps can be taken from model (iii) or from (ii) by using either or both of Brown's tests. Stepwise (or stagewise) algorithms allow additional flexibility by permitting bidirectional steps so that interactions previously added to a simple model can be eliminated or interactions previously eliminated from a complicated model can be added. In view of the discussion of the power of sequences of G^2 or X^2 tests, backward or stepwise algorithms are to be preferred.

The fourth and final element of any model building algorithm is a terminal decision rule. Virtually all algorithms stop when goodness-of-fit tests employed by the algorithm for the addition of a term are nonsignificant or tests for the elimination of a term are significant.

The next few paragraphs summarize some of the features of specific model selection algorithms that have appeared in the literature.

Goodman (1971a) formulates a multidirectional procedure for model selection. His starting model is obtained by performing a test for order. Forward or backward steps based on conditional tests-of-fit are performed. Goodman (1971a, 1973) also discusses a "guided" version of the procedure which uses standardized estimates of the λ-terms from the fully saturated model to guide the steps. The guided procedure first fits the saturated model M_d, estimates the variances of the MLEs $\hat{\boldsymbol{\lambda}}^*$ for various interactions $*$, and computes the standardized variates $\hat{\boldsymbol{\lambda}}^*_s$ which are the $\hat{\boldsymbol{\lambda}}^*$ divided by their estimated standard deviations. Standardized variates large in magnitude suggest significant model terms. The variance of $\hat{\boldsymbol{\lambda}}^*$ is easily estimated using the delta method because $\hat{\boldsymbol{\lambda}}^*$ is a linear combination of the logs of the observed counts $Y_{\mathbf{i}}$ for the saturated model. See Fienberg (1980, pp. 70–76) for more details on the method and Aitkin (1979) for a discussion of its limitations.

Wermuth (1976) considers a backward elimination procedure within the class of direct models which starts with the saturated model and uses the zpa test (4.5.2) to remove terms in its first step and conditional versions of these tests (assuming certain pairs of variables have zpa) in subsequent steps. Benedetti and Brown (1978) summarize and illustrate several of these model selection procedures.

More recently Edwards and Kreiner (1983) develop three variants of a backward elimination procedure for graphical models. In a slightly different spirit, Havranek (1984) and Edwards and Havranek (1985) consider selection methods for graphical and for hierarchical models which obey the coherence principle. In their scheme a submodel of a rejected model is said to be "weakly" rejected and a supermodel of an accepted model is said to be "weakly" accepted. Forward and backward stepwise procedures are described which terminate when all models are classified as accepted, weakly accepted, rejected, or weakly rejected. The coherence principle is not necessarily satisfied by an algorithm based on separate statistical tests for each model; i.e., if model $\mathcal{M}'$ is a submodel of $\mathcal{M}$ then $G^2(\mathcal{M}')$ $(X^2(\mathcal{M}'))$ may be significant when $G^2(\mathcal{M})$ $(X^2(\mathcal{M}))$ is not.

For all the stepwise methods described above, the selected model depends, of course, on whether stepping is forward, backward, or multidirectional. Even with a fixed type of stepping, the resulting model need not be unique if the *order* in which certain interactions are tested is arbitrary. In addition, the usual problems associated with overall significance levels for multiple tests apply here. Aitkin (1979) develops a formal simultaneous test procedure for fitting models to cross-classified data. His procedure, discussed also in Whittaker and Aitkin (1978) and Aitkin (1980), provides control of the overall error rate. It is a backward elimination scheme based on a partitioning of the G^2 statistic.

Case Study

The remainder of this section is devoted to a case study in which the Madsen data of Example 4.1.3 is analyzed using Goodman's (1971a) multidirectional algorithm for choosing a model within the class of HLLMs, Wermuth's (1976) algorithm for choosing a model within the class of direct models, and the all-possible models plot proposed by Fowlkes et al. (1988). The data are repeated in Table 4.5.2 for ease of reference. The Madsen data were selected for this study because they allow illustration of the modifications required by variable selection algorithms to accommodate product multinomial sampling. Recall that the data were collected by fixing a priori the number of tenants to be interviewed from each of the four housing types. In the ensuing discussion, the following mnemonic notation for the four classification variables is adopted: H = housing type, I = feeling of influence with apartment management, C = degree of contact with other residents, and S = satisfaction with housing. Only HLLMs containing the housing main effect $\boldsymbol{\lambda}^H$ will be considered so that the fitted counts $\hat{Y}_{\text{chis}}$ satisfy the sampling constraints $\hat{Y}_{+h++} = Y_{+h++}$ for $h = 1(1)4$.

Table 4.5.3 lists the unconditioned (H_0: M_k versus H_A: (not M_k)) and conditional (H_0: M_{k-1} versus H_A: M_k) tests of model order. At any reasonable significance level, both tests for order give the same result: neither the 4-way nor the 3-way interactions are required but at least some of the 2-way interactions are needed.

Table 4.5.4 gives the results of marginal association, partial association, and Birch zpa tests for the six 2-way interactions. Brown's marginal association tests for IS, CI, and CS compare $\mathcal{M}' = [I, S]$ with $\mathcal{M} = [IS]$, $\mathcal{M}' = [C, I]$ with $\mathcal{M} = [CI]$, and $\mathcal{M}' = [C, S]$ with $\mathcal{M} = [CS]$, respectively. However, none of these six models contains the $\boldsymbol{\lambda}^H$ main effect. Thus Table 4.5.4 lists modified versions of each of the models which adds the generator set $\{H\}$.

Table 4.5.2. 1,681 Persons Classified According to Satisfaction, Contact, Influence, and Type of Housing (Reprinted with permission from M. Madsen: "Statistical Analysis of Multiple Contingency Tables: Two Examples," *Scandinavian Journal of Statistics,* **3**. The Almquist & Wiksell Periodical Company, Stockholm, Sweden, 1976.)

Contact...		Low			High		
Satisfaction		Low	Medium	High	Low	Medium	High
Housing	Influence						
Tower blocks	Low	21	21	28	14	19	37
	Medium	34	22	36	17	23	40
	High	10	11	36	3	5	23
Apartments	Low	61	23	17	78	46	43
	Medium	43	35	40	48	45	86
	High	26	18	54	15	25	62
Atrium houses	Low	13	9	10	20	23	20
	Medium	8	8	12	10	22	24
	High	6	7	9	7	10	21
Terraced houses	Low	18	6	7	57	23	13
	Medium	15	13	13	31	21	13
	High	7	5	11	5	6	13

Table 4.5.3. (a) Tests of H_0: M_k versus H_A: (not M_k)

k	*df*	G^2	P-value	X^2	P-value
3	12	5.94	.92	5.97	.92
2	40	43.95	.31	44.18	.30
1	63	295.35	$< 10^{-4}$	305.93	$< 10^{-4}$
0	71	833.66	$< 10^{-4}$	949.08	$< 10^{-4}$

(b) Tests of H_0: M_{k-1} versus H_A: M_k

k	*df*	G^2	P-value	X^2	P-value
4	12	5.94	.92	5.97	.92
3	28	38.01	.10	38.21	.09
2	23	251.40	$< 10^{-4}$	261.75	$< 10^{-4}$
1	8	538.31	$< 10^{-4}$	643.15	$< 10^{-4}$

Table 4.5.4. Marginal Association (M), Partial Association (P), and Birch zpa (Z) Tests for 2-Way Interactions in Madson Data of Table 4.5.2

Interaction	Test	*df*	$\mathcal{M}'$	$\mathcal{M}$	$G^2(\mathcal{M}'\|\mathcal{M})$	*P*-value
	M	6	$[H,S]$	$[HS]$	60.69	$< 10^{-4}$
HS	P	6	$[HI,CH,IS,CI,CS]$	M_2	62.18	$< 10^{-4}$
	Z	6	$[HS]^d$	$[HS,CHI,CIS]$	61.62	$< 10^{-4}$
	M	6	$[H,I]$	$[HI]$	16.89	$< 10^{-2}$
HI	P	6	$[HS,CH,IS,CI,CS]$	M_2	13.69	.03
	Z	6	$[HI]^d$	$[HI,CHS,CIS]$	13.43	.037
	M	3	$[H,C]$	$[CH]$	39.06	$< 10^{-4}$
CH	P	3	$[HS,HI,IS,CI,CS]$	M_2	44.04	$< 10^{-3}$
	Z	3	$[CH]^d$	$[CH,HIS,CIS]$	43.30	$< 10^{-4}$
	M	4	$[S,I,H]$	$[IS,H]$	106.37	$< 10^{-4}$
IS	P	4	$[HS,HI,CH,CI,CS]$	M_2	109.08	$< 10^{-4}$
	Z	4	$[IS]^d$	$[IS,CHI,CHS]$	107.43	$< 10^{-4}$
	M	2	$[C,I,H]$	$[CI,H]$	17.54	$< 10^{-3}$
CI	P	2	$[HS,HI,CH,IS,CS]$	M_2	23.71	$< 10^{-3}$
	Z	2	$[CI]^d$	$[CI,HIS,CHS]$	16.84	$< 10^{-3}$
	M	2	$[C,S,H]$	$[CS,H]$	5.13	.04
CS	P	2	$[HS,HI,CH,IS,CI]$	M_2	16.02	$< 10^{-3}$
	Z	2	$[CS]^d$	$[CS,HIS,CHI]$	16.71	$< 10^{-3}$

Goodman's backward elimination scheme starting with the model M_2 uses the partial association tests since they compare the starting model M_2 with the model in which a single 2-way interaction is deleted from M_2. For $\alpha = .01$, the conclusion is that $\boldsymbol{\lambda}^{HI}$ is not required in the model. The remaining 2-way interactions are tested by deleting them one at a time from $\mathcal{M} = [HS, CH, IS, CI, CS]$. The results presented in Table 4.5.5 show that none of these terms can be deleted. The final model selected by this algorithm is $\mathcal{M} = [HS, CH, IS, CI, CS]$; $\mathcal{M}$ is easily seen to not be graphical (and hence not direct) by drawing its interaction graph, displayed in Figure 4.5.1, and recognizing that the corresponding graphical model is $[CHS, CIS]$. The same model $[HS, CH, IS, CI, CS]$ is also identified by a forward selection algorithm starting with the main-effects-only model M_1. The 2-way interactions enter in the order IS, HS, CH, CI, and CS which is in accordance with the significance levels of the marginal and partial association tests and the Birch zpa test as given in Table 4.5.4.

Table 4.5.5. Tests in Step 2 of Goodman's Algorithm

Interaction	*df*	$\mathcal{M}'$	$G^2(\mathcal{M}' \mid \mathcal{M})$	P-value
HS	6	$[CH, IS, CI, CS]$	66.55	$< 10^{-4}$
CH	3	$[HS, IS, CI, CS]$	42.94	$< 10^{-4}$
IS	4	$[HS, CH, CI, CS]$	111.41	$< 10^{-4}$
CI	2	$[HS, CH, IS, CS]$	22.61	$< 10^{-4}$
CS	2	$[HS, CH, IS, CI]$	15.76	$< 10^{-3}$

Figure 4.5.1. Interaction graph of $[HS, CH, IS, CI, CS]$.

Next Wermuth's (1976) backward selection algorithm in the class of direct models is illustrated. Wermuth's starting model is the saturated model with interaction graph in Figure 4.5.2. Step 1 calculates the statistics for the tests of zpa of the six 2-way interactions. Geometrically the models $\mathcal{M}'$ have interaction graphs obtained by deleting edges one at a time from the graph in Figure 4.5.2 (see Table 4.5.6). The largest P-value among these six tests is .18 and corresponds to H_0: (H and I are conditionally independent given C and S); the edge HI is deleted and the algorithm goes to Step 2.

Table 4.5.6. Tests in Step 1 of Wermuth Algorithm

Edge	$\mathcal{M}'$	*df*	$G^2(\mathcal{M}')$	*P*-value
HI	$[HI]^d$	36	43. 76	.18
HS	$[HS]^d$	36	99.09	$< 10^{-4}$
CH	$[CH]^d$	27	64.35	$< 10^{-4}$
IS	$[IS]^d$	32	135.69	$< 10^{-4}$
CI	$[CI]^d$	24	33.72	.09
CS	$[CS]^d$	24	32.87	.11

H · · I
S · · C

Figure 4.5.2. Interaction graph of the saturated model $[CHIS]$.

The interaction graph of $[HI]^d = [CHS, CIS]$ is shown in Figure 4.5.3. Using this graph it is easy to see that only four of the five models obtained by deleting one additional edge are direct; eliminating the edge CS yields a graph with cycle HS, IS, CI, CH which has no chords and hence is not direct (Proposition 4.3.1). The tests for the remaining four models relative to $\mathcal{M} = [HI]^d$ are listed in Table 4.5.7. All four tests are rejected and thus Wermuth's algorithm terminates with the model $[HI]^d = [CHS, CIS]$.

Table 4.5.7. Tests in Step 2 of Wermuth Algorithm

Edge	$\mathcal{M}'$	*df*	$G^2(\mathcal{M}')$	*P*-value
HS	$[HI, HS]^d$	12	76.64	$< 10^{-4}$
CH	$[HI, CH]^d$	9	55.03	$< 10^{-4}$
IS	$[HI, IS]^d$	8	113.25	$< 10^{-4}$
CI	$[HI, CI]^d$	6	24.40	$< 10^{-4}$

Figure 4.5.3. Interaction graph of the model $[HI]^d$.

It is interesting to compare these two models with other ones by means of the likelihood ratio plot suggested by Fowlkes et al. (1988). Figure 4.5.4 plots G^2 versus degrees of freedom for the 166 possible HLLMs for this table. There is a dichotomy in the goodness-of-fit statistic between the models which contain the main effect of H (upper cloud of points) and those which do not. The lower line is $y = x$ and the upper line is the upper 5% critical point for the χ^2 distribution with x degrees of freedom. Models with plotting positions above this line are rejected by an $\alpha = .05$ level test. The lower right-hand part of this scatterplot ($G^2 \leq 90$ and degrees of freedom between 30 and 50) is expanded in Figure 4.5.5. The models selected by Goodman's algorithm and Wermuth's algorithm are Models 8 and 4, respectively. Other parsimonious models with reasonably good overall fit are also readily identified from Figure 4.5.5 and listed in Table 4.5.8. The models in this table differ from the selected models 4 and 8 by their inclusion of the 2-way interaction HI (Models 1, 2, 3, and 5) and their inclusion of a single 3-way interaction (Models 1, 2, 3, 6, and 7). The 3-way interactions CHS and CIS are each present in two of these six models. Of the eight models only one is direct (model 4) and the remaining seven are not even graphical. However, models 5–8 all have the *same* interaction graph as Model 4.

Table 4.5.8. Parsimonious Models Exhibiting No Lack-of-Fit for Madsen Data of Table 4.5.2

Number	Model	df	$G^2(\mathcal{M})$
1	$[HI, IS, CI, CHS]$	34	31.78
2	$[HS, CS, IS, CHI]$	34	38.66
3	$[HS, CH, HI, CIS]$	36	42.29
4	$[CHS, CIS]$	36	43.76
5	$[HS, CH, CS, HI]$	40	43.95
6	$[IS, CI, CHS]$	40	45.55
7	$[HS, CH, CIS]$	42	55.85
8	$[HS, CH, IS, CI, CS]$	46	57.64

This case study is concluded with residual analyses and interpretations for the more simply stated of our two final models; i,e., $[HS, CH, IS, CI, CS]$ or

$$\ell_{\text{chis}} = \lambda + \lambda_h^H + \lambda_i^I + \lambda_c^C + \lambda_s^S + \lambda_{hs}^{HS} + \lambda_{ch}^{CH} + \lambda_{is}^{IS} + \lambda_{ci}^{CI} + \lambda_{cs}^{CS}. \quad (4.5.4)$$

Table 4.5.9 lists the estimated expected cell means $\{\hat{\mu}_{\text{chis}}\}$ based on model (4.5.4). Table 4.5.10 lists the adjusted residuals (recommended) and, for comparison, Table 4.5.11 presents the Pearson residuals. There are no obvious patterns; only one Pearson residual and four adjusted residuals exceed ± 2 in magnitude. The adjusted residuals vary between 1.08 and 1.62 times the Pearson residuals and the insights provided by the two sets of residuals are basically the same. Figure 4.5.6 is a probability plot of the ordered adjusted residuals. The plot shows a very slightly heavy right tail (there are 2 adjusted residuals greater than 2.5), but overall the fit to the model is acceptable.

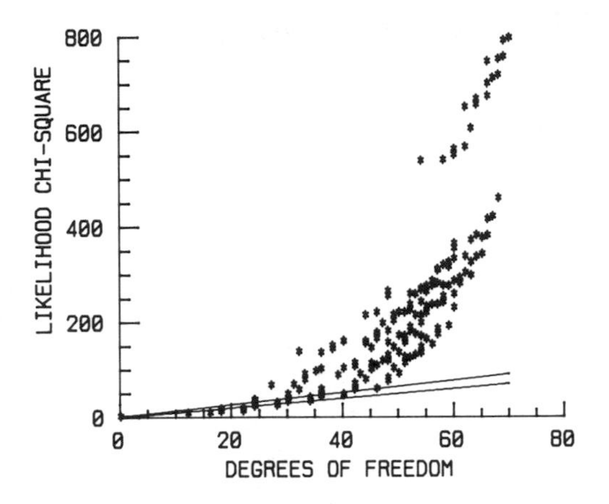

Figure 4.5.4. Scatterplot of (df = degrees of freedom, G^2); the line $y = df$; and $y = \chi^2_{.05,df}$ for various HLLMs.

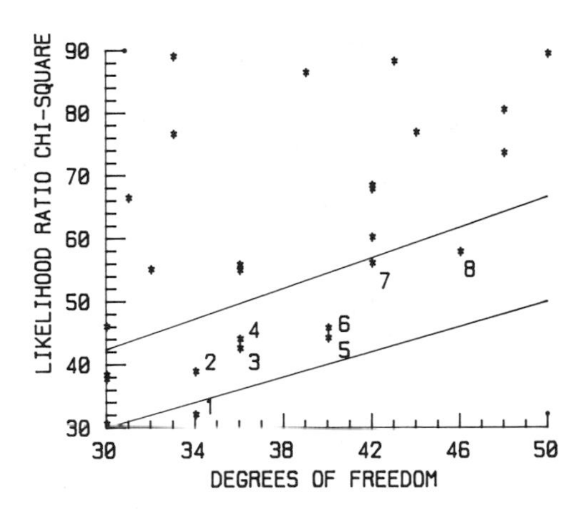

Figure 4.5.5. Expanded view of Figure 4.5.4 for $G^2 \leq 90$ and $30 \leq$ degrees of freedom ≤ 50.

Table 4.5.9. Expected Counts Based on Model (4.5.4)

Contact...		Low			High		
Satisfaction		Low	Medium	High	Low	Medium	High
Housing	Influence						
Tower blocks	Low	26.94	17.66	22.38	20.99	19.63	28.09
	Medium	22.72	23.11	39.86	13.64	19.81	38. 56
	High	10.42	13.11	42.80	4.28	7.68	28.31
Apartments	Low	56.49	24.63	24.69	78.22	48.68	55.06
	Medium	47.63	32.24	43.96	50.85	49.11	75.59
	High	21.86	18.29	47.21	15.95	19.05	55.50
Atrium houses	Low	11.39	8.48	6.55	20.83	22.14	19.30
	Medium	9.60	11.10	11.67	13.54	22.33	26.50
	High	4.40	6.29	12.53	4.25	8.66	19.46
Terraced houses	Low	22.66	7.57	4.55	44.48	21.21	14.39
	Medium	19.11	9.91	8.10	28.91	21.40	19.75
	High	8.77	5.62	8.70	9.07	8.30	14.50

Table 4.5.10. Adjusted Residuals for Model (4.5.4)

Contact...		Low			High		
Satisfaction		Low	Medium	High	Low	Medium	High
Housing	Influence						
Tower blocks	Low	−1.51	0.99	1.47	−1.94	0.18	2.12
	Medium	3.01	−0.30	−0.84	1.07	0.90	0.31
	High	−0.15	−0.70	−1.48	−0.67	−1.09	−1.26
Apartments	Low	0.89	−0.43	−1.94	−0.04	−0.56	−2.33
	Medium	−0.96	0.66	−0.83	−0.57	−0.86	1.83
	High	1.15	−0.08	1.41	−0.29	1.72	1.26
Atrium houses	Low	0.58	0.21	1.50	−0.24	0.24	0.19
	Medium	−0.61	−1.12	0.12	−1.16	−0.09	−0.63
	High	0.83	0.32	−1.20	1.45	0.52	0.43
Terraced houses	Low	−1.27	−0.65	1.24	2.79	0.51	−0.44
	Medium	−1.18	1.16	1.97	0.51	−0.11	−1.93
	High	−0.68	−0.29	0.90	−1.54	−0.91	−0.47

Table 4.5.11. Pearson Residuals for Model (4.5.4)

Contact...		Low			High		
Satisfaction		Low	Medium	High	Low	Medium	High
Housing	Influence						
Tower blocks	Low	−1.15	0.80	1.19	−1.53	−0.14	1.68
	Medium	2.37	−0.23	−0.61	0.91	0.72	0.23
	High	−0.13	−0.58	−1.04	−0.62	−0.97	−1.00
Apartments	Low	0.60	−0.33	−1.55	−0.02	−0.38	−1.62
	Medium	−0.67	0.49	−0.60	−0.40	−0.59	1.20
	High	0.89	−0.07	0.99	0.24	1.36	0.87
Atrium houses	Low	0.48	0.18	1.35	−0.18	0.18	0.16
	Medium	−0.52	−0.93	0.10	−0.96	−0.07	−0.49
	High	0.76	0.28	−1.00	1.34	0.45	0.35
Terraced houses	Low	−0.98	−0.57	1.15	1.88	0.39	−0.37
	Medium	−0.94	0.98	1.72	0.39	−0.09	−1.52
	High	−0.60	−0.26	0.78	−1.35	−0.80	−0.39

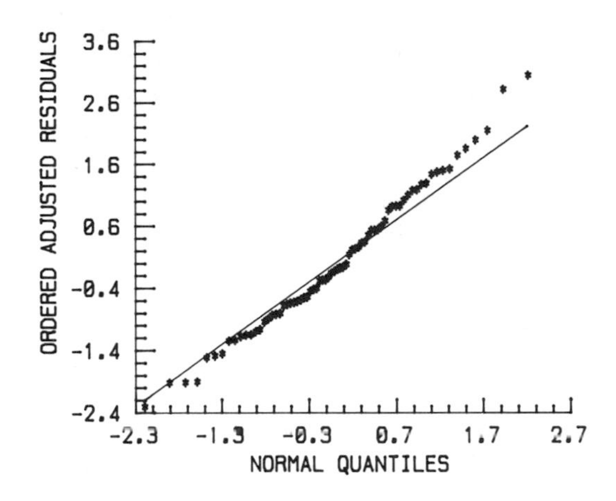

Figure 4.5.6. Normal probability plot of ordered adjusted residuals in Table 4.5.10.

Table 4.5.12 lists the estimated 2-way interactions from the final model. S.K. Lee (1977) gives expressions for the asymptotic variances of λ-terms in the case of direct models. In a direct model each λ-term is a linear combination of the natural logarithms of the cell means and it is thus straightforward but tedious to determine asymptotic variances for them. However Model (4.5.4) is not direct and hence no standardization is performed in Table 4.5.12.

Table 4.5.12. Estimated Two-Factor λ Parameters in Model (4.5.4)

Satisfact.	Low	Medium	High
Housing			
Tower blocks	−0.28	−0.04	0.33
Apartments	0.07	−0.10	0.03
Atrium houses	−0.20	0.16	0.04
Terraced houses	0.42	−0.02	−0.40

Contact	Low	High
Housing		
Tower blocks	0.29	−0.29
Apartments	0.01	−0.01
Atrium houses	−0.13	0.13
Terraced houses	−0.17	0.17

Satisfact.	Low	Medium	High
Influence			
Low	0.38	0.02	−0.40
Medium	−0.01	0.06	−0.05
High	−0.37	−0.08	0.45

Influence	Low	Medium	High
Contact			
Low	−0.15	−0.02	0.17
High	0.15	0.02	−0.17

Satisfact.	Low	Medium	High
Contact			
Low	0.14	−0.04	−0.10
High	−0.14	0.04	0.10

Table 4.5.12 suggests the following conclusions.

(a) The degree of satisfaction is weakly though positively related to the degree of contact with the landlord.

(b) The degree of satisfaction is strongly related to the apartment dwellers' feelings of influence with the landlord.

(c) The feeling of influence with the landlord is negatively related to the degree of contact.

(d) The degree of satisfaction is highest in high-rise blocks and worst in terraced houses.

(e) The association between contact and housing is (nearly) the reverse of that between satisfaction and housing.

The final section of this chapter uses the models developed in Section 4.3 to introduce the topic of collapsing tables "without losing information."

4.6 Collapsing Tables

Simpson's Paradox

This section introduces the problem of collapsing a table "without losing information." Attention is restricted to collapsing an $R \times C \times L$ table over the layer variable. The problem is illustrated by two examples.

Example 4.6.1. Consider a two-arm clinical trial (Treatments 1 and 2) in which the outcome is classified as a success or failure. Table 4.6.1 shows hypothetical data for such a trial stratified by patient sex. For men the estimated odds ratio of success for Treatment 1 to Treatment 2 is 100(95)/800(5) = 2.37 and the Pearson chi-squared statistic for testing equality of the success probabilities for the two treatments is $X^2 = 3.58$ ($P = .059$). For women the estimated odds ratio is 195(400)/4(500) = 39 and the Pearson chi-squared statistic is $X^2 = 149.808$ ($P < 10^{-4}$). Thus Treatment 1 is preferable to Treatment 2 for both men and women as both estimated odds ratios are greater than unity. (The chi-squared statistics can be made arbitrarily large by increasing the sample sizes in each cell by the same proportion.) However in the marginal table collapsed over sex, Table 4.6.2, the estimated odds of success for Treatment 1 to Treatment 2 is 295(495)/805(405) = .447 with a chi-square of 71.93 ($P < 10^{-4}$). Thus, presented only with the collapsed Table 4.6.2 a physician would erroneously conclude that success is strongly related to treatment and *Treatment 2 is preferable to Treatment 1,* despite the fact that the opposite holds in both sex-specific estimates.

Table 4.6.1. Numbers of Successes and Failures Using Treatments 1 and 2 Cross-Classified by Sex

	Sex			
	Male		Female	
	Treatment		Treatment	
	1	2	1	2
Success	100	5	195	400
Failure	800	95	5	400

Table 4.6.2. Numbers of Successes and Failures for Treatments 1 and 2

	Treatment	
	1	2
Success	295	405
Failure	805	495

Example 4.6.1 shows that collapsing tables can lead to loss of information. Simpson (1951) is usually credited with first recognizing this paradox which is named after him. Park (1985) presents a graphical interpretation of the paradox. Simpson's paradox is of more than hypothetical interest as the set of real-world examples collected by Wagner (1982) show. Example 4.6.2 below is due to him.

Example 4.6.2. Between the years 1974 and 1978 the federal personal income tax rate overall increased from 14.1% to 15.2%. However, Table 4.6.3 shows the tax rate for each adjusted gross income category actually decreased. Because of inflation, in 1978 there were more persons (and consequently more taxable dollars) assigned to the higher income brackets.

Table 4.6.3. Total Income, Total Tax (in thousands of dollars) and Tax Rate for Taxable Income Classified by Income Category and Year (Reprinted with permission from C.H. Wagner: "Simpson's Paradox in Real Life," *The American Statistician,* **36**, #1, 1982. American Statistical Association.)

Adj. Gross Income	Year	Income	Tax	Tax Rate
$< \$5,000$	1974	41,651,643	2,244,467	.054
	1978	19,879,622	689,318	.035
$[\$5,000, \$9,999)$	1974	146,400,740	13,646,348	.093
	1978	122,853,315	8,819,461	.072
$[\$10,000, \$14,999)$	1974	192,688,922	21,449,597	.111
	1978	171,858,024	17,155,758	.100
$[\$15,000, \$99,999)$	1974	470,010,790	75,038,230	.160
	1978	865,037,814	137,860,951	.159
$\geq \$100,000$	1974	29,427,152	11,311,672	.384
	1978	62,806,159	24,051,698	.383
Totals & Overall	1974	880,179,247	123,690,314	.141
Tax Rates	1978	1,242,434,934	188,577,186	.152

Collapsibility

Suppose the cell probabilities for an observed $R \times C \times L$ table $\{Y_{ijk}: 1 \leq i \leq R,\ 1 \leq j \leq C,\ 1 \leq k \leq L\}$ based on m multinomial trials are $\{p_{ijk}\}$. Then the cell probabilities for the $R \times C$ table collapsed over layers are $\{p_{ij+}\}$. The discussion that follows considers loglinear models for the means of both the original table $\{\ell_{ijk} = \ln(mp_{ijk})\}$ and the collapsed table $\{\ell_{ij} = \ln(mp_{ij+})\}$. The symbols ${}^3\boldsymbol{\lambda}$ and ${}^2\boldsymbol{\lambda}$ denote the λ-terms for the 3 and 2 dimensional tables, respectively.

One mathematical formulation of collapsibility, discussed in Bishop, Fienberg, and Holland (1975), is to define the table to be collapsible over the layer variable if the 2-way interactions between the row and column variables are the same; i.e.,

$$ {}^2\lambda_{ij}^{12} = {}^3\lambda_{ij}^{12} \tag{4.6.1} $$

for $i = 1(1)R$ and $j = 1(1)C$. This implies that certain functions of the $\{p_{ijk}\}$ are the same as when computed using $\{p_{ij+}\}$. Theorem 2.4-1 of Bishop, Fienberg and Holland (1975) shows that (4.6.1) is implied by the two conditions

$$ {}^3\boldsymbol{\lambda}^{123} = \mathbf{0} \tag{4.6.2} $$

and

$$ {}^3\boldsymbol{\lambda}^{13} = \mathbf{0} \quad \text{or} \quad {}^3\boldsymbol{\lambda}^{23} = \mathbf{0}. \tag{4.6.3} $$

However, (4.6.2) and (4.6.3) are only sufficient and not necessary for (4.6.1) as the following example due to Whittemore (1978) shows.

Example 4.6.3. Suppose the values in Table 4.6.4 are the means μ_{ijk} for a $2 \times 2 \times 2$ table. The odds ratio for layer 1 is 4 while that for layer 2 is .25. It is easy to calculate that

$$ {}^3\lambda_{11}^{12} = \left(\frac{1}{2}\right)\left(\frac{1}{4}\right)\left[\ln\left(\frac{36}{9}\right) + \ln\left(\frac{9}{36}\right)\right] = 0 $$

which implies ${}^3\boldsymbol{\lambda}^{12} = \mathbf{0}$. Further,

$$ {}^3\lambda_{111}^{123} = \left(\frac{1}{2}\right)\left(\frac{1}{4}\right)\left[\ln(4) - \ln\left(\frac{1}{4}\right)\right] = \frac{\ln(4)}{4} \neq 0, $$

so equation (4.6.2) fails. Collapsed over layers, Table 4.6.4 becomes Table 4.6.5. The collapsed Table 4.6.5 has ${}^2\lambda_{11}^{12} = \frac{1}{4}\ln(\frac{9 \cdot 9}{9 \cdot 9}) = 0$ so that ${}^2\boldsymbol{\lambda}^{12} = \mathbf{0} = {}^3\boldsymbol{\lambda}^{12}$. Hence (4.6.1) *holds* and the table would be defined as "collapsible" over the two layers, despite the fact that the layers behave *differently* as the failure of (4.6.2) shows.

Table 4.6.4. Mean Values for a $2\times2\times2$ Table with ${}^2\boldsymbol{\lambda}^{12} = {}^3\boldsymbol{\lambda}^{12}$ (Reprinted with permission from A.S. Whittemore: "Collapsibility of Multidimensional Contingency Tables," *Journal of the Royal Statistical Society,* **40**, #3, 1978. Royal Statistical Society, London, England.)

	Layer			
	1		2	
Column	1	2	1	2
Row 1	6	3	3	6
Row 2	3	6	6	3

Table 4.6.5. Table 4.6.4 Collapsed over Layer Variable (Reprinted with permission from A.S. Whittemore: "Collapsibility of Multidimensional Contingency Tables," *Journal of the Royal Statistical Society,* **40**, #3, 1978. Royal Statistical Society, London, England.)

	Column 1	Column 2
Row 1	9	9
Row 2	9	9

Example 4.6.3 shows the inadequacy of equation (4.6.1) as a definition of collapsibility. Whittemore (1978) gives additional examples in which none of ${}^3\boldsymbol{\lambda}^{123}$, ${}^3\boldsymbol{\lambda}^{13}$ or ${}^3\boldsymbol{\lambda}^{23}$ is identically zero, and yet ${}^3\boldsymbol{\lambda}^{12} = {}^2\boldsymbol{\lambda}^{12}$ holds. The preceding example and discussion motivate her formulation of a definition of "strict collapsibility."

Definition. A cross-classified table with cell probabilities $\{p_{ijk}\}$ is *strictly collapsible* with respect to the layer variable means (i) ${}^3\boldsymbol{\lambda}^{123} = \mathbf{0}$ and (ii) ${}^2\boldsymbol{\lambda}^{12} = {}^3\boldsymbol{\lambda}^{12}$.

This definition requires both that the interactions be *constant* across layers (4.6.2 holds), and that the common interaction structure in each layer be the *same* as in the collapsed table (4.6.1 holds). In a $2 \times 2 \times L$ table with layer specific odds ratios $\psi_k := (p_{11k}p_{22k})/(p_{12k}p_{21k})$, $1 \leq k \leq L$, and collapsed table odds ratio $\psi^C := (p_{11+}p_{22+})/(p_{12+}p_{21+})$, strict collapsibility holds if and only if

$$\psi_1 = \ldots = \psi_L = \psi^C. \tag{4.6.4}$$

In $R \times C \times L$ tables, strictly collapsibility requires that a set of $(R-1)(C-1)$ odds ratio statements similar to (4.6.4) hold. Shapiro (1982) gives a geometric interpretation of strict collapsibility by plotting various models in odds coordinate-axes. Whittemore (1978) provides the following characterization of the condition.

Proposition 4.6.1. *A table having cell probabilities* $\{p_{ijk}\}$ *is strictly collapsible if and only if there exist functions* $g(\cdot,\cdot)$ *and* $h(\cdot,\cdot)$ *such that* $p_{ijk} \equiv p_{ij+}g(j,k)h(i,k)$ for all (i,j,k).

It can be shown that Bishop, Fienberg, and Holland's conditions (4.6.2) and (4.6.3) are sufficient, but not necessary, for strict collapsibility to occur. Thus if a test of either

$${}^3\boldsymbol{\lambda}^{13} = \mathbf{0} = {}^3\boldsymbol{\lambda}^{123} \tag{4.6.5}$$

or

$${}^3\boldsymbol{\lambda}^{23} = \mathbf{0} = {}^3\boldsymbol{\lambda}^{123} \tag{4.6.6}$$

accepts the respective null hypothesis, then the data suggest that

$$H_0\colon \{p_{ijk}\} \text{ satisfies strict collapsibility}$$

may be true. Unfortunately, this reasoning relies on believing a null hypothesis which is accepted by a test whose power is typically uncontrolled. If tests of both (4.6.5) and (4.6.6) reject the null hypotheses, nothing can be concluded. Thus the only advantage of testing (4.6.5) or (4.6.6) is that the LR tests are readily computed from many statistical packages.

The likelihood ratio test of strict collapsibility is more difficult to compute than (4.6.5) or (4.6.6). However it can be carried out using special software available from Whittemore. Alternatively, Ducharme and Lepage (1986) present a closed-form asymptotic test of strict collapsibility for $2 \times 2 \times L$ tables; their test is a quadratic form in the layer specific log odds ratio estimators $\hat{\omega}_k = \ln[(Y_{11k}Y_{22k})/(Y_{12k}Y_{21k})]$, $1 \le k \le L$, and collapsed table log odds ratio estimator $\hat{\omega}^C = \ln[(Y_{11+}Y_{22+})/(Y_{12+}Y_{21+})]$. Their test rejects strict collapsibility when

$$m\left\{\sum_{k=1}^{L}\frac{(\hat{\omega}_k - \hat{\omega}^C)^2}{\hat{V}_k} + (\{\hat{V}^+\}^{-1} - \sum_{k=1}^{L}\hat{V}_k^{-1})^{-1}\left(\sum_{k=1}^{L}\frac{\hat{\omega}_k - \hat{\omega}^C}{\hat{V}_k}\right)^2\right\} \ge \chi^2_{\alpha,L}$$

where $m = Y_{+++}$, $\hat{V}_k = \sum_{i,j} m/Y_{ijk}$, and $\hat{V}^+ = \sum_{i,j} m/Y_{ij+}$. They describe how to extend their test to arbitrary $R \times C \times L$ tables.

Davis (1988a) derives a third test of strict collapsibility by applying the union intersection principle. For a parametric problem with parameter space Ω, the union intersection test of a null hypothesis of the form $H_0\colon \cup_q \omega_q$ versus $H_A\colon \Omega - (\cup_q \omega_q)$ where $\omega_q \subset \Omega$ for all q is defined as follows. If A_q is the acceptance region of a test of the individual null hypothesis $H_{0q}\colon \theta \in \omega_q$ versus $H_{Aq}\colon \theta \in \Omega - \omega_q$, then the union intersection test of the union null hypothesis H_0 has acceptance region $\cap_q A_q$; equivalently it rejects H_0 if *any* of the component tests reject the corresponding component H_{0q}. Davis shows how to write the null hypothesis of strict collapsibility as a finite union of pairwise distinct models. She shows her test of strict collapsibility is at least as powerful as the LR test.

The following example illustrates data where *neither* sufficient model (4.6.5) nor (4.6.6) fits the data and yet the table is strictly collapsible.

Example 4.6.4. (Combining case-control studies of the association between smoking and lung cancer.) Dorn (1954) reports the data in Table 4.6.6 from eight separate case-control studies of the association between smoking and lung cancer. One weakness of the data is that the amount of smoking is not quantified for smokers. The question of interest is whether the data can be combined over the 8 studies.

Fitting the no 3-factor interaction model ${}^3\boldsymbol{\lambda}^{123} = \mathbf{0}$ shows no lack-of-fit ($G^2 = 5.95$ with 7 degrees of freedom, $P-$value $= .55$). Thus the interaction structure (odds of contracting lung cancer in the smoking group to that of contracting lung cancer in the control group) appears constant across the 8 studies. However, neither of the models ${}^3\boldsymbol{\lambda}^{123} = \mathbf{0} = {}^3\boldsymbol{\lambda}^{13}$ nor ${}^3\boldsymbol{\lambda}^{123} = \mathbf{0} = {}^3\boldsymbol{\lambda}^{23}$ yields a good fit ($G^2 = 105.39$ and $G^2 = 136.15$, respectively, both with 14 degrees of freedom and both with P-values $< .001$). Thus the question of the collapsibility of these 8 studies is unresolved.

Table 4.6.6. Results of 8 Case-Control Studies on the Association Between Smoking and Lung Cancer (Reprinted with permission from H.F. Dorn: "The Relationship of Cancer of the Lung and the Use of Tobacco," *American Statistician,* **8**, 1954. American Statistical Association.)

	Control patients		Lung cancer patients	
Study	Non-smokers	Smokers	Non-smokers	Smokers
1	14	72	3	83
2	43	227	3	90
3	19	81	7	129
4	81	534	18	459
5	61	1296	7	1350
6	27	106	3	60
7	56	462	19	499
8	28	259	5	260
Total	329	3037	65	2930

Table 4.6.7. Summary of 8 Case-Control Studies on the Association Between Smoking and Lung Cancer

	Smokers	Non-smokers
Lung Cancer	2930	65
Controls	3037	329

The likelihood ratio and Ducharme and Lepage tests of strict collapsibility are 9.23 and 13.21, respectively, both with 8 degrees of freedom (P-values .32 and .11, respectively). Thus both indicate the data can be collapsed over studies as in Table 4.6.7. The estimated odds of lung cancer for smokers to non-smokers from Table 4.6.7 is $\hat{\psi}^C = (2930(329))/(3037(65)) = 4.88$. A 95% confidence interval based on the normal approximation to $\ln(\hat{\psi}^C)$ is (3.71, 6.36). Thus these studies suggest smokers have roughly 4 to 6 times the odds of developing lung cancer as non-smokers.

Ducharme and Lepage (1986) introduce a third definition of collapsibility, called strong collapsibility, which can be motivated from the idea of partially collapsing an $R \times C \times L$ table by pooling some of the layer categories.

Definition. A cross-classified table with cell probabilities $\{p_{ijk}\}$ is *strongly collapsible* with respect to the layer variable provided every $R \times C \times L'$ table, $1 \leq L' \leq L$, obtained by pooling layer categories of the original table is strictly collapsible.

When $L = 2$, strict and strong collapsibility are the same. For $L > 2$, strong collapsibility implies many statements of strict collapsibility—one for each of the possible ways of partitioning the L layer categories into two or more nonempty sets. The Ducharme and Lepage characterization of strong collapsibility is equivalent to Bishop, Fienberg, and Holland's sufficient conditions (4.6.1) and (4.6.2) as the following Proposition states. Thus a table is strongly collapsible if and only if either rows and layers are conditionally independent given the columns or the columns and layers are conditionally independent given the rows (Problem 4.11).

Proposition 4.6.2. *A table having cell probabilities* $\{p_{ijk}\}$ *is strongly collapsible if and only if either*

(i) ${}^{3}\boldsymbol{\lambda}^{123} = \mathbf{0} = {}^{3}\boldsymbol{\lambda}^{13}$ *or*

(ii) ${}^{3}\boldsymbol{\lambda}^{123} = \mathbf{0} = {}^{3}\boldsymbol{\lambda}^{23}$.

Example 4.6.4 (continued). The calculations performed earlier show that the 8 tables in the Dorn study are not strongly collapsible. Studies 1, 2, 3, and 5 were performed in European countries and the remainder elsewhere. Table 4.6.8 lists the $2 \times 2 \times 2$ table obtained by pooling the European and Non-European studies. The estimated odds ratio for lung cancer in smokers to non-smokers is 6.75 for the European studies with a 95% confidence interval of (4.29, 11.09), while that for the non-European studies is 4.01 with a 95% confidence interval of (2.90, 5.65). Since the original table is *not* strongly collapsible, there exist one or more subtables formed by partial pooling of the studies which are not strictly collapsible. Consider whether this particular pooled table is strictly collapsible. Let Variable 1 be Disease State (Case/Control), Variable 2 be Smoking Status (Yes/No), and Variable 3 by Country Group (European/Other). Testing H_0: ${}^{3}\boldsymbol{\lambda}^{123} = \mathbf{0}$ gives $G^2 = 3.24$ (P-value = .08) and $X^2 = 3.16$ (P-value = .08) providing some evidence against a common odds ratio. However, in this case $L = 2$ and hence Proposition 4.6.2 gives necessary and sufficient conditions for strict collapsibility. Testing Proposition 4.6.2(i) (H_0: ${}^{3}\boldsymbol{\lambda}^{123} = \mathbf{0} = {}^{3}\boldsymbol{\lambda}^{13}$) gives $G^2 = 3.61$ (P-value = .17) and $X^2 = 3.54$ (P-value = .17) which viewed alone would suggest the table is strictly collapsible. In light of the earlier test, it is inadvisable to combine the two groups of studies. (For completeness, testing Proposition 4.6.2(ii) (H_0: ${}^3\boldsymbol{\lambda}^{123} = \mathbf{0} = {}^3\boldsymbol{\lambda}^{23}$), yields $G^2 = 38.84$ (P-value $< .001$) and $X^2 = 38.83$ (P-value $< .001$).)

This discussion is specific to collapsing a three-dimensional table of cross-classified counts over a single variable. Whittemore (1978) gives a general definition of strict collapsibility of an arbitrary d-dimensional cross-classified table over an arbitrary subset of variables; similarly Ducharme and Lepage give a general definition of strong collapsibility. Asmussen and Edwards (1983) and Davis (1986a, 1988b) consider other notions of collapsibility based on likelihood ratio tests for "response models"; i.e., models appropriate for tables in which some of the variables are explanatory and others are responses.

Table 4.6.8. Summary of European and Non-European Case-Control Studies on the Association Between Smoking and Lung Cancer

	Controls		Cases	
	Nonsmokers	Smokers	Nonsmokers	Smokers
Europeans[1]	137	1676	20	1652
Non-Europeans[2]	192	1361	45	1278

[1](Studies 1,2,3,5).
[2](Studies 4,6,7,8).

Problems

4.1. Derive the equations (4.2.1) for λ, λ_i^1, λ_j^2, λ_{ij}^{12}.

4.2. Derive equations (4.3.3) which state the relationships between the λ-terms in (4.3.1) and the ν-terms in (4.3.2).

4.3. (a) State the graphical models corresponding to the graphs (a)–(e) in Figure 4.P.3 using both the generator set and the dual model notations.

(b) Draw the interaction graph for each of the following models, state the associated graphical model, and interpret the graphical model.

(i) $[14, 15, 34, 45]^d$
(ii) [21, 23, 24, 25, 13]
(iii) $[16, 35, 234, 134, 36, 135, 25, 26]^d$
(iv) $[126, 253, 264, 143, 1245, 1356, 1456, 3456]^d$
(v) $[21436, 351, 542, 1256]^d$
(vi) [21436, 351, 542, 1256].

4.4. Apply the IPF algorithm to find the MLE of $\boldsymbol{\mu}$ under the model [12, 13] for independent Poisson data $\{Y_{ijk}: i = 1(1)R,\ j = 1(1)C,\ k = 1(1)L\}$. Show that the algorithm converges to $\hat{p}_{ijk} = Y_{i+k}Y_{ij+}/Y_{i++}$ in one cycle.

4.5. In a 2-way table with logmean model $\ell_{ij} = \lambda + \lambda_i^1 + \lambda_j^2$, $1 \le i \le R$ and $1 \le j \le C$, suppose that the IPF algorithm is initialized by setting $\boldsymbol{\mu}^0 = \mathbf{Y}/m$ where $m = Y_{++}$ instead of $\boldsymbol{\mu}^0 = \mathbf{1}_{RC}$. Show that with this alternate initialization, the algorithm preserves all the observed odds ratios; that is, at each step g, the odds ratios in terms of components of $\boldsymbol{\mu}^g$ are identical to those in terms of components of $\mathbf{Y}$. Given an arbitrary initialization, what does the IPF algorithm do? Why does it calculate the MLE when initialized to $\boldsymbol{\mu}^0 = \mathbf{1}_{RC}$?

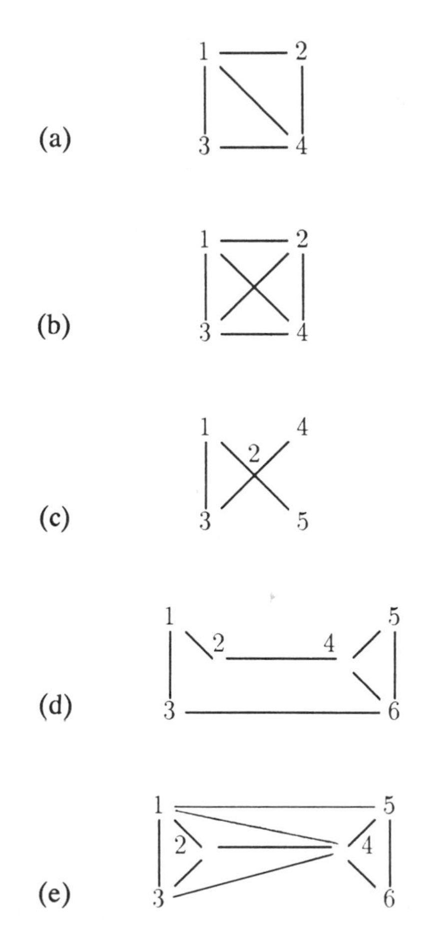

Figure 4.P.3. Five Interaction Graphs for Problem 4.3.

4.6. Fit the following HLLMs to the food poisoning data of Example 4.1.2: [12,23,13], [12,13], [13,23], [1,23], [2,13], [3,12] and [1,2,3]. What conclusions can you make?

4.7. Ceder, Thorngren, and Wallden (1980) present the data in Table 4.P.7 which cross-classify 103 elderly hip fracture patients according to sex, age group, and type of fracture. Fit the loglinear model [12,13,23] to this data. Which 2-way interactions can be eliminated? Use your model to determine whether both types of fractures are equally likely among men and women. Does your answer depend on the patient's age group?

Table 4.P.7. Cross-Classification of 103 Hip Fractures by Fracture Mode, Patient's Age, and Patient's Sex (Reprinted with permission from L. Ceder, K.-G. Thorngren, B. Wallden: "Prognostic Indicators and Early Home Rehabilitation in Elderly Patients with Hip Fractures," *Clinical Orthopedics,* 152. J.P. Lippincott, Philadelphia, PA, 1980.)

Sex	Age (years)	Type of Fracture Cervical	Trochanteric
Male	50–64	3	2
	65–79	6	7
	> 80	8	2
Female	50–64	9	5
	65–79	24	9
	> 80	15	13

4.8. Table 4.P.8 from Sewell and Shah (1968) is a $4 \times 4 \times 2 \times 2 \times 2$ table based on a survey of 10,318 high school seniors in Wisconsin. The table gives information on their socioeconomic status (low/lower middle/upper middle/high), intelligence (low/lower middle/upper middle/high), sex (M/F), parental encouragement (high/low), and college plans (yes/no). Sewell and Shah (1968) contains more detailed discussion about these data. What are the key relationships among these five factors?

4.9. The data in Table 4.P.9 also deal with the college plans of high school seniors. Besides indicating whether they intend to go to college or to get a job, students responded as to their academic preferences (liberal arts or math-science), whether they agree or disagree with the statement that "I'll need mathematics in my future work," and whether or not they attended a Women and Mathematics (WAM) Secondary School Lectureship Program lecture. The sex of the students and the location of their high school (suburban or urban) is also indicated. The WAM program is designed to encourage the study of mathematics by women; it sponsors lectures in high schools by women who use mathematics. More details about WAM and this data set can be found in Lacampagne (1979). These data form a 6-dimensional contingency table with variables: (1) Future Plans, (2) Academic Preference, (3) School, (4) Sex, (5) Lecture Attendence, and (6) Need for Mathematics. Start with the model containing all 3-way interactions and use partial and marginal tests of association to eliminate interactions. Calculate the estimates λ-terms based on your final model. What interpretation does your model have? Fowlkes, Freeny and Landwehr (1988) fit the model [23456, 135, 146]; contrast the fit of your model with theirs.

Table 4.P.8.

Males; Low Parental Encouragement

Plan College					Do Not Plan College				
Socioeconomic	Intelligence				Socioeconomic	Intelligence			
Status	L	LM	UM	H	Status	L	LM	UM	H
L	4	9	12	10	L	349	207	126	67
LM	2	7	12	17	LM	232	201	115	79
UM	8	6	17	6	UM	164	120	92	42
H	4	5	9	8	H	48	47	41	17

Males; High Parental Encouragement

Plan College					Do Not Plan College				
Socioeconomic	Intelligence				Socioeconimic	Intelligence			
Status	L	LM	UM	H	Status	L	LM	UM	H
L	13	33	38	49	L	64	72	54	43
LM	27	64	93	119	LM	84	95	92	59
UM	47	74	148	198	UM	91	110	100	73
H	39	123	224	414	H	57	90	65	54

Females; Low Parental Encouragement

Plan College					Do Not Plan College				
Socioeconomic	Intelligence				Socioeconomic	Intelligence			
Status	L	LM	UM	H	Status	L	LM	UM	H
L	5	5	8	13	L	455	312	216	96
LM	11	19	12	15	LM	285	236	164	113
UM	7	13	12	20	UM	163	193	174	81
H	6	5	13	13	H	50	70	47	49

Females; High Parental Encouragement

Plan College					Do Not Plan College				
Socioeconomic	Intelligence				Socioeconomic	Intelligence			
Status	L	LM	UM	H	Status	L	LM	UM	H
L	9	14	20	28	L	44	47	35	24
LM	29	47	62	72	LM	61	88	85	50
UM	36	75	91	142	UM	72	90	100	77
H	36	110	230	360	H	58	76	81	98

Table 4.P.9. Women and Mathematics Secondary School Lectureship Program Data (Reprinted with permission from *An Evaluation of the Women and Mathematics (WAM) Program and Associated Sex-Related Differences in the Teaching, Learning, and Counseling of Mathematics* by Carole Baker Lacampagne, Ph.D. dissertation, Teachers College, Columbia University, 1979.)

plans: college
preference: math-science

school:	suburban				urban			
sex:	female		male		female		male	
lecture:	yes	no	yes	no	yes	no	yes	no
response:								
agree:	37	27	51	48	51	55	109	86
disagree:	16	11	10	19	24	28	21	25

plans: college
preference: liberal arts

school:	suburban				urban			
sex:	female		male		female		male	
lecture:	yes	no	yes	no	yes	no	yes	no
response:								
agree:	16	15	7	6	32	34	30	31
disagree:	12	24	13	7	55	39	26	19

plans: job
preference: math-science

school:	suburban				urban			
sex:	female		male		female		male	
lecture:	yes	no	yes	no	yes	no	yes	no
response:								
agree:	10	8	12	15	2	1	9	5
disagree:	9	4	8	9	8	9	4	5

plans: job
preference: liberal arts

school:	suburban				urban			
sex:	female		male		female		male	
lecture:	yes	no	yes	no	yes	no	yes	no
response:								
agree:	7	10	7	3	5	2	1	3
disagree:	8	4	6	4	10	9	3	6

4.10. (a) Table 4.P.10 (a) from Knoke (1976) lists survey data on the political affiliation of 6026 individuals and that of their fathers. It is clearly not meaningful to ask whether the row and column classifications are independent as the bulk of the data is on the diagonal. Consider Model (1) below which adds one parameter to the independence model to account for the attraction of the diagonal:

$$\ln(p_{ij}) = \lambda + \lambda_i^1 + \lambda_j^2 + \delta I[i = j] \tag{1}$$

with $\lambda_+^1 = 0 = \lambda_+^2$. A more restrictive formulation assumes that party affiliation is symmetric with respect to respondents and their fathers; i.e., $p_{ij} = p_{ji}$, $1 \leq i, j \leq 3$. Model (2) below forces symmetry by choice of the main effect parameters

$$\ln(p_{ij}) = \lambda + \lambda_i^1 + \lambda_j^1 + \delta I[i = j] \tag{2}$$

with $\lambda_+^1 = 0$. Fit Models (1) and (2) to Knoke's data and interpret your results.

Table 4.P.10(a). Knoke's Survey of Father and Son Party Affiliations (Reprinted with permission from *Change and Continuity in American Politics* by David Knoke. The Johns Hopkins University Press, Baltimore, MD, 1976.)

Father's Party	Respondent's Party: Democrat	Independent	Republican	
Democrat	2335	860	431	3626
Independent	108	247	76	431
Republican	367	501	1101	1969
	2810	1608	1608	6026

(b) Table 4.P.10(b) contains data similar to that of part (a). These data, originally reported by Stuart (1953), concern the unaided distance vision measured in each eye separately of 7477 women, aged between 30 and 39 years, who were working in Royal ordinance factories. Distance vision was graded (1/2/3/4) with grade 1 denoting the best acuity. Because the classifying variable is *ordered,* the following modifications of Models (1) and (2) are reasonable:

$$\ln(p_{ij}) = \lambda + \alpha i + \beta j + \delta I[i = j] \tag{1'}$$

and

$$\ln(p_{ij}) = \lambda + \alpha i + \alpha j + \delta I[i = j]. \tag{2'}$$

Models (1′) and (2′) place *no* restrictions on the parameters α or β. Fit (1′) and (2′) to Stuart's data and interpret your results.

Table 4.P.10(b). Unaided Distance Vision of 7477 Women (Reprinted with permission from A. Stuart: "The Estimation and Comparison of Strengths of Association in Contingency Tables," *Biometrika,* **40**. Biometrika Trust, London, England, 1953.)

		Left Eye Grade			
		1	2	3	4
Right	1	1520	266	124	66
Eye	2	234	1512	432	78
Grade	3	117	362	1772	205
	4	36	82	179	492

4.11. Consider a $2 \times 2 \times L$ contingency table with cell probabilities $\{p_{ijk}\}$. Prove that if conditions (i) and (ii) of Proposition 4.6.2 hold, then the table is strongly collapsible.

[Ducharme and Lepage (1986)]

4.12. Consider a 2-way table of counts Z_{ij}, $1 \leq i \leq R$ and $1 \leq j \leq C$, and assume that the true probabilities satisfy $p_{ij} > 0$ for all i, j. Let $K = \min\{R-1, C-1\}$. Then there exist ("canonical scores") x_{ik} and y_{jk}, $1 \leq i \leq R$, $1 \leq j \leq C$, $1 \leq k \leq K$ and ("canonical correlations") ρ_k, $1 \leq k \leq K$, such that

$$p_{ij} = p_{i+}p_{+j}\left[1 + \sum_{k=1}^{K} \rho_k x_{ik} y_{jk}\right] \tag{1}$$

for $1 \leq i \leq R$, $1 \leq j \leq C$, and $1 \geq \rho_1 \geq \rho_2 \geq \cdots \geq \rho_K$, $\sum_{i=1}^{R} p_{i+}x_{ik} = \sum_{j=1}^{C} p_{+j}y_{jk} = 0$ for $1 \leq k \leq K$, and $\sum_{i=1}^{R} p_{i+}x_{ik}x_{i\ell} = \sum_{j=1}^{C} p_{+j}y_{jk}y_{j\ell} = \delta_{k\ell}$ for $1 \leq k, \ell \leq K$ with $\delta_{k\ell} = 1$ if $k = \ell$ and 0 otherwise.

Representation (1) dates to Fisher (1940). It can be obtained from a singular value decomposition of the $R \times C$ matrix $\mathbf{S}$ with $S_{ij} = p_{ij}/(p_{i+}p_{+j})^{1/2}$. The singular values of $\mathbf{S}$ satisfy $1 = \rho_0 \geq \rho_1 \geq \cdots \geq \rho_K$; $\{\mathbf{x}_k = (x_{1k}, \ldots, x_{Rk})'\colon 1 \leq k \leq K\}$ are the eigenvectors of $\mathbf{SS}'$; and $\{\mathbf{y}_k = (y_{1k}, \ldots, y_{Ck})'\colon 1 \leq k \leq K\}$ are the eigenvectors of $\mathbf{S}'\mathbf{S}$. The quantities in (1) have the following interpretation. Assigning the scores x_{ik} and y_{jk} to the row and column categories, respectively, yields (unobserved) canonical random variables X_k and Y_k with zero means and unit variances. The correlation ρ_k between X_k and Y_k is maximized subject to the constraints that X_k (Y_k) be uncorrelated with X_ℓ (Y_ℓ) for all $1 \leq \ell < k$. The independence model corresponds to $\rho_1 = 0$. Models inbetween the independence model and the fully saturated model have the property that $\rho_k = 0$ for some $k \geq 2$.

The sample analog of Equation (1) replaces p_{ij} with $\hat{p}_{ij} = Z_{ij}/n$, p_{i+} with $\hat{p}_{i+} = Z_{i+}/n$, p_{+j} with $\hat{p}_{+j} = Z_{+j}/n$, and the canonical scores and correlations with their sample counterparts $\hat{\rho}_k$, $\hat{x}_{ik}$, and $\hat{y}_{jk}$. Here $n = Z_{++}$ is the total number of counts, and the $\hat{\rho}_k$, $\hat{x}_{ik}$, and $\hat{y}_{jk}$ are derived from a singular value decomposition of the matrix $\hat{\mathbf{S}}$ with $\hat{S}_{ij} = \hat{p}_{ij}/(\hat{p}_{i+}\hat{p}_{+j})^{1/2}$ in a completely analogous fashion. Sample values of $\hat{\rho}_k$ "appreciably" different from zero are evidence against the independence model. In fact, it can be shown that the sample canonical correlations satisfy

$$n \sum_{k=1}^{K} \hat{\rho}_k^2 = X^2$$

where X^2 is the Pearson chi-squared statistic for testing independence. (Kettenring (1982) contains further discussion about canonical analysis of contingency tables and, in particular, the connections to classical canonical analysis for sets of continuous variables.)

This problem explores the use of canonical analysis to partially collapse contingency tables by combining homogeneous columns (or rows). Columns j and q are called *homogeneous* if $p_{ij}/p_{+j} = p_{iq}/p_{+q}$ for all $1 \leq i \leq R$; i.e., if the conditional probability of falling in row i given that an observation is in column j is the same as the conditional probability of falling in row i given column q, for all $1 \leq i \leq R$. Prove that columns j and q are homogeneous if and only if $y_{jk} = y_{qk}$ for all $1 \leq k \leq K$. How can this representation be used if one is interested in simplifying a 2-way table by combining homogeneous columns?

[Gilula (1986)]

5

Univariate Discrete Data with Covariates

5.1 Introduction

Polychotomous response regression data has the form $(\mathbf{Y}_i, m_i, \mathbf{x}_i)$, $1 \leq i \leq T$, where, without loss of generality, each response takes one of the values $\{0, \ldots, g\}$. Each vector $\mathbf{Y}_i = (Y_{i0}, \ldots, Y_{ig})'$, giving the number of times the $(g+1)$ outcomes occur at the ith design point, is assumed to follow an independent multinomial distribution with m_i trials ($\sum_{j=0}^{g} Y_{ij} = m_i$). The vector $\mathbf{x}_i = (x_{i1}, \ldots, x_{ik})'$ contains covariate values affecting the cell probabilities of $\mathbf{Y}_i$. Example 1.2.7 on the severity of nausea in cancer patients undergoing chemotherapy is typical of such data. There are six possible outcomes ($g = 5$), two design points ($T = 2$), a scalar covariate ($k = 1$) which is 1 or 0 according as the chemotherapy includes cisplatinum or not, and the numbers of patients are $m_1 = 58$ and $m_2 = 161$ for the cisplatinum and no-cisplatinum groups, respectively.

An important special type of discrete response regression data is *binary* regression data meaning there are only two possible outcomes ($g = 1$). The response in such a case will simply be denoted by the *scalar* Y_i which is the number of 1's ("successes") among the m_i trials. Examples 1.2.4, 1.2.5, 1.2.6, 1.2.8, and 1.2.9 have binary responses; some of the covariates are qualitative (e.g., treatment or block effects) and others are quantitative (e.g., temperature).

The development of appropriate models for general polychotomous responses has received much attention particularly in the econometric literature. (See Maddala, 1983; Amemiya, 1985; and Dalal and Klein, 1988 for recent summaries.) This chapter will almost exclusively consider binary regression data although occasional comments will be made about polychotomous responses. The following stochastic model will be assumed throughout. The covariates $\mathbf{x}_i$ are constants measured without error, $Y_1, \ldots, Y_T$ are mutually independent random variables with $Y_i \sim B(m_i, p_i)$, and $p_i = p(\mathbf{x}_i)$ is the conditional (success) probability of $Y_i = 1$ given $\mathbf{x}_i$.

The precise assumptions made about $p(\mathbf{x})$ lead to different models. The following is one possible hierarchy of models:

$$p(\mathbf{x}) = F(\mathbf{x}) \quad \text{where} \quad F(\cdot) \in \mathcal{F}_U, \tag{5.1.1}$$

and $\mathcal{F}_U \subset \{F : \mathbb{R}^k \rightarrow [0,1]\}$;

$$p(\mathbf{x}) = F(\mathbf{x}'\boldsymbol{\beta}) \quad \text{where} \quad F(\cdot) \in \mathcal{F}_D,\ \boldsymbol{\beta} \in \mathbb{R}^k, \tag{5.1.2}$$

and $\mathcal{F}_D \subset \{F : \mathbb{R} \rightarrow [0,1]\}$ is a set of cumulative distribution functions (cdfs); and

$$p(\mathbf{x}) = F_0(\mathbf{x}'\boldsymbol{\beta}) \quad \text{where } F_0(\cdot) \text{ is a specified cdf, and } \boldsymbol{\beta} \in \mathbb{R}^k. \tag{5.1.3}$$

Model (5.1.1) gives the most flexible class of models for $p(\cdot)$ while (5.1.3) is the most restrictive. Each of (5.1.1)–(5.1.3) forces $p(\mathbf{x}) \in [0,1]$ for all $\mathbf{x}$. Problem 5.1 motivates the classes (5.1.2) and (5.1.3). In the first two models, some care has to be taken to insure identifiability. To illustrate the problem, consider model (5.1.2) with $\mathcal{F}_D$ containing distributions $F_1(w) = F(w)$ and $F_2(\cdot) = F(\alpha w)$ which differ only by a scale factor $\alpha > 0$. Then

$$F_1(\mathbf{x}'\boldsymbol{\beta}) = F(\mathbf{x}'\boldsymbol{\beta}) = F_2(\mathbf{x}'\boldsymbol{\beta}/\alpha)$$

so that $(F_1, \boldsymbol{\beta})$ and $(F_2, \boldsymbol{\beta}/\alpha)$ yield identical success probabilities and are thus not identifiable.

When $\mathcal{F}_U$ is unrestricted model (5.1.1) does not permit a simple interpretation of individual covariates and display is difficult when $k > 2$. In this case, O'Sullivan, Yandell, and Raynor (1986) use smoothing methods to estimate $p(\mathbf{x})$. Hastie and Tibshirani (1987) derive estimators of $p(\mathbf{x})$ when $\mathcal{F}_U$ is the set of functions of the form $\text{logit}(p(\mathbf{x})) = \beta_0 + \sum_{j=1}^{k} q_j(x_j)$ where the $q_j(\cdot)$ are unknown "smooth" functions.

By introducing the linear term $\mathbf{x}'\boldsymbol{\beta}$, Models (5.1.2) and (5.1.3) permit easy interpretation of the individual components of the covariate vector. In some problems the restriction to distribution functions can be justified by an appeal to a threshold model for Y_i (Problem 5.1). Ostensibly, (5.1.2) implies monotonicity of the covariate effect over its entire range although, as with other regression models, the use of quadratic and other non-linear terms in continuous variables allows great flexibility. Model (5.1.2) reduces the estimation of success probabilities to that of estimating the distribution (link) function $F(\cdot)$ and the parameter vector $\boldsymbol{\beta}$ while Model (5.1.3) only requires estimation of $\boldsymbol{\beta}$. Klein and Spady (1988) consider estimation under (5.1.2) when $\mathcal{F}_D$ is a class of unknown distributions subject to identifiability conditions; they also consider estimation under the more general model $p(\mathbf{x}) = F(\nu(\mathbf{x}, \boldsymbol{\beta}))$ where $\nu(\cdot,\cdot)$ is any given function.

Several parametric families have been proposed in the literature as link families $\mathcal{F}_D$ in Model (5.1.2). Prentice (1976) considers the family of distributions of natural logarithms of F_{ν_1,ν_2} random variables; Pregibon (1980) considers a generalization of the "Tukey lambda" family; Aranda-Ordaz (1981) considers families of symmetric and asymmetric distributions; Guerrero and Johnson (1982) consider Box–Cox transforms of $\ln(p/[1-p])$; and Czado (1988) considers the Burr family of transformations. Lastly, Stukel

(1988) proposes a two parameter family which is quite flexible and appears very useful. All of these families include the (important case) of the logistic link $F_0(w) = (1 + \exp(-w))^{-1}$ (Problem 5.2).

Nonidentifiability is not a problem in Model (5.1.3). Three popular choices of link function $F_0(\cdot)$ are:

(i) $F_0(t) = \Phi(t)$ which is called the probit model,

(ii) $F_0(t) = \exp\{t\}/(1 + \exp\{t\})$ which is called the logistic model, and

(iii) $F_0(t) = 1 - \exp\{-\exp(t)\}$ which is called the complementary log-log model.

Historically the probit model was advocated for bioassay work (Finney (1971)). The logit model has received more attention recently for reasons sketched below.

This chapter focuses on the analysis of binary regression data using the logistic link function so that

$$p(\mathbf{x}) = \frac{\exp\{\mathbf{x}'\boldsymbol{\beta}\}}{1 + \exp\{\mathbf{x}'\boldsymbol{\beta}\}} \tag{5.1.4}$$

or, equivalently,

$$\text{logit}[p(\mathbf{x})] = \mathbf{x}'\boldsymbol{\beta} \tag{5.1.5}$$

where $\text{logit}[w] := \ln(w/(1 - w))$.

Four reasons for emphasizing the logistic model are: (1) the formal connection of the logistic model with loglinear model theory, (2) the applicability of the model to the analysis of retrospective as well as prospective data, (3) the ease of interpretation of the regression coefficients β_j, and (4) theoretical statistical considerations including the availability of "exact" analyses of individual parameters.

To see the connection between the logistic model and loglinear models, recall from Example 3.1.3 that, under the logistic model, if $Z_{i1} = Y_i$ and $Z_{i2} = m_i - Y_i$, $1 \le i \le T$, then

$$\begin{aligned} \ln(E[Z_{i1}]) &= \ln(m_i p_i) \\ &= \gamma_i + \mathbf{x}_i'\boldsymbol{\beta} \end{aligned}$$

where $\gamma_i := \ln(E[Z_{i2}])$. Hence the random variables $\{Z_{ij}: 1 \le i \le T,\ j = 1, 2\}$ follow a loglinear model. One practical aspect of this observation is that the existence of the MLE of $\boldsymbol{\beta}$ under the logistic model for a given data set can be determined by studying that of $(\boldsymbol{\beta}, \boldsymbol{\gamma})$ under Poisson sampling for the $\{Z_{ij}\}$ using Propositions 3.2.1 or 3.2.2.

A second motivation for studying the logistic model is its versatility in analyzing retrospective as well as prospective data. The following calculation is suggestive of a result that holds more generally. Consider a prospective experiment in which subjects from "exposed" (E) and "non-exposed"

$(\sim E)$ groups are sampled. A binary response is then observed for each subject which will be called "diseased" (D) or "non-diseased" $(\sim D)$. If $P(D \mid E)$ and $P(D \mid \sim E)$ are the probabilities of disease in the exposed and nonexposed groups, respectively, then the (prospective) ratio of odds of disease for the exposed group to odds of disease for the nonexposed group is

$$\psi_P = \frac{P(D \mid E)}{1 - P(D \mid E)} \times \frac{1 - P(D \mid \sim E)}{P(D \mid \sim E)}.$$

Furthermore, the same odds ratio ψ_P is of primary interest even if the study is conducted using retrospective sampling. In a retrospective study separate samples of diseased and non-diseased subjects are collected; for each subject it is determined whether they were exposed or not. In this case the probability of exposure among the diseased group, $P(E \mid D)$, and the probability of exposure among the healthy group, $P(E \mid \sim D)$, can be estimated as well as any functions of these two probabilities. In particular the "retrospective" odds ratio

$$\psi_R := \frac{P(E \mid D)}{1 - P(E \mid D)} \times \frac{1 - P(E \mid \sim D)}{P(E \mid \sim D)}$$

can be estimated. However a little algebra shows $\psi_P = \psi_R$ so that the retrospective study can be used to make inferences about ψ_P. More generally, Breslow and Day (1980, Sec. 6.2 and 6.3) and Prentice and Pike (1979) show that valid prospective odds ratio estimates can be obtained from retrospective data by fitting a prospective model to the data.

The third reason for considering the logistic model is the ease of interpreting the model's coefficients. If $\mathbf{x}$ and $\mathbf{x}^*$ are valid covariate vectors for which $x_j = x_j^* - 1$ and $x_i = x_i^*$ for $1 \leq i \leq k$ and $i \neq j$, then

$$\beta_j = \ln\left[\frac{p(\mathbf{x})}{1 - p(\mathbf{x})} \times \frac{1 - p(\mathbf{x}^*)}{p(\mathbf{x}^*)}\right]. \tag{5.1.6}$$

Thus $\exp(\beta_j)$ is the odds that $Y = 1$ corresponding to a unit increase in x_j when all other covariates are held fixed. In cases where there do not exist valid $\mathbf{x}$ and $\mathbf{x}^*$ which differ only in the jth component, the interpretation of β_j is modified. For example, suppose the model contains an intercept so that

$$\text{logit}[p(\mathbf{x})] = \beta_0 + \beta_1 x_1 + \cdots + \beta_k x_k.$$

Instead of being an odds *ratio*, $\exp(\beta_0)$ is simply the odds of $Y = 1$ when all covariates are zero. As another example suppose x is a scalar covariate and

$$\text{logit}[p(x)] = \beta_0 + \beta_1 x + \beta_2 x^2.$$

Then the odds ratio of the probability $Y = 1$ corresponding to a unit increase in x starting at x_0 is $\exp\{\beta_1 + \beta_2[(x_0+1)^2 - x_0^2]\} = \exp\{\beta_1 + \beta_2[2x_0 + 1]\}$. The fact that the value of the odds ratio corresponding to a unit increase from x_0 to $x_0 + 1$ depends on x_0 greatly complicates the model's interpretation compared to the simpler (5.1.6). As a final caveat concerning coefficient interpretation, note that even if β_j admits the interpretation (5.1.6) as a change in the log-odds ratio, its physical interpretation depends on whether x_j is quantitative or qualitative. The three following examples illustrate the remarks above concerning coefficient interpretation.

Example 1.2.4 (continued). In this two treatment problem suppose the covariate value is $x = 0$ (1) for subjects given the standard treatment (vitamin plus standard treatment). Consider a model of the form

$$p(x) = F(\beta_0 + \beta_1 x), \quad \beta \in \mathbb{R}^2$$

for a cdf $F(\cdot)\colon \mathbb{R} \xrightarrow{\text{into}} (0,1)$. For the logistic distribution function $F(\cdot)$, $\exp\{\beta_0\}$ is the odds of clinical improvement for the control group and $\exp\{\beta_1\}$ is the odds ratio of improvement for the vitamin C group to the control group. However if $F(\cdot)\colon \mathbb{R} \xrightarrow{\text{onto}} (0,1)$ then the model is equivalently and more simply parameterized by $p_1 = F(\beta_0)$ and $p_2 = F(\beta_0 + \beta_1)$ with parameter space $(0,1)^2$. Similarly, the parameterization $p_1, \ldots, p_T$ can be used in the T treatment problem.

Example 1.2.6 (continued). Define the indicator variables

$$\begin{aligned} x_{i1} &= I[\text{Sex} = \text{male}], \\ x_{i2} &= I[\text{Strain} = X], \text{ and} \\ x_{i3} &= I[\text{Treatment} = \text{avadex}] \end{aligned}$$

and let $p(x_{i1}, x_{i2}, x_{i3})$ be the probability of tumor for the ith animal. Suppose the following block-by-treatment model holds

$$\text{logit}(p(x_{i1}, x_{i2}, x_{i3})) = \beta_0 + \beta_1 x_{i1} + \beta_2 x_{i2} + \beta_3 x_{i1} x_{i2} + \beta_4 x_{i3}.$$

For any of the four sex-by-strain combinations, $\exp\{\beta_4\}$ is the odds ratio of tumor for the avadex animals to the control animals. The parameters β_1, β_2, and β_3 represent baseline changes to the odds of tumor for different sex-by-strain groups. The parameter $\exp\{\beta_0\}$ is the odds of tumor for a control diet-female-strain Y animal.

Example 5.1.1. The data in Table 5.1.1 is a summary of an experiment originally reported and analyzed in Hoblyn and Palmer (1934) and re-analyzed in Bartlett (1935). The response is the number of plum root stock (Y_i) surviving out of 240 root stocks transplanted under each of 4 experimental conditions corresponding to a 2^2 experiment. There are two discrete explanatory variables: (i) the time of planting (at once/in the spring) and (ii) the length of stock (short/long). Let the variable

$x_{i1} = I$[Length = long] be an indicator variable denoting long root stock and $x_{i2} = I$[Planting Time = at once] be an indicator denoting planting at once. Statistics is not required to see that the most favorable set of conditions for successful transplantation is to plant long root stock at once. Conversely the worst treatment is to plant short cuttings in the spring. The question investigated here is to quantify the effect of the two explanatory variables on the probability $p(x_{i1}, x_{i2})$ that a transplant survives and, in particular, to determine if there is any "interaction" between them. One method of quantifying the effect of x_{i1} and x_{i2} is to consider the model

$$\text{logit}[p(x_{i1}, x_{i2})] = \beta_0 + \beta_1 x_{i1} + \beta_2 x_{i2}, \quad 1 \leq i \leq 4.$$

Then $\exp\{\beta_0\}$ is the odds of survival for short root stock planted in the spring and $\exp\{\beta_1\}$ is the ratio of the odds of survival for long root stock to short root stock (when both are planted either in the spring or at once). Similarly, $\exp\{\beta_2\}$ is the ratio of the odds of survival for root stock planted at once to that planted in the spring (both either short or long). Problem 5.26 presents the complete data set from which Table 5.1.1 was collapsed together with some suggestions for its analysis.

Table 5.1.1. Number of Plum Root Stock Surviving under Four Experimental Conditions (Reprinted with permission from Hoblyn and Palmer: "A Complex Experiment in the Propagation of Plum Rootstocks from Root Cuttings," *Journal of Pomology and Horticultural Science,* 1934. Headley Brothers Ltd., Invicta Press, Ashford, Kent, England.)

Y_i	m_i	x_{i1}	x_{i2}
156	240	1	1
84	240	1	0
107	240	0	1
31	240	0	0

The final motivation for using the logistic model is that many general results for exponential families apply directly to it. The reason is that the logit of p_i is the "natural parameter" in the exponential family representation of the probability mass function of $\mathbf{Y}$. Among the general exponential family results applicable to the logistic model are Fahrmeir and Kaufmann (1985, 1986) who give conditions under which the MLE of $\boldsymbol{\beta}$ is consistent, asymptotically normally distributed, and efficient. These results do not apply to the closely related probit model. Another immediate result is that $T(\mathbf{Y}) = \mathbf{X}'\mathbf{Y}$ is a complete sufficient statistic for $\boldsymbol{\beta}$. Furthermore the conditional distribution of $\mathbf{Y}$ given a subset of the elements of $\mathbf{X}'\mathbf{Y}$ can be used to perform small sample ("exact") inference for specific coefficients β_j. Problem 5.17 gives a simple example of forming a confidence interval for a single β_j and Sections 5.2–5.5 give additional applications of this idea and references.

Despite the four reasons cited above for using the logistic model, it should be stressed that the particular subject matter application should dictate the choice of model. Section 5.4 will describe several diagnostics for checking the adequacy of the logistic model.

This section concludes with a few remarks about several topics of considerable practical importance which are beyond the scope of this chapter. First, the logistic model (5.1.5) can be extended to accommodate polychotomous responses with values $0, 1, \ldots, g$ by assigning jth cell probability

$$p_j(\mathbf{x}) = \frac{\exp\{\mathbf{x}'\boldsymbol{\beta}_j\}}{1 + \sum_{i=1}^{g} \exp\{\mathbf{x}'\boldsymbol{\beta}_i\}}$$

for $j = 1(1)g$ where $\boldsymbol{\beta}_j = (\beta_{j1}, \ldots, \beta_{jk})'$ and

$$p_0(\mathbf{x}) = 1 - \sum_{j=1}^{g} p_j(\mathbf{x}) = \frac{1}{1 + \sum_{i=1}^{g} \exp\{\mathbf{x}'\boldsymbol{\beta}_i\}}. \tag{5.1.7}$$

This model involves $k \times g$ parameters. The likelihood equations based on model (5.1.7) are analogous to those developed in Section 5.4 (Problem 5.3).

A second important extension of the methods discussed in this chapter is to the analysis of *ordinal* response data as in Example 1.2.7. Latent variable models are one important tool for analyzing such data. These models assume that there exists an unobserved continuous variable Z and cut points $-\infty = a_{-1} < a_0 < \ldots < a_g = +\infty$ so that for each multinomial trial, outcome j is observed if and only if $Z \in (a_{j-1}, a_j)$, $0 \leq j \leq g$. McCullagh (1980) and Agresti (1984, Chapters 6 and 7) discuss the analysis of ordinal response data in detail.

A third generalization considers data that are "clustered" meaning that not all of the $\{(Y_i, m_i, \mathbf{x}_i)\}_{i=1}^{T}$ are independent. As an example, suppose a binary response is measured on the same subject at several points in time. The observations on the same subject are not independent. Such data come from so-called "longitudinal," "panel," or "repeated measures" studies. Similarly, the responses of littermates in a teratology experiment (i.e., an animal study of the causes of malformations in offspring), or of members of the same family in a genetic study of cancer incidence should not be treated as independent observations. Both nonparametric moment models and specific parametric models incorporating dependence have been used to analyze dependent discrete regression data. The book by McCullagh and Nelder (1983) and the papers by Liang and Zeger (1986) and Prentice (1988) discuss the former. Haseman and Kupper (1979), Jewell (1984), Stiratelli et al. (1984), Kupper et al. (1986), and Wypij (1986, Chapter 1) give surveys of specific parametric models.

Section 2 of this chapter considers simple binary regression with one discrete covariate; i.e., the T-sample problem. Sections 3 and 4 study logistic regression. Section 3 covers standard likelihood-based inference and omnibus tests-of-fit while Section 4 reviews alternative methods of estimation and graphical techniques for assessing fit. The final section considers specialized techniques for models with many nuisance parameters.

5.2 Two by T Tables

A. Introduction

Suppose $Y_1, \ldots, Y_T$ are mutually independent responses with $Y_i \sim B(m_i, p_i)$ where m_i is a known number of trials and p_i is an unknown success probability with $0 < p_i < 1$ for $i = 1(1)T$. The symbol Π_i is used to denote the ith binomial distribution. Such data are a special case of the binary regression model in which $k = 1$ and there is one *discrete* covariate denoting the population of the response. The data $\mathbf{Y} = (Y_1, \ldots, Y_T)'$ are usually presented as in Table 5.2.1.

Table 5.2.1. Notation for $2 \times T$ Tables

	Π_1	...	Π_T	Total
Success	Y_1	...	Y_T	Y_+
Failure	$m_1 - Y_1$	...	$m_T - Y_T$	$m_+ - Y_+$
Total	m_1	...	m_T	m_+

The T populations will either have a natural ordering or not according as the underlying covariate is ordinal or nominal. For example, the populations corresponding to breast cancer patients of different menopausal *ages* are ordered, but the populations corresponding to breast cancer patients receiving different *treatments* are not ordered. The case of a nominal covariate can be thought of as the binary response analog of the one-way layout.

More generally, the responses might be polychotomous (i.e., multinomial) with R possible values measured on either a nominal or ordinal scale. With polychotomous responses the data can be arranged in an $R \times T$ table with fixed columns totals. Example 1.2.7 concerning the use of cisplatinum as part of a combination chemotherapy illustrates the polychotomous ordinal response case. Only statistical methods for binary responses are discussed in this section; Example 5.2.1 is typical. (Problem 5.4 illustrates an analysis of the ordinal data of Example 1.2.7.)

Example 5.2.1. The data of Table 5.2.2, extracted from Knoke (1976), concern the proportions of Southern Protestants identifying themselves as

Democrats in the Presidential election years between 1952 and 1972, inclusive. The goal of this study is to quantify the pattern of change during this 20 year period. Note that the Π_i (which correspond to the election years) are ordered.

Table 5.2.2. Numbers of Southern Protestants Identifying Themselves as Democrats (Y_i) in the Presidential Election Years 1952–1972 (Reprinted with permission from David Knoke: *Change and Continuity in American Politics,* 1976, The Johns Hopkins University Press, Baltimore, Maryland.)

	1952	1956	1960	1964	1968	1972
Y_i	240	253	287	209	169	266
m_i	334	381	460	325	347	593
$\hat{p}_i$	.719	.664	.624	.643	.487	.449

Because of the independence of the $Y_1, \ldots, Y_T$, most of the results discussed in Section 2.1 concerning point estimation for a single binomial success probability extend directly to the T-sample problem. The MLE of $\mathbf{p} := (p_1, \ldots, p_T)'$ is $\hat{\mathbf{p}} = (\hat{p}_1, \ldots, \hat{p}_T)' := (Y_1/m_1, \ldots, Y_T/m_T)'$ which is also the UMVUE of $\mathbf{p}$. Further, the admissibility of the $\hat{p}_i$ with respect to any component loss function $L_i(p_i, a_i)$ is inherited by $\hat{\mathbf{p}}$ with respect to the ensemble loss $L_+(\mathbf{p}, \mathbf{a})$ defined by *summing* the component losses $L_i(p_i, a_i)$; i.e., $L_+(\mathbf{p}, \mathbf{a}) = \sum_{i=1}^{T} L_i(p_i, a_i)$ (Gutmann (1982)). In particular, $\hat{\mathbf{p}}$ is admissible with respect to summed squared error loss $L_S(\mathbf{p}, \mathbf{a}) = \sum_{i=1}^{T} m_i(p_i - a_i)^2$ and summed relative squared error loss $L_R(\mathbf{p}, \mathbf{a}) = \sum_{i=1}^{T} m_i(p_i - a_i)^2/p_i(1-p_i)$.

Generalizations of the Bayes and related estimators introduced in Section 2.1 are quite appealing in the $2 \times T$ case when there is reason to believe that the unknown p_i are related. For example, Leonard (1972), Novick, Lewis, and Jackson (1973), and Berry and Christensen (1979) study hierarchical Bayes estimators of $\mathbf{p}$ under the key modeling assumption that the prior distribution of $p_1, \ldots, p_T$ is exchangeable. Leonard's formulation is parametric with $\ln(\mathbf{p}[\mathbf{1}_T - \mathbf{p}]) \sim N_T(\mu \mathbf{1}_T, \sigma^2 \mathbf{I}_T)$ in the first stage. In the second stage, μ and σ^2 are modeled as independent with μ uniformly distributed over $\mathbb{R}$ and σ^2 inversely chi-squared distributed (see Section 2.1). Novick, Lewis, and Jackson's (1973) formulation is also parametric, but based on the Arcsine transformation of p_i rather than the logistic. Berry and Christensen (1979) adopt a nonparametric hierarchical Bayes approach in which the first stage specifies $p_1, \ldots, p_T$ are iid with completely unknown distribution $G(\cdot)$ having support on $[0,1]$. In the second stage, $G(\cdot)$ is modeled by a Dirichlet distribution.

The empirical Bayes approach can be employed to estimate the prior parameters rather than using a second stage prior to model them. Brier, Zacks, and Marlow (1986) give an example of the empirical Bayes approach based on the Arcsine transform. The smoothing methods of Section 2.2 can also be used to derive estimators which "pull" the p_i toward some structural model.

This section will concentrate on the use of testing and interval estimation to determine whether, or to what extent, the populations $\Pi_1, \ldots, \Pi_T$ are similar. Special attention will be given to the important case of $T = 2$ populations. In addition, the problem of selecting populations with large p_i's will be discussed, which is appropriate when $T \geq 3$.

Before considering hypothesis tests, the following conditional distribution, required for later use, is reviewed. When $T = 2$, let $t := Y_+$, $L := \max\{0, t - m_2\}$ and $U := \min\{t, m_1\}$. Then for $L \leq y_1 \leq U$

$$P[Y_1 = y_1 \,|\, Y_+ = t] = \frac{P[Y_1 = y_1, Y_2 = t - y_1]}{P[Y_+ = t]}$$

$$= \frac{\binom{m_1}{y_1}\binom{m_2}{t-y_1}\psi^{y_1}}{\sum_{j=L}^{U}\binom{m_1}{j}\binom{m_2}{t-j}\psi^j} =: f(y_1 \,|\, t; \psi) \qquad (5.2.1)$$

where $\psi = \dfrac{p_1(1-p_2)}{(1-p_1)p_2}$ is the odds ratio. The conditional probability (5.2.1) depends on p_1 and p_2 only through ψ as the notation for the mass function indicates. The symbol $E_\psi[\cdot \,|\, Y_+ = t]$ denotes expectation taken with respect to the conditional distribution (5.2.1), and the symbol $E_{\mathbf{p}}[\cdot]$ denotes expectation with respect to the unconditional distribution of $\mathbf{Y}$. In the special case $\psi = 1$ (equivalently $p_1 = p_2$),

$$f(y_1 \,|\, t; 1) = \binom{m_1}{y_1}\binom{m_2}{y_2} \Big/ \binom{m_+}{t} \qquad (5.2.2)$$

which is the hypergeometric distribution. This calculation shows Y_+ is a sufficient statistic for the family of distributions of $\mathbf{Y}$ when $0 < p_1 = p_2 < 1$ since (5.2.2) is independent of p_1 and p_2. Alternatively, exponential family theory shows that Y_+ is not only sufficient but also complete in this case.

B. Hypothesis Tests

The Two-Sample Problem

Consider first the problem of testing homogeneity H_0: $p_1 = p_2$. The most commonly described large sample tests of H_0 versus the global alternative $H_{\neq}$: $p_1 \neq p_2$ are the score and Wald tests. To define these tests let $\hat{p}_i = Y_i/m_i$ denote the MLE of p_i for $i = 1, 2$ and $\hat{p} = Y_+/m_+$ be the pooled MLE of $p_1 = p_2$ under H_0. Problem 5.5 shows that the score test of H_0 versus $H_{\neq}$ rejects H_0 if and only if

$$X^2 = \frac{(\hat{p}_1 - \hat{p}_2)^2}{\hat{p}(1-\hat{p})\left(\frac{1}{m_1} + \frac{1}{m_2}\right)} \geq \chi^2_{\alpha,1} \qquad (5.2.3)$$

and the Wald test rejects H_0 if and only if

$$W = \frac{(\hat{p}_1 - \hat{p}_2)^2}{\frac{\hat{p}_1(1-\hat{p}_1)}{m_1} + \frac{\hat{p}_2(1-\hat{p}_2)}{m_2}} \geq \chi^2_{\alpha,1}. \tag{5.2.4}$$

The intuition behind both tests is that the null hypothesis $p_1 = p_2$ is rejected when $\hat{p}_1 - \hat{p}_2$ is large in absolute value. The difference between the tests is that X^2 calibrates $\hat{p}_1 - \hat{p}_2$ by an estimate of its standard error which is valid only under H_0 while W standardizes $\hat{p}_1 - \hat{p}_2$ by an estimated standard error valid under H_0 or $H_{\neq}$.

The score statistic is presented more frequently in textbooks as a large sample test of H_0 than the Wald test. Problem 5.6 shows that the score statistic is Pearson's chi-squared statistic as well as an analog of the one-way ANOVA F-statistic in that it rejects homogeneity when the individual $\hat{p}_1$ and $\hat{p}_2$ vary too much.

Several comparisons of the size and power of X^2, W, and continuity corrected versions of X^2 have appeared in the literature. If $m_1 = m_2 = m$, algebra shows that

$$W = 2m(X^2/(2m - X^2)) \geq X^2$$

with strict inequality unless $Y_1 = Y_2$. Thus W will be more powerful than X^2 but also have larger size; a fair comparison between the two tests must adjust for the latter. Eberhardt and Fligner (1977) and Berengut and Petkau (1979) compare the score and the Wald tests; they conclude that X^2 more nearly achieves nominal size, but W has higher asymptotic power over a larger portion of (p_1, p_2)-space. Berengut and Petkau recommend that X^2 be used if $\min\{m_1, m_2\} \leq 30$ and W be used if m_1 and m_2 are both large (say, both ≥ 40). In another study of large sample tests, Storer and Kim (1988) also compute the exact achieved size and power of X^2; they compare X^2 with several continuity corrected alternatives (but not the Wald test) for $m_1 = m_2 = 1(1)100$. They find that X^2 achieves approximately nominal size once $(m_1 = m_2)$ is about 20 although it can be slightly liberal. Their calculations also show that X^2 has smaller achieved size than nominal for $(m_1 = m_2) \leq 15$.

In sum, for m_1 and m_2 nearly equal and between 20 and 30, X^2 has size close to nominal and reasonable power characterizations. For m_1 and m_2 both large (≥ 40), W is preferable. For intermediate cases the comparison is not clear-cut; the operating characteristics of the two tests can be compared in the Berengut and Petkau technical report.

Example 5.2.2. Consider the data in Table 5.2.3 on the incidence of colds in a French study discussed by Pauling (1971). One hundred and thirty-nine skiers received a supplement of one gram of vitamin C per day in their diets and 140 received no supplement. Seventeen skiers in the supplemented group and 31 skiers in the control group caught colds during the course of the study. The score statistic is $X^2 = 4.81$ (P-value $= .028$) and the Wald statistic is $W = 4.90$ (P-value $= .027$). Since $m_1 = 140$ and $m_2 = 139$ are nearly equal and $Y_1 \neq Y_2$, one anticipates the observed relationship $W > X^2$ should hold. These data provide substantial though not overwhelming evidence that Vitamin C reduces cold incidence.

Table 5.2.3. Cold Incidence Among 279 French Skiers (Reprinted with permission from L. Pauling: "The Significance of the Evidence About Ascorbic Acid and the Common Cold," *Proceedings of the National Academy of Sciences,* vol. 68, 1971, National Academy of Sciences, Washington, D.C.)

	Vitamin C	No Vitamin C
Cold	17	31
No Cold	122	109
Total	139	140

Consider testing $H_0: p_1 = p_2$ when the large sample tests described in the previous paragraphs are inappropriate and when it is desired that the test have size no more than α. The most famous small sample test of H_0 is the Fisher–Irwin test. This test is randomized; it is the uniformly most powerful unbiased test (UMPU) of H_0. The exact form of the test depends on the alternative. The UMPU test of $H_0: p_1 = p_2$ versus $H_>: p_1 > p_2$ is

$$\phi_>^R = \phi_>^R(Y_1, Y_2) = \begin{cases} 1 & \text{if } Y_1 > c \\ \gamma & \text{if } Y_1 = c, \\ 0 & \text{if } Y_1 < c \end{cases} \tag{5.2.5}$$

where $c = c(y_+)$ and $\gamma = \gamma(y_+)$ are chosen to satisfy

$$E_{\psi=1}[\phi_>^R(Y_1, Y_2) \,|\, Y_+ = t] = P[Y_1 > c \,|\, Y_+ = t; \psi = 1] + \gamma P[Y_1 = c \,|\, Y_+ = t; \psi = 1] = \alpha.$$

Here $\phi_>^R(y_1, y_2)$ is the probability of rejecting H_0 given data (y_1, y_2). The constants c and γ are chosen to make the test conditionally (given $Y_+ = t$) size α which implies that it is also unconditionally size α. Tests $\phi(\cdot)$ which satisfy $E_{\mathbf{p}}[\phi(\mathbf{Y})] = \alpha$ for $0 < p_1 = p_2 < 1$ are called similar (on the boundary) since they have exactly α probability of rejecting H_0 on the boundary between $H_0: p_1 = p_2$ and $H_>: p_1 > p_2$.

The Fisher–Irwin UMPU tests of H_0 versus $H_<: p_1 < p_2$ and versus $H_{\neq}: p_1 \neq p_2$ are also randomized. Despite its theoretical optimality the Fisher–Irwin test is seldom, if ever, used in practice because of the randomization.

Numerous authors have studied nonrandomized small sample tests of H_0 versus $H_>$, $H_<$, and $H_{\neq}$. One of the earliest proposals, called the "Fisher exact test" by some authors, is to use a truncated nonrandomized Fisher–Irwin test. For testing H_0 versus $H_>$ this test is defined by

$$\phi_>^{NR} = \begin{cases} 1 & \text{if } Y_1 > c \\ 0 & \text{if } Y_1 \leq c \end{cases} \tag{5.2.6}$$

where $c = c(y_+)$ is the same as in (5.2.5). Note that

$$E_{\psi=1}[\phi_>^{NR}(\mathbf{Y}) \,|\, Y_+ = t] \leq \alpha$$

which implies the unconditional probability of rejection satisfies $E_{\mathbf{p}}[\phi_>^{NR}(\mathbf{Y})] \leq \alpha$ for all $p_1 = p_2$. Furthermore, $E_{\mathbf{p}}[\phi_>^{NR}(\mathbf{Y})] < \alpha$ is almost always the case for $p_1 = p_2$ and, even worse, $\max_{\mathbf{p}:p_1=p_2} E_{\mathbf{p}}[\phi_>^{NR}(\mathbf{Y})]$ can be substantially less than α (Garside and Mack, 1976; Storer and Kim, 1988). Thus (5.2.6) is ordinarily quite conservative and consequently has low power.

Although the Fisher exact test is very conservative, it has been widely used. Tables of the critical constant c required to implement $\phi_>^{NR}$ can be found in Finney et al. (1963) and Bennett and Horst (1966). Tables of the power function $\beta_>^{NR}(p_1, p_2) :=\mathrm{E}_{\mathbf{p}}[\phi_>^{NR}(\mathbf{Y})]$ can be found in several places. Bennett and Hsu (1960) plot isopower curves for various choices of m_1 and m_2. Casagrande, Pike, and Smith (1978) provide a FORTRAN program for calculating $\beta_>^{NR}(p_1, p_2)$. Gail and Gart (1973) table the smallest common sample size $m_1 = m_2$ required for $\phi_>^{NR}$ to achieve power .50 (.80, .90) against a given alternative $p_1 > p_2$; these tables can be used in designing comparative binomial experiments. Haseman (1978) corrects some approximation errors in Gail and Gart.

In general, nonrandomized tests formed by *truncating* randomized Neyman–Pearson tests need not have optimal power properties in the class of nonrandomized tests. In particular, such a test need *not* be the most powerful nonrandomized test of size $\leq \alpha$, even when testing a simple null versus a simple alternative as Example 5.2.3 shows.

Example 5.2.3. Suppose a random variable W takes values $1, 2, \ldots, 10$ and has true probability mass function $f(\cdot)$ which is either $f_0(\cdot)$ or $f_1(\cdot)$. Table 5.2.4 lists $f_0(\cdot)$, $f_1(\cdot)$, and their likelihood ratio. Consider testing H_0: $f = f_0$ versus H_1: $f = f_1$. The uniformly most powerful (UMP) test of H_0 versus H_1 is randomized and rejects for large values of $f_1(k)/f_0(k)$. When $\alpha = .05$ the UMP test ϕ^U rejects with probability one for $k = 4$ or 9 and with probability 1/2 for $k = 7$; it has size .05 and power .45. Consider the conservative nonrandomized test ϕ^T obtained by truncating ϕ^U. For $\alpha = .05$, ϕ^T rejects with probability one for $k = 4$ or 9 and has achieved size .04 and power .38. The alternative nonrandomized test ϕ^N which rejects with probability one when $k = 4$ or 7 has achieved size .05 and power .41; thus ϕ^N has greater power than ϕ^T. The reason, of course, is that ϕ^N also has a greater chance of rejecting H_0. In particular, ϕ^N attains the target size while ϕ^T is conservative.

Table 5.2.4. Probability Mass Functions f_0, f_1 and Their Likelihood Ratio

k	1	2	3	4	5	6	7	8	9	10
$f_0(k)$	.02	.18	.07	.03	.17	.24	.02	.08	.01	.18
$f_1(k)$	.10	.13	.04	.27	.01	.05	.14	.03	.11	.12
$f_1(k)/f_0(k)$	5	.72	.57	9	.06	.21	7	.38	11	.67

Several approaches have been investigated for deriving nonrandomized tests which are not as conservative as Fisher's exact test. Boschloo (1970) proposes using $\phi_{>}^{NR}(\cdot,\cdot)$ with size $K\alpha$ where K (> 1) is selected so that $\max_{p_1=p_2} \beta_{>}^{NR}(p_1,p_2) \simeq \alpha$. Armesan (1955) and McDonald, Davis and Milliken (1977) study Sterne-like critical regions.

A third approach, considered by Suissa and Shuster (1985) and Haber (1987), is to use one of the large sample test statistics, such as X^2 or W, with a critical point calculated to guarantee that the exact size of the test is $\leq \alpha$. For testing $H_0\colon p_1 = p_2$ versus $H_{>}\colon p_1 > p_2$, Suissa and Shuster use this approach based on the one-sided versions of the score statistic

$$S_{>} = \frac{\hat{p}_1 - \hat{p}_2}{\left\{\hat{p}(1-\hat{p})\left(\frac{1}{m_1}+\frac{1}{m_2}\right)\right\}^{1/2}} \tag{5.2.7}$$

and the Wald statistic

$$W_{>} = \frac{\hat{p}_1 - \hat{p}_2}{\left\{\frac{\hat{p}_1(1-\hat{p}_1)}{m_1}+\frac{\hat{p}_2(1-\hat{p}_2)}{m_2}\right\}^{1/2}}. \tag{5.2.8}$$

As in the two sample case, $W_{>}$ is a monotone function of $S_{>}$ and thus either statistic leads to the same small sample critical region. Suissa and Shuster table the small sample critical points for $\alpha \in \{01, .025, .05\}$ and $m_1 = m_2 = m = 10(1)150$. Their calculations show this test is *uniformly more powerful* than the Fisher exact test for these (α, m) combinations. Suissa and Shuster also give tables of the common sample size required by $W_{>}$ to achieve power .80 for a size $(\leq)$ .05.

In a similar way Haber (1987) compares the X^2, W, LR, and other test statistics for H_0: $p_1 = p_2$ versus $H_{\neq}\colon p_1 \neq p_2$ when the critical points for each test are chosen to make the tests have achieved size $\leq \alpha$; both unequal and equal sample sizes are studied. While no single test uniformly dominates in all his comparisons, the score statistic with small sample critical point performs well overall.

In summary, for the 2 population problem with small sample sizes, if guaranteed size is desired, the score statistic with a specifically computed small sample critical point is recommended. The Fisher exact test (5.2.6) is extremely conservative and has low power; it should not be used. If the α

level is not prespecified, a test's P-value can be calculated. Davis (1986b) investigates the calculation of P-values for a variety of tests including the score, Wald, and Fisher exact tests (Problems 5.11 and 5.12).

The T-Sample Problem

When $T \geq 3$, there are more alternatives possible to the homogeneity hypothesis H_0: $p_1 = \ldots = p_T$. The discussion below considers only the global alternative $H_{\neq}$: (not H_0) and the trend alternative $H_{\geq}$: $p_1 \geq \ldots \geq p_T$ (with at least one inequality holding strictly). Also only large sample tests are described.

The LR, score, and Wald statistics, as well as other members of the Cressie–Read family, can be used to construct large sample tests of H_0 versus $H_{\neq}$. In light of the comparative studies described earlier for $T = 2$, only the score test will be stated here. The analog of (5.2.3) rejects H_0 if and only if

$$X^2 = \sum_{i=1}^{T} \frac{m_i(\hat{p}_i - \hat{p})^2}{\hat{p}(1 - \hat{p})} \geq \chi^2_{\alpha, T-1}. \tag{5.2.9}$$

The score statistic is easy to compute; its asymptotic power can be calculated under contiguous alternatives by standard analytical techniques.

When the T populations differ with respect to an ordinal explanatory variable (e.g., age of individuals in 10-year ranges or the dose of a drug), it is often of interest to test H_0 versus the trend alternative $H_{\geq}$. The LRT requires calculation of the MLE of $\mathbf{p}$ under both H_0 (which is easy) and $H_{\geq}$ (which is iterative). Robertson, Wright, and Dykstra (1988) describe algorithms for the latter. The LR statistic is asymptotically distributed as a chi-bar-squared random variable. The chi-bar-squared distribution is the distribution of a convex combination of chi-squared random variables with different degrees of freedom. The reason the limiting distribution is so complicated has to do with the boundary of the alternative hypothesis; this same difficulty occurs in most order restricted inference problems. Robertson, Wright, and Dykstra (1988) contain some tables of critical points of the chi-bar-squared distribution as well as several approximations to such critical points.

Perhaps the most commonly used test for H_0 versus $H_{\geq}$ is the Cochran–Armitage (CA) test. This test is closed form but requires additional information in the form of scores x_i for each population Π_i, $1 \leq i \leq T$. Cochran (1954) and Armitage (1955) proposed regressing $\hat{\mathbf{p}}$ on the score vector $\mathbf{x} = (x_1, \ldots, x_T)'$ with weights $m_i/\hat{p}(1-\hat{p})$ where $\hat{p} = Y_+/m_+$ is the MLE of the common success probability under H_0. The test rejects H_0 if and only if

$$X^2_{CA} := \frac{\left(\sum_{i=1}^{T} y_i x_i - \hat{p} \sum_{i=1}^{T} m_i x_i\right)^2}{\hat{p}(1 - \hat{p}) \sum_{i=1}^{T} m_i (x_i - \overline{x}_w)^2} \geq \chi^2_{\alpha,1} \tag{5.2.10}$$

where $\bar{x}_w := (\sum_{i=1}^{T} m_i x_i)/m_+$ is the weighted average of the scores. The CA test is robust in the following two senses. First, X_{CA}^2 is invariant with respect to linear transformations of the scores which is important when obvious scores, such as drug dose, are not available. In the latter case Armitage recommends that equally spaced scores centered around zero be used. Second, X_{CA}^2 is the score statistic for testing $H_0: \beta = 0$ in the model $p_i = G(\alpha + \beta x_i)$ for *any* monotone twice differentiable function $G(\cdot)$ (Problem 5.13). Chapman and Nam (1968) compute the noncentrality parameter of the CA test under contiguous alternatives and study the asymptotic power. Nam (1987) provides a simple power approximation which allows calculation of sample sizes. Thus, despite the arbitrariness of assigning scores in some cases, the CA test's strong robustness with respect to misspecification of the functional dependence of the success probabilities p_i on the scores, and its known power approximation make it attractive.

Before illustrating (5.2.9) and (5.2.10), it should be noted that Robertson, Wright, and Dykstra (1988) describe several other approximate tests for H_0 versus $H_{\geq}$ including one based on the Arcsine transformation. They also formulate the LRT for $H_{\geq}$: $p_1 \geq \ldots \geq p_T$ versus the global alternative H_G: (not $H_{\geq}$); this test also requires iterative calculation and has a limiting chi-bar-squared distribution.

Example 5.2.1 (continued). The data of Table 5.2.2 suggest an erosion of Democratic support among Southern Protestants over the 1952–1972 time period. The score statistic for testing H_0 versus $H_{\neq}$ is $X^2 = 100.80$ (P-value $< 10^{-6}$, 5 degrees of freedom). If instead, the trend alternative $H_{\geq}$ is considered and the scores $\mathbf{x} = (-2.5, -1.5, -.5, .5, 1.5, 2.5)'$ are used, then the CA statistic is $X_{CA}^2 = 91.50$ (P-value $< 10^{-6}$, 1 degree of freedom). This value would be unchanged if the election years $1952, \ldots, 1972$ were used as scores instead. The LR statistic for testing H_0 versus $H_{\geq}$ has value 101.02. A conservative critical point based on the χ_5^2 distribution gives a highly significant result. These test statistics strongly support the trend observed above, namely, the probability of Democratic identification is decreasing among Southern Protestants over this 20-year period.

C. Interval Estimation

The Two-Sample Problem

Many scalar measures have been suggested in the epidemiological literature for comparing Π_1 and Π_2. Walter (1976) and Gart (1971) define and interpret a number of these quantities. Three important measures are considered below.

(1) *The Attributable Risk.* The attributable risk Δ is defined as $\Delta := p_1 - p_2$; Δ ranges over $(-1, +1)$. The attributable risk is easy to interpret.

For a group of N subjects, $N\Delta$ is the difference in the expected number of successes if the subjects come from Π_1 versus Π_2.

(2) *The Relative Risk.* The relative risk ρ is defined as $\rho := p_1/p_2$; ρ ranges over $(0, \infty)$. The relative risk is less easy to interpret than Δ. Consider $(p_1, p_2) = (.6, .3)$ and $(p_1^*, p_2^*) = (2 \times 10^{-6}, 10^{-6})$. Both have relative risk $\rho = 2$ although their associated success probabilities (and Δ values) differ greatly in magnitude.

(3) *The Odds Ratio.* The odds ratio ψ is defined as $\psi := \dfrac{p_1(1-p_2)}{(1-p_1)p_2}$; ψ ranges over $(0, \infty)$. Like the relative risk, ψ is a ratio but it is a ratio of *odds* of success rather than of success *probabilities.* The odds ratio is approximately equal to the relative risk for small p_1 and p_2. For example, if $p_1 = .05$ and $p_2 = .10$ then $\rho = .5$ while $\psi = .473$. This observation is important since it is much easier to determine small sample confidence intervals for ψ than ρ, and ψ may be an adequate measure of treatment comparison in many problems.

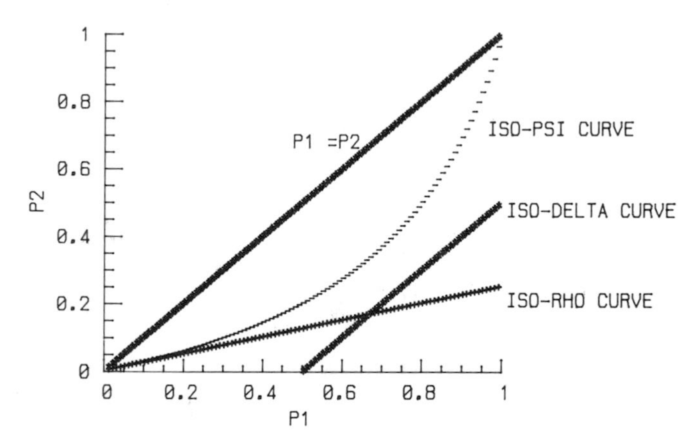

Figure 5.2.1. The line $p_1 = p_2$ ($\Delta = 0$, $\rho = 1$, $\psi = 1$), the iso-Δ curve for $\Delta = .5$, the iso-ρ curve for $\rho = 4$, and the iso-ψ curve for $\psi = 4$.

Geometrically, the three measures can be contrasted by viewing their isocurves in (p_1, p_2) space. Figure 5.2.1 contrasts isocurves for Δ, ρ, and ψ. The iso-Δ curves are a family of parallel lines with p_2-axis intercepts ranging over $(-1, +1)$. The iso-ρ curves are a family of lines through the origin with slope ranging over $(0, \infty)$. The iso-ψ curves are a family of curves each joining $(0, 0)$ to $(1, 1)$. Figure 5.2.1 shows that the three measures are not in 1–1 correspondence with one another so that knowing one gives little information about the other two.

Both large and small sample confidence intervals are described for Δ, ρ, and ψ. Some of these intervals require iterative methods to calculate while others are closed-form expressions of the data $\mathbf{Y}$. References are given to both Monte Carlo and exact comparative studies of various systems of intervals and to published software for implementing the iterative methods. The point estimators about which some of these intervals are constructed contain bias corrections. See Gart, Pettigrew, and Thomas (1985) and the references therein for details about the bias and higher order moments of these estimators.

The basis for the large sample Δ confidence intervals considered below is the approximate distribution.

$$\hat{\Delta} \overset{app}{\sim} N\left(\Delta, \sigma_{\hat{\Delta}}^2 := \left[\frac{p_1(1-p_1)}{m_1} + \frac{p_2(1-p_2)}{m_2}\right]\right) \tag{5.2.11}$$

where $\hat{\Delta} := \hat{p}_1 - \hat{p}_2$. The normal distribution becomes inaccurate when either p_1 or p_2 is near its extremes.

One class of confidence intervals has the form

$$\hat{\Delta} \pm z_{\alpha/2}\hat{\sigma}_{\hat{\Delta}} \tag{5.2.12}$$

where $\hat{\sigma}_{\hat{\Delta}}$ is an estimate of $\sigma_{\hat{\Delta}}$. These intervals can be thought of as analogs of the naive intervals I_N for a single binomial success probability discussed in Section 2.1. Hauck and Anderson (1985) report a simulation study comparing the coverage probabilities of (5.2.12) for eight estimators $\hat{\sigma}_{\hat{\Delta}}$. They recommend using (5.2.12) with

$$\hat{\sigma}_{\hat{\Delta}} := \left\{\frac{\hat{p}_1(1-\hat{p}_1)}{m_1} + \frac{\hat{p}_2(1-\hat{p}_2)}{m_2} + \frac{1}{2\max\{m_1, m_2\}}\right\}^{1/2} \tag{5.2.13}$$

which is adequate when

$$\min\{m_1 p_1, m_1(1-p_1), m_2 p_2, m_2(1-p_2)\} \geq \begin{cases} 3, & \alpha = .10, .05 \\ 5, & \alpha = .01. \end{cases}$$

There is no guarantee that this interval is consistent with the Pearson chi-square test (5.2.3) in the sense that the confidence interval covers $\Delta = 0$ if and only if (5.2.3) rejects $H_0\colon \Delta = 0$.

A second method of interval construction based on (5.2.11) is the analog of the sophisticated binomial interval I_S discussed in Section 2.1. Express $\sigma_{\hat{\Delta}}^2$ as a function of Δ and a nuisance parameter, say p_1, and then solve

$$\left(\frac{\hat{\Delta} - \Delta}{\sigma_{\hat{\Delta}}}\right)^2 \leq \chi_{\alpha,1}^2$$

for Δ after substituting $\hat{p}_1$ for p_1 (Anbar (1983); Mee (1984); Meittinen and Nurminen (1985)). These intervals depend on the choice of nuisance parameter and must be iteratively calculated. Beal (1987) compares several systems of iteratively calculated intervals with a more complicated though closed-form version of (5.2.12)–(5.2.13) called the "Jeffreys–Perks interval" (his Equation 2) and concludes the Mee and Meittinen–Nurminen intervals are best. However the Jeffreys–Perks interval also has reasonable coverage and expected length properties.

Bedrick (1987), Gart and Nam (1988), and the references therein review the literature on approximate confidence intervals for ρ. One simple non-iterative interval is based on the approximation distribution

$$\ln(\hat{\rho}) \overset{app}{\sim} N\left(\ln(\rho), \sigma^2_{\ln(\hat{\rho})} := \left[\frac{1-p_1}{m_1 p_1} + \frac{1-p_2}{m_2 p_2}\right]\right) \tag{5.2.14}$$

where $\ln(\hat{\rho}) := \ln\left[\frac{(Y_1+.5)(m_2+.5)}{(m_1+.5)(Y_2+.5)}\right]$. Walter (1975) shows $\ln(\hat{\rho})$ is a nearly unbiased point estimate of $\ln(\rho)$. Gart, Pettigrew, and Thomas (1986) show that a nearly unbiased estimator of the variance $\sigma^2_{\ln(\hat{\rho})}$ is

$$\hat{\sigma}^2_{\ln(\hat{\rho})} := (Y_1+.5)^{-1} - (m_1+.5)^{-1} + (Y_2+.5)^{-1} - (m_2+.5)^{-1}. \tag{5.2.15}$$

Combining (5.2.14) and (5.2.15) yields the $100(1-\alpha)\%$ confidence interval

$$\hat{\rho}\exp\{\pm z_{\alpha/2}\hat{\sigma}_{\ln(\hat{\rho})}\}. \tag{5.2.16}$$

It should be noted that the final interval (5.2.16) is not symmetric about the point estimate $\hat{\rho}$ because of the inversion. The same is true for the ψ interval (5.2.19) below.

Although (5.2.16) has coverage reasonably close to the nominal level, improved intervals are possible by (iteratively) inverting either continuity corrected score tests or Cressie–Read tests of $H_0: \rho = \rho_0$ versus $H_A: \rho \neq \rho_0$. The latter can be used for m_1 and m_2 as small as 25 (Bedrick, 1987; Gart and Nam, 1988).

Fleiss (1979) and Gart and Thomas (1982) study interval estimation of the odds ratio in 2 by 2 tables. Gart and Thomas compute the exact coverage probabilities of several large-sample methods. The best non-iterative method they discuss is based on the approximate distribution

$$\ln(\hat{\psi}) \overset{app}{\sim} N\left(\ln(\psi), \sigma^2_{\ln(\hat{\psi})} := \left[\frac{1}{m_1 p_1(1-p_1)} + \frac{1}{m_2 p_2(1-p_2)}\right]\right) \tag{5.2.17}$$

where $\hat{\psi} := \frac{(Y_1+.5)(m_2-Y_2+.5)}{(m_1-Y_1+.5)(Y_2+.5)}$. Using the estimated variance

$$\begin{aligned}\hat{\sigma}^2_{\ln(\hat{\psi})} &:= (Y_1+.5)^{-1} + (m_1-Y_1+.5)^{-1}\\ &\quad + (Y_2+.5)^{-1} + (m_2-Y_2+.5)^{-1}\end{aligned} \tag{5.2.18}$$

and (5.2.17) yields the interval

$$\hat{\psi}\exp\{\pm z_{\alpha/2}\hat{\sigma}_{\ln(\hat{\psi})}\}. \tag{5.2.19}$$

Their calculations also show that an iterative system of intervals due to Cornfield (1956) has achieved coverage very close to nominal levels for a wide range of m_1, m_2, p_1, and p_2.

Example 5.2.2 (continued). Point estimates and 90%-confidence intervals are computed for Δ, ρ, and ψ based on (5.2.12)–(5.2.13), (5.2.16), and (5.2.19), respectively, for the cold incidence of skiers receiving supplemental vitamin C compared with those who do not. The results are listed in Table 5.2.5. The width of these intervals tempers the conclusions obtained from the tests calculated earlier. They offer only marginal support of the claim that use of vitamin C supplements is effective in reducing cold incidence.

The large sample confidence intervals for Δ, ρ, or ψ described above are suspect when used on small samples such as those in Example 5.2.4 below or in settings where the p_i are extreme.

Table 5.2.5. Point and 90% Confidence Intervals for Three Measures of Association Comparing Cold Incidence in French Skiers Taking Vitamin C and Placebo

Measure of Association	Point Estimate	90% Confidence Interval
Δ	.099	(−0.047, .245)
ρ	1.787	(1.046, 3.053)
ψ	2.014	(1.063, 3.813)

Example 5.2.4. The data in Table 5.2.6 from Fisher (1935) concern the criminal status (convicted or not convicted) of same sex twins of criminals. There were 30 individuals in the study each of whom had a twin of the same sex who *was* a criminal. The study group is classified into two populations according as they are monozygotic (identical) or dizygotic (fraternal) twins. The question of interest is whether monozygotic same sex twins of criminals are more likely to be convicted of crimes than dizygotic same sex twins of criminals.

Table 5.2.6. Convictions of Same Sex Monozygotic and Dizygotic Twins of Criminals (Reprinted with permission from R. Fisher: "The Logic of Inductive Inference," *Journal of the Royal Statistical Society*, vol. 48, 1935, pp. 39–54. Royal Statistical Society, London, England.)

	Dizygotic	Monozygotic	Total
Convicted	2	10	12
Not Convicted	15	3	18
Total	17	13	30

Small sample confidence intervals for ψ are discussed first due to their simplicity compared to the intervals for Δ and ρ. Suppose that lower and upper interval limits $\underline{\psi}(\mathbf{Y})$ and $\overline{\psi}(\mathbf{Y})$ satisfy

$$P[\underline{\psi}(\mathbf{Y}) < \psi < \overline{\psi}(\mathbf{Y}) \,|\, Y_+ = t, \psi] \geq 1 - \alpha \tag{5.2.20}$$

for all t and $0 < \psi < \infty$. Equation (5.2.20) implies that unconditionally

$$P_{\mathbf{p}}[\underline{\psi}(\mathbf{Y}) < \psi < \overline{\psi}(\mathbf{Y})] \geq 1 - \alpha \text{ for all } \mathbf{p} \in (0,1)^2, \tag{5.2.21}$$

so that $(\underline{\psi}(\mathbf{Y}), \overline{\psi}(\mathbf{Y}))$ forms a $100(1-\alpha)\%$ confidence interval for ψ. Two methods for constructing $\underline{\psi}(\mathbf{Y})$ and $\overline{\psi}(\mathbf{Y})$ satisfying the conditional requirement (5.2.20) have been considered in the literature: the tail and Sterne methods.

Cornfield (1956) was the first to observe that the tail method could be applied to the conditional distribution $f(y \,|\, t; \psi)$ of (5.2.1). Recall that the tail method replaces (5.2.20) with the stronger requirement

$$\begin{aligned} &P_{\mathbf{p}}[\psi \geq \overline{\psi}(\mathbf{Y}) \,|\, Y_+ = t, \psi] \leq \alpha/2 \\ \text{and } &P_{\mathbf{p}}[\psi \leq \underline{\psi}(\mathbf{Y}) \,|\, Y_+ = t, \psi] \leq \alpha/2. \end{aligned} \tag{5.2.22}$$

As shown in Section 2.1 for the single binomial success probability, (5.2.22) defines upper and lower tail limits $\underline{\psi}(\mathbf{Y})$ and $\overline{\psi}(\mathbf{Y})$ as

$$\underline{\psi}(i,j) = \begin{cases} 0 & , \ i = \max\{0, i+j-m_2\} \\ P[Y_1 \geq i \,|\, Y_+ = i+j, \underline{\psi}] \overset{set}{=} \alpha/2 \ , & \text{otherwise} \end{cases}$$

$$\overline{\psi}(i,j) = \begin{cases} +\infty & , \ i = \min\{m_1, i+j\} \\ P[Y_1 \leq i \,|\, Y_+ = i+j, \overline{\psi}] \overset{set}{=} \alpha/2 \ , & \text{otherwise,} \end{cases} \tag{5.2.23}$$

where $P[E \,|\, Y_+ = t, \psi] \overset{set}{=} \alpha/2$ for an event E means that ψ is set equal to that value which solves the equation. StatXact contains software which solves (5.2.23) for $\underline{\psi}(\cdot)$ and $\overline{\psi}(\cdot)$.

Sterne intervals are constructed by inverting a family of acceptance regions, $A(\psi_0)$, $0 < \psi_0 < +\infty$, corresponding to tests of $H_0\colon \psi = \psi_0$ versus $H_A\colon \psi \neq \psi_0$. The Sterne region $A(\psi_0)$ is composed of the minimal number of most likely outcomes under H_0. To apply this method to the ψ problem, fix $\psi_0 \in (0,\infty)$ and an integer t, $0 \leq t \leq m_1 + m_2$, and order the possible outcomes for Y_1 according to the conditional distribution $f(\cdot \,|\, t; \psi_0)$ as $j_1, \ldots, j_R$ where $f(j_1 \,|\, t; \psi_0) \geq \ldots \geq f(j_R \,|\, t; \psi_0)$ with $R = U + L - 1$. Then $A(\psi_0)$ is defined as

$$A(\psi_0) = \left\{ j_1, \ldots, j_J : \sum_{q=1}^{J} f(j_q \,|\, t; \psi_0) \geq 1 - \alpha > \sum_{q=1}^{J-1} f(j_q \,|\, t; \psi_0) \right\}.$$

Inversion of $A(\psi_0)$ for $0 \leq \psi_0 < \infty$ produces the conditional Sterne interval. Baptista and Pike (1977) provide a FORTRAN subroutine which calculates conditional Sterne ψ intervals.

Example 5.2.4 (continued). A point estimate of the odds ratio of conviction for monozygotic to dyzygotic twins for the data of Table 5.2.6 is

$$\hat{\psi} = \left(\frac{\text{odds of conviction} \mid \text{dizygotic}}{\text{odds of conviction} \mid \text{monozygotic}}\right) = \frac{2.5(3.5)}{10.5(15.5)} = .05;$$

95% conditional ψ tail limits are (.0033, .3632) and 95% conditional ψ Sterne limits are (.0050, .3617). The Sterne interval is wholly contained within the tail interval. This phenomenon has been observed in earlier comparisons of tail and Sterne intervals; tail intervals are often excessively wide. These data strongly support the notion that dizygotic same sex twins of criminals have much lower odds of conviction than monozygotic same sex twins of criminals.

Small sample confidence intervals for ρ and Δ are very difficult to compute with one exception. If one is allowed to design a binomial experiment based on an a priori fixed total number T of binomial trials, then Buhrman (1977) shows that if m_1 is chosen randomly according to a certain binomial distribution (and $m_2 = T - m_1$), then confidence intervals for ρ and Δ can be easily calculated.

Santner and Snell (1980) construct tail intervals for Δ. As this is a nuisance parameter problem, care must be taken defining the tails. Problems 5.7 and 5.8 describe some issues involved in forming confidence intervals for problems containing nuisance parameters. Tail intervals could be similarly devised for ρ.

Sterne–Crow intervals for either Δ or ρ are considerably more complex to implement than tail intervals and will not be discussed in detail here. Santner and Yamagami (1988) provide an algorithm and code for computing Sterne–Crow intervals for Δ; they assess the advantages of these intervals over tail intervals in terms of expected interval length and coverage probability.

The T-Sample Problem

When $T \geq 3$, simultaneous confidence intervals for several different families of functions involving $p_1, \ldots, p_T$ are of interest. Three examples of such families are (1) the T individual success probabilities $\{p_i : 1 \leq i \leq T\}$, (2) all pairwise differences $\{\Delta_{ij} = p_i - p_j\colon 1 \leq i \neq j \leq T\}$, and (3) all pairwise differences with a standard (known success probability) or control (unknown success probability) $\{\Delta_{i0} = p_i - p_0\colon 1 \leq i \leq T\}$ where p_0 is the success probability of the control or standard. Obviously, pairwise comparisons might also be of interest in terms of relative risks $\rho_{ij} := p_i/p_j$, or odds ratios $\psi_{ij} = p_i(1-p_j)/(1-p_i)p_j$ rather than differences.

In large samples, approximate simultaneous confidence intervals can be constructed for all these problems by applying normal theory procedures, appealing to the fact that when $\min\{m_i\}$ is large, $\hat{\mathbf{p}}$ is approximately distributed as $N_T[\mathbf{p}, \mathbf{\Sigma}(\hat{\mathbf{p}})]$ where $\mathbf{\Sigma}(\mathbf{p}) := \text{Diag}\{\ldots, p_i(1-p_i)/m_i, \ldots\}$. For example, if p_0 is a known standard then

$$p_i - p_0 \in \hat{p}_i - p_0 \pm z_{\alpha/2T}(\hat{p}_i(1-\hat{p}_i)/m_i)^{1/2} \tag{5.2.24}$$

are obvious asymptotic $(1-\alpha)$-simultaneous confidence intervals for $\{p_i\}$. As another example applying the Tukey–Kramer procedure yields,

$$\Delta_{ij} \in \hat{p}_i - \hat{p}_j \pm C\left(\{\hat{p}_i(1-\hat{p}_i)/m_i + \hat{p}_j(1-\hat{p}_j)/m_j\}/2\right)^{1/2} \tag{5.2.25}$$

as simultaneous confidence intervals for all pairwise differences Δ_{ij} where C is the upper α percentile of the distribution of the range of T iid $N(0,1)$ random variables (Hochberg and Tamhane, 1987, Chapter 10). Intervals for families of more complicated contrasts among the $\{p_i\}$ can be formed by applying Scheffe's projection technique (Problem 5.14). Similarly, simultaneous confidence intervals can be constructed for the family of individual success probabilities using the maximum modulus distribution, and for all comparisons with a standard by applying Dunnett's procedure. Asymptotic intervals for more complicated families such as $\{\rho_{ij}\}$ and $\{\psi_{ij}\}$ can be derived from Scheffe's projection method and the variance formulae (5.2.15) and (5.2.18). Hochberg and Tamhane (1987) give the derivations of these normal theory intervals.

When forming large sample intervals based on normal approximations, several points should be noted. First, applying the (normality enhancing) Arcsine transformation is helpful in some cases. In particular since $\text{Arcsine}(\sqrt{p_i})$ is monotone in p_i, this transformation is appropriate for forming simultaneous confidence intervals for $\{p_i\}$; it is not useful in forming confidence intervals for $\{\Delta_{ij}\}$ since the difference $(\text{Arcsine}\sqrt{p_i} - \text{Arcsine}\sqrt{p_j})$ is not monotonically related to Δ_{ij}. Second, more sophisticated variance estimators of $\hat{\Delta}_{ij}$ such as that of Hauck and Anderson given in (5.2.13) presumably yield intervals with achieved coverage closer to their nominal values.

With one exception, small sample simultaneous confidence intervals are much more difficult to construct for the families above. The exception is that simultaneous $100(1-\alpha)\%$ confidence intervals for the family $\{p_i\}$ can be formed from individual $100(1-\alpha/T)\%$ Sterne–Crow intervals for each p_i because of the independence of the $\hat{p}_i$. Otherwise the only general method guaranteed to achieve a given confidence level is Bonferroni's method. For example, constructing confidence intervals for all treatment comparisons ψ_{ij} requires forming individual $100(1-\alpha/\binom{T}{2})\%$ small sample confidence intervals from (5.2.23).

Example 5.2.5. The data in Table 5.2.7 are extracted from Carp and Rowland's (1983) study of 27,772 Federal District Court opinions divided according to the type of case. Focusing only on decisions handed down between 1969 and 1977 and on 5 of the 20 case categories, the number of liberal decisions and the total number of decisions are given. For purposes of illustration, these data will be used to determine simultaneous confidence intervals for the proportions of liberal decisions in the 5 types of cases. In fact, a critical examination of the data may suggest that the binomial assumption may not hold as the cases in each category are accumulated over time and space and the probability of a liberal decision may vary with these factors as well as type of case. Nevertheless Table 5.2.8 lists the 95% simultaneous confidence intervals (5.2.25) for all pairwise comparisons between case categories of the proportions of liberal decisions. Five of the 10 intervals do not cover zero indicating a significant difference in the proportion of liberal opinions for these comparisons. Three of these significant differences involve cases on convictions for a criminal offense; they indicate that liberal decisions are less likely in such cases than in cases of types 2, 4, or 5.

Table 5.2.7. Numbers of Liberal Decisions by Federal District Courts During the Years 1969–1977 (Reprinted with permission from Robert A. Carp and C.K. Rowland: *Policymaking and Politics in the Federal District Courts,* 1983, The University of Tennessee Press, Knoxville, Tennessee.)

i	Type of Case	Liberal Decisions	Total Decisions	$\hat{p}_i$
1	Conviction for a Criminal Offense	104	267	.39
2	Fair Labor Standard Act	224	373	.60
3	Union and Company	202	440	.46
4	Women's Rights	136	256	.53
5	Freedom of Religion	222	353	.63

Table 5.2.8. Simultaneous Confidence Intervals (5.2.25) for All Pairwise Differences Δ_{ij} of Proportions of Liberal Decisions by Federal District Courts During 1969–1977

i	j	$\hat{\Delta}$	$\underline{\Delta}(i,j)$	$\overline{\Delta}(i,j)$
1	2	−.210	−.317	−.103*
1	3	−.070	−.174	.034
1	4	−.140	−.258	−.022*
1	5	−.240	−.348	−.133*
2	3	.140	.045	.235*
2	4	.070	−.040	.180
2	5	−.030	−.129	.069
3	4	−.070	−.177	.037
3	5	−.170	−.266	−.074*
4	5	−.100	−.210	.010

*Interval does not cover zero.

D. Selection and Ranking

In some applications such as the design and analysis of clinical trials, selecting the treatment with the highest probability of curing the disease is a very relevant formulation. Both the indifference zone and the subset selection approaches have been applied to the selection of binomial populations with large p_i's. The symmetric problem of selecting Π_i's with small p_i's will not be discussed. Because many of the applications are in medical experiments, ethical considerations have motivated the development of sequential selection procedures which minimize the criteria (i) the expected total number of patients given the worst treatments or (ii) the expected total number of failures over the trial. The paragraphs that follow give a brief introduction to this literature and provide references where details are available.

Sobel and Huyett (1957) introduced the indifference zone formulation for the binomial selection problem. Let

$$p_{[1]} \leq \cdots \leq p_{[T]}$$

denote the ordered true success probabilities. Given $\delta > 0$ and $0 < \alpha < 1$, their goal was to select the population associated with $p_{[T]}$ subject to the requirement:

$$P_{\mathbf{p}}[\text{Correct Selection}] \geq 1 - \alpha \tag{5.2.26}$$

whenever the true $\mathbf{p} \in \Omega(\delta) := \{\mathbf{p}: p_{[T]} - p_{[T-1]} \geq \delta\}$. They considered the following intuitive procedure.

Procedure $\mathcal{P}_{SH}$. Observe a common number of Bernoulli trials $m_1 = \ldots = m_T = m$, say, from each population. Select the Π_i associated with $\max\{\hat{p}_i\}$

as the best population, using randomization to break ties among populations Π_j for which $\hat{p}_j = \max\{\hat{p}_i\}$.

Sobel and Huyett prove that for given m, the lowest probability of correct selection over vectors $\mathbf{p} \in \Omega(\delta)$ is for configurations satisfying

$$p_{[1]} = \ldots = p_{[T-1]} = p_{[T]} - \delta. \tag{5.2.27}$$

However, unlike the multinomial case where the extra constraint $\Sigma_i p_i = 1$ holds, (5.2.27) does not uniquely determine the least favorable configuration in $\Omega(\delta)$. Sobel and Huyett table both exact and asymptotic sample sizes m achieving (5.2.26). When the probabilities p_i are bounded a priori, the sample sizes required to satisfy (5.2.26) can be decreased.

Chapter 4 of Gupta and Panchapakesan (1979) summarizes other work on binomial indifference zone approaches including alternative procedures, modified probability requirements, curtailment, and adaptive sampling. Bechhofer (1985) and Kulkarni and Kulkarni (1987) survey recent enhancements to the basic Sobel and Huyett procedure.

Gupta and Sobel (1960) proposed a single-stage procedure for selecting a *subset* of $\Pi_1, \ldots, \Pi_T$ containing the Π_i associated with $p_{[T]}$ and satisfying the requirement

$$P_{\mathbf{p}}[\text{Correct Selection}] \geq 1 - \alpha \tag{5.2.28}$$

for all $\mathbf{p} \in \Omega := X_{i=1}^{T}(0,1)$. The event "Correct Selection" means that the subset contains the population associated with $p_{[T]}$.

Procedure $\mathcal{P}_{GS}$. Observe a common number m of Bernoulli trials from each Π_i. Place Π_i in the selected subset if and only if

$$Y_i \geq \max\{Y_j : 1 \leq j \leq T\} - d.$$

Assuming the common sample size m is given, Gupta and Sobel consider the determination of d to satisfy (5.2.28). They prove that $\inf_{\Omega} P_{\mathbf{p}}$ [Correct Selection] occurs when

$$p_{[1]} = \ldots = p_{[T]}. \tag{5.2.29}$$

Thus, as with the indifference zone approach, the least favorable configuration in Ω must be determined numerically among those satisfying (5.2.29). Gupta and Panchapakesan (1979) describe other subset selection procedures. Sanchez (1987) uses curtailment and adaptive sampling to derive a sequential procedure which chooses the same subset as $\mathcal{P}_{GS}$, achieves the same probability requirement (5.2.28), and uses uniformly fewer total observations.

5.3 Logistic Regression: Basic Techniques

A. Introduction

This section and the next consider the analysis of data of the form $\{(Y_i, m_i, \mathbf{x}_i)\colon 1 \leq i \leq T\}$ where the Y_i are mutually independent $B(m_i, p_i)$ random variables, $\mathbf{x}_i' = (x_{i1}, \ldots, x_{ik})$ is a vector of nonstochastic covariates, and $p_i = p(\mathbf{x}_i, \boldsymbol{\beta})$ satisfies the logistic model

$$p(\mathbf{x}, \boldsymbol{\beta}) = \exp\{\mathbf{x}'\boldsymbol{\beta}\}/(1 + \exp\{\mathbf{x}'\boldsymbol{\beta}). \tag{5.3.1}$$

Let $N := \sum_{i=1}^{T} m_i$ denote the total number of Bernoulli trials, $\mathbf{m} := (m_1, \ldots, m_T)'$ be the vector of the numbers of trials, and $\mathbf{X}$ denote the $T \times k$ design matrix with rows $\mathbf{x}_i'$. It is assumed that $\mathbf{X}$ has full column rank k.

One subtle change in notation should be emphasized at the outset. In Chapters 3 and 4, the symbol $\mathbf{X}$ was used to denote a matrix whose columns spanned the linear space of logmeans of the responses. Recall from Example 3.1.3 that if the *enlarged data* are defined as the set $\{(Y_i, m_i - Y_i)\}_{i=1}^{T}$ of $2T$ successes *and* failures, then assuming a logistic model for the success data $\{Y_i\}_{i=1}^{T}$ yields a loglinear model for the enlarged data. The relationship between the $T \times k$ design matrix $\mathbf{X}$ in this chapter and the corresponding $2T \times (T + k)$ matrix for the enlarged data is given in Example 3.1.3. The columns of the latter matrix span the linear space of the logmeans of the enlarged data set.

Section 5.1 gives examples of logistic models for the data in Examples 1.2.4, 1.2.6, and 5.1.1. Examples 5.1.1 and 5.3.1 will be used to illustrate the material of this section.

Example 5.3.1. Table 5.3.1 from Lee (1974) lists the remission STATUS (1 = in remission/0 = relapsed) of 27 acute myeloblastic leukemia (AML) patients together with two covariates culled from a larger set of potential prognostic variables. The covariate LI is the "percent labeling index;" it is a pretreatment measure of the "proliferation fraction" (the proportion of cells undergoing DNA synthesis in the presence of chemotheraphy times 10). The covariate TEMP is also available prior to treatment; TEMP is the highest recorded patient temperature (in °F divided by 100) prior to start of chemotherapy. In contrast to Example 5.1.1, where the data come from a designed experiment, these data are observational. Lee (1974) contains a definitive analysis of the full data set of which these are a subset.

Table 5.3.1. Remission Status of Twenty-Seven AML Patients (Reprinted with permission from Elisa T. Lee: "A Computer Program for Linear Logistic Regression Analysis," *Computer Programs in Biomedicine,* 1974, North-Holland Publishing Co., Amsterdam, The Netherlands.)

STATUS	LI	TEMP
1	1.9	.996
1	1.4	.992
0	.8	.982
0	.7	.986
1	1.3	.980
0	.6	.982
1	1.0	.992
0	1.9	1.020
0	.8	.990
0	.5	1.038
0	.7	.988
0	1.2	.982
0	.4	1.006
0	.8	.990
0	1.1	.990
1	1.9	1.020
0	.5	1.014
0	1.0	1.004
0	.6	.990
1	1.1	.986
0	.4	1.010
0	.6	1.020
1	1.0	1.002
0	1.6	.988
1	1.7	.990
1	.9	.986
0	.7	.986

One other aspect about notation deserves comment. No assumption has been made about the uniqueness of the covariates $\mathbf{x}_i$; i.e., about whether there are T distinct design points. The data of Example 5.1.1 have unique covariates whereas that of Example 5.3.1 do not. Even if the raw data have distinct covariates, logistic models formed from a subset of the data can lead to nondistinct $\mathbf{x}_i$. The likelihood function (see Equation 5.3.4) is, aside from constants, invariant with respect to grouping of responses with common covariate vectors. However, the validity of many techniques based on the likelihood function depends on the number of distinct $\mathbf{x}_i$ and the size of the m_i. In those cases where methods assume large m_i it is important to group responses if appropriate.

The asymptotic properties of the procedures discussed in this section have been studied in two different settings which are outlined below; the regularity assumptions required by each formulation are given in the appropriate references. Haberman (1973b, 1977b) and Fahrmeir and Kaufmann (1985, 1986) consider *standard asymptotics* in which $N = \sum_{i=1}^{T} m_i \to \infty$ and the number k of parameters is fixed. Standard asymptotics is usually further subdivided into two cases: (i) the *large strata model* in which T is fixed and $m_i/N \to c_i \in (0,1)$ for $i = 1(1)T$, and (ii) the *many strata model* in which $T \to \infty$. In the many strata model, the sequence of m_i can be bounded. Thinking of the enlarged data set and the corresponding loglinear model induced by (5.3.1), the many strata model reflects the notion that the number of nuisance parameters (T) grows but the number of parameters of interest is fixed (see also Section 5.5). Haberman (1977b) and Lindsay (1983) address *large sparse asymptotics* in which both $N \to \infty$ *and* $k \to \infty$; implicit in large sparse asymptotics is $T \to \infty$ since k (= rank of $\mathbf{X}$) $\leq T$. Large sparse asymptotics reflect the notion that additional covariates are investigated as the quantity of data increases. In the sequel when large sample results are given, they will hold for both models of standard asymptotics but not for large sparse asymptotics unless otherwise noted.

The remainder of this section is organized as follows. First, two commonly-used likelihood based methods for analyzing logistic regression data are described. Estimation, testing, confidence band construction, and omnibus goodness-of-fit tests are considered. The methods are illustrated using Examples 5.1.1 and 5.3.1. Section 5.4 discusses recent developments for analyzing logistic regression data such as alternative estimation methods and case diagnostics.

B. Likelihood-Based Inference

The two most frequently used methods for making inferences about β based on logistic regression data are *weighted least squares* (WLS) (Berkson, 1944; Grizzle, Starmer, and Koch, 1969; and Cox, 1970, Chapter 3 and Sections 6.1–6.3) and *maximum likelihood* (ML) (Cox, 1970, Sections 6.4–6.6).

Weighted Least Squares Estimation

The idea of the WLS approach is as follows. Define empiric logits

$$W_i := \ln\left[\frac{Y_i + \frac{1}{2}}{m_i - Y_i + \frac{1}{2}}\right], \quad 1 \leq i \leq T$$

for which

$$\mathbf{W} = (W_1, \ldots, W_T)' \overset{app}{\sim} N[\mathbf{X}\beta, \hat{\Sigma}] \tag{5.3.2}$$

where

$$\hat{\Sigma} = \text{Diag}\left(\ldots, \frac{1}{m_i \hat{p}_i(1-\hat{p}_i)}, \ldots\right) \tag{5.3.3}$$

with $\hat{p}_i = Y_i/m_i$, $1 \le i \le T$ (Chapter 2.1). Standard weighted linear regression techniques are then used to compare models and make inferences about the $\boldsymbol{\beta}$ vector. The WLS estimate of $\boldsymbol{\beta}$ minimizes

$$\sum_{i=1}^{T} m_i \hat{p}_i (1-\hat{p}_i)[W_i - \mathbf{x}_i'\boldsymbol{\beta}]^2;$$

hence Berkson (1944) called this estimator the *minimum logit chi-squared estimator.* Since the usual unknown scale parameter σ is unity in this WLS model, output from most normal theory regression software will not give correct estimates of standard errors and test statistics.

The WLS method has traditionally been justified on the basis of the large strata model ($N \to \infty$, T and k fixed) in which it is consistent, asymptotically normal, and efficient. Davis (1985) studies the behavior of the WLS (and related) estimator(s) under the many strata standard asymptotic model and determines that (i) the estimator need not be consistent and (ii) there is no simple correction to make it consistent. Discussion of the small sample bias and related properties of the WLS estimator of $\boldsymbol{\beta}$ is given in Gart, Pettigrew, and Thomas (1985, 1986) and the references therein. In practice this method is recommended only for situations with reasonably large values of m_i.

Maximum Likelihood Estimation

The maximum likelihood approach is based on the likelihood function

$$\begin{aligned} L(\boldsymbol{\beta}) &= \prod_{i=1}^{T} \binom{m_i}{y_i} p_i^{y_i}(1-p_i)^{m_i - y_i} \\ &= \left\{ \prod_{i=1}^{T} \binom{m_i}{y_i} \right\} \exp\left(\sum_{i=1}^{T} y_i \mathbf{x}_i'\boldsymbol{\beta} \right) \Big/ \prod_{i=1}^{T} \{1+\exp(\mathbf{x}_i'\boldsymbol{\beta})\}^{m_i}. \qquad (5.3.4) \end{aligned}$$

The loglikelihood and its derivatives are easily calculated from Equation (5.3.4) using the results in Appendix A.2 as

$$\ln L = \mathbf{Y}'\mathbf{X}\boldsymbol{\beta} - \sum_{i=1}^{T} \left\{ m_i \ln[1+\exp(\mathbf{x}_i'\boldsymbol{\beta})] + \ln \binom{m_i}{y_i} \right\},$$

$$\begin{aligned} \nabla \ln L &= \mathbf{X}'\mathbf{Y} - \sum_{i=1}^{T} \frac{m_i \exp(\mathbf{x}_i'\boldsymbol{\beta})}{1+\exp(\mathbf{x}_i'\boldsymbol{\beta})} \mathbf{x}_i \\ &= \mathbf{X}'\mathbf{Y} - \sum_{i=1}^{T} m_i p_i \mathbf{x}_i \\ &= \mathbf{X}'\mathbf{Y} - \mathbf{X}' E_{\boldsymbol{\beta}}[\mathbf{Y}], \text{ and} \end{aligned}$$

$$\nabla^2 \ln L = -\mathbf{X}'\mathbf{D}\mathbf{X} \tag{5.3.5}$$

where $\mathbf{D} := \text{Diag}(\ldots, m_i p_i(1-p_i), \ldots)$ is a $T \times T$ diagonal matrix.

It is easy to check that $\ln L(\boldsymbol{\beta})$ is strictly concave in $\boldsymbol{\beta}$ since $\nabla^2 \ln L(\boldsymbol{\beta})$ is positive definite for every $\boldsymbol{\beta} \in \mathbb{R}^k$. Hence if the MLE exists, it must satisfy

$$\nabla \ln L(\hat{\boldsymbol{\beta}}) = \mathbf{0}_k \Longleftrightarrow \mathbf{X}'(\mathbf{Y} - \hat{\mathbf{Y}}) = \mathbf{0}_k \tag{5.3.6}$$

where $\hat{\mathbf{Y}} = (\ldots, \hat{Y}_i := m_i \exp(\mathbf{x}_i'\hat{\boldsymbol{\beta}})/\{1 + \exp(\mathbf{x}_i'\hat{\boldsymbol{\beta}})\}, \ldots)'$ is the $T \times 1$ estimated mean vector of $\mathbf{Y}$. Equation (5.3.6) can also be derived from Proposition 3.3.1 by starting with the enlarged set of $2T$ successes and failures.

The general discussion in Section 3.2 shows that the likelihood equations (5.3.6) need not have a solution $\hat{\boldsymbol{\beta}}$ as the log likelihood need not achieve a maximum. Silvapulle (1981), Albert and Anderson (1984), and Santner and Duffy (1986) give necessary and sufficient conditions directly in terms of $\{(Y_i, m_i, \mathbf{x}_i)\colon 1 \le i \le T\}$ for the existence of $\hat{\boldsymbol{\beta}}$. These conditions have an easy intuitive description. Suppose first that $Y_i = m_i$ or 0 for all $1 \le i \le T$. For example, in the not uncommon situation of $m_i = 1$ for $1 \le i \le T$ (no replication), the latter assumption is true. Suppose further that there exists a $\boldsymbol{\beta}^* \in \mathbb{R}^k$ such that

$$\mathbf{x}_i'\boldsymbol{\beta}^* \left\{ \begin{matrix} > \\ < \end{matrix} \right\} 0 \quad \text{as} \quad Y_i = \left\{ \begin{matrix} m_i \\ 0 \end{matrix} \right\} \tag{5.3.7}$$

for $i = 1(1)T$. Geometrically this means there exists a hyperplane in covariate space such that the covariates corresponding to $Y_i = 0$ all lie on one side of the hyperplane while covariates corresponding to $Y_i = m_i$ all lie on the opposite side. Consider the classification rule which guesses Y_i is m_i (0) when the probability $p_i^* := \exp(\mathbf{x}_i'\boldsymbol{\beta}^*)/\{1 + \exp(\mathbf{x}_i'\boldsymbol{\beta}^*)\}$ corresponding to $\boldsymbol{\beta}^*$ satisfies $p_i^* > (<)\ 1/2$ for $i = 1(1)T$. By (5.3.7), this rule gives a perfect classification of the Y_i as m_i or 0. It can be shown that the likelihood (5.3.4) approaches its global supremum (of unity) along the ray $\{\alpha\boldsymbol{\beta}^* : \alpha > 0\}$ and hence the MLE of $\boldsymbol{\beta}$ does not exist.

A similar phenomenon occurs if a weaker version of (5.3.7) holds. In particular suppose that $Y_i = 0$ or m_i for some $1 \le i \le T$ and that there exists $\boldsymbol{\beta}^* \in \mathbb{R}^k$ satisfying

$$\mathbf{x}_i'\boldsymbol{\beta}^* \left\{ \begin{matrix} \ge \\ \le \\ = \end{matrix} \right\} 0 \quad \text{for} \quad \left\{ \begin{matrix} Y_i = m_i \\ Y_i = 0 \\ 0 \le Y_i \le m_i \end{matrix} \right\} \tag{5.3.8}$$

with strict inequality for some i, $1 \le i \le T$. Then one or more of the factors of the likelihood (5.3.4) can be driven to unity and again the MLE of $\boldsymbol{\beta}$ does not exist.

If neither (5.3.7) nor (5.3.8) holds for any $\boldsymbol{\beta}^*$ then the MLE of $\boldsymbol{\beta}$, denoted $\hat{\boldsymbol{\beta}}$, exists and $\hat{\mathbf{p}} = \mathbf{p}(\hat{\boldsymbol{\beta}})$ is the MLE of $\mathbf{p}$. The problem of determining whether $\hat{\beta}$ exists can be formulated as a linear programming problem similar to that described in Section 3.2. Silvapulle (1981) discusses the existence of the MLE of β for other link functions $F(\cdot)$ in the general binary regression model (5.1.2) (Problem 5.23).

Although the Iterative Proportional Fitting algorithm (Section 4.4) can be used to calculate $\hat{\mathbf{Y}}$ (apply it to the enlarged table of successes and failures), the Newton–Raphson method is usually employed to compute $\hat{\beta}$ directly. In this case Newton–Raphson is a sequence of weighted regressions; each update is similar to that in (4.4.1) where the "working" response vector and weights depend on the previous guess of β. The initial iteration uses the response vector $\mathbf{W}$ and weights $[m_i\hat{p}_i(1-\hat{p}_i)]^{-1}$ given in Equations (5.3.2) and (5.3.3) (see Problem 5.18).

Study of the small sample properties of $\hat{\beta}$ is hampered by its lack of a closed-form expression. Equation (5.3.6) shows that $\hat{\beta}$ has several intuitive types of invariance. If successes and failures are relabeled so that $\mathbf{Y} \to \mathbf{m} - \mathbf{Y}$ then $\hat{\beta} \to -\hat{\beta}$ (hence the estimated vector of success probabilities and means are correspondingly switched $\hat{\mathbf{p}} \to 1-\hat{\mathbf{p}}$ and $\hat{\mathbf{Y}} \to \mathbf{m}-\hat{\mathbf{Y}}$). If any single covariate is rescaled; i.e., if $x_{ij} \to \alpha x_{ij}$ for all observations $i = 1(1)T$ for some $\alpha \neq 0$, then $\hat{\beta}_j \to \hat{\beta}_j/\alpha$ (hence $\hat{\mathbf{p}}$ and $\hat{\mathbf{Y}}$ remain unchanged). The MLE is biased for any fixed sample size. Duffy and Santner (1987b) study the bias of $\hat{\beta}$ (given $\hat{\beta}$ exists) for a series of 45 test problems with $m_i \equiv 1$ and values of T and k ranging from $(T,k) = (30,2)$ to $(T,k) = (200,5)$. Their results show that $\hat{\beta}_j$ generally has positive (negative) bias for $\beta_j >$ $(<)\ 0$. In words, $\hat{\beta}$ systematically "pulls" too far from the origin.

A number of authors have studied the summed mean squared error (MSE) of $\hat{\mathbf{p}}$: $R(\beta,\hat{\mathbf{p}}) := \sum_{i=1}^{T} E[\{\hat{p}_i - p(\mathbf{x}_i,\beta)\}^2 \mid \beta]$, and $\hat{\beta}$: $R_{\beta}(\beta,\hat{\beta}) := \sum_{j=1}^{k} E[\{\hat{\beta}_j - \beta_j\}^2 \mid \beta]$. The following example typifies the results found in Duffy and Santner (1987a and 1989).

Example 5.3.2. Suppose $T = 5$, $k = 2$, $m_i = 1$ for $1 \leq i \leq 5$, and

$$\mathbf{X} = \begin{bmatrix} 1 & -1 \\ 1 & 0 \\ 0 & 2 \\ -2 & 1 \\ 1 & 1 \end{bmatrix}.$$

Figure 5.3.1 is the linear interpolation of a 3-dimensional plot of $(\beta_1, \beta_2, R(\beta,\hat{\mathbf{p}}))$ for β_1, β_2 each in the range $-4(1/3)4$. The MSE $R(\beta,\hat{\mathbf{p}})$ is symmetric with respect to reflections through the origin. Further $R(\mathbf{0}_2,\hat{\mathbf{p}})$ is the global maximum and the MSE decreases as the norm $||\beta|| = \{\sum_{i=1}^{2} \beta_i^2\}^{1/2}$ increases. More precisely, it can be shown that for all but a finite set of directions β on the unit circle ($||\beta|| = 1$), the risk $R(c\beta,\hat{\mathbf{p}}) \to 0$ as $c \to \infty$.

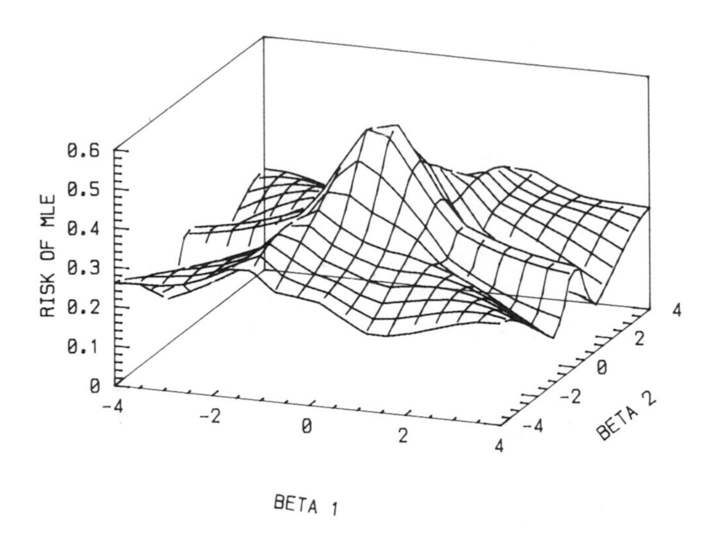

Figure 5.3.1. Mean squared error for estimating $\mathbf{p}$ in Example 5.3.2. (Reprinted with permission from S. Gupta and J. Berger: *Statistical Decision Theory and Related Topics IV*, 1988, Springer-Verlag, New York.)

In general, Duffy and Santner (1987a) show that the MSE is invariant with respect to reflections through the origin and goes to zero along all but a finite number of directions through the origin (Problem 5.19). This phenomenon of the MLE of $\mathbf{p}$ having greatest MSE for "central" $\mathbf{p}$ values and small MSE for "extreme" $\mathbf{p}$ values was previously observed in Sections 2.1 and 2.2 in the study of the binomial and multinomial distributions. In the latter two settings the MLE is in fact admissible; whether this holds in the logistic case is an open question.

Asymptotically much more is known about $\hat{\boldsymbol{\beta}}$. One class of results are moment expansions of $\hat{\boldsymbol{\beta}}$. For example, Schaefer (1983) gives a T^{-1}-order bias expression b_T

$$E[\hat{\boldsymbol{\beta}}] = \boldsymbol{\beta} + b_T + o(1/T). \tag{5.3.9}$$

under large strata model asymptotics. Another example is Amemiya (1980) who derives expressions for the T^{-2}-order MSE of both the MLE and WLS estimators of $\boldsymbol{\beta}$ in the same case. Amemiya (1980) and Davis (1984) give examples in which the WLS estimator has smaller T^{-2}-order MSE than the MLE and conversely, showing that in large strata cases neither method uniformly dominates the other with respect to this criterion.

A second class of results concerns the asymptotic distribution and efficiency of $\hat{\boldsymbol{\beta}}$. While many authors have studied this, Fahrmeir and Kaufmann (1985) subsumes most of the previous work. Fahrmeir and Kaufmann (1985) consider standard asymptotics for regression problems with arbitrary exponential family responses. Their conditions are easily stated and interpreted in the logistic regression case. Consider a sequence of problems the Nth of which is obtained from the $(N-1)$st by adding a single binary trial with covariate vector $\mathbf{x}_N$ (so that $m_i = 1$). In this formulation $\mathbf{x}_N$ is permitted to be a repeat of a previous $\mathbf{x}_i$. Let $\mathbf{X}_N$ denote the $N \times k$ design matrix for the Nth problem and $\mathbf{D}_N$ the $N \times N$ diagonal matrix with elements $(\ldots, p(\mathbf{x}_i)\{1-p(\mathbf{x}_i)\}, \ldots)$; then $\mathbf{X}'_N\mathbf{D}_N\mathbf{X}_N$ is the Fisher information about $\boldsymbol{\beta}$ in the Nth problem. If

(i) the minimum eigenvalue of $\mathbf{X}'_N\mathbf{D}_N\mathbf{X}_N \to \infty$ as $N \to \infty$, and (5.3.10)

(ii) $\mathbf{x}'_N(\mathbf{X}'_N\mathbf{D}_N\mathbf{X}_N)^{-1}\mathbf{x}_N \to 0$ as $N \to \infty$ (5.3.11)

then $\hat{\boldsymbol{\beta}}$ is strongly consistent, asymptotically normal, and efficient. Intuitively, (5.3.10) guarantees that the data contains sufficient information to estimate $\boldsymbol{\ell}'\hat{\boldsymbol{\beta}}$ efficiently for the "hardest" direction $\boldsymbol{\ell}$; i.e., for $\boldsymbol{\ell}$ proportional to an eigenvector corresponding to the *minimum* eigenvalue of $\mathbf{X}'_N\mathbf{D}_N\mathbf{X}_N$. The expression on the left-hand side of (5.3.11) is the asymptotic variance of $\mathbf{x}'_N\hat{\boldsymbol{\beta}}$; this condition assures that no single $\mathbf{x}_N$ plays too dominant a role in estimating $\boldsymbol{\beta}$. Furthermore since

$$\mathbf{x}'_N(\mathbf{X}'_N\mathbf{D}_N\mathbf{X}_N)^{-1}\mathbf{x}_N \leq ||\mathbf{x}_N||^2 \text{ (min eigenvalue of } \mathbf{X}'_N\mathbf{D}_N\mathbf{X}_N)^{-1},$$

(5.3.11) automatically holds under (5.3.10) when the regressors $\mathbf{x}_N$ come from a bounded set. The general condition (5.3.11) allows regressors to become unbounded. Under mild conditions, the negative of the Hessian matrix evaluated at $\hat{\boldsymbol{\beta}}$ is a consistent estimate of the inverse of the asymptotic covariance matrix of $\hat{\boldsymbol{\beta}}$ so that approximately for large samples

$$\hat{\boldsymbol{\beta}} \sim N[\boldsymbol{\beta}, \mathbf{V}(\hat{\boldsymbol{\beta}})] \tag{5.3.12}$$

where $\mathbf{V}(\boldsymbol{\beta}) = [\mathbf{X}'\mathbf{D}(\boldsymbol{\beta})\mathbf{X}]^{-1}$.

Equation (5.3.12) forms the basis of one large-sample test of H_0: $\beta_j = 0$. Using the fact that the square roots of the diagonal elements of $\mathbf{V}(\hat{\boldsymbol{\beta}})$ are approximate standard errors of the estimated coefficients $\hat{\beta}_j$, the Wald test compares the ratio of $\hat{\beta}_j$ and its standard error to a standard normal critical point. Two other asymptotically equivalent chi-square test statistics are the likelihood ratio statistic

$$2\{\ln L(\hat{\boldsymbol{\beta}}) - \ln L(\hat{\boldsymbol{\beta}}^0)\} = 2\sum_{i=1}^{T}\left[Y_i \ln\left\{\frac{\hat{p}_i}{\hat{p}_i^0}\right\} + (m_i - Y_i)\ln\left\{\frac{1-\hat{p}_i}{1-\hat{p}_i^0}\right\}\right]$$

where $\hat{\boldsymbol{\beta}}^0 = (\hat{\beta}_1^0, \ldots, \hat{\beta}_{j-1}^0, 0, \hat{\beta}_{j+1}^0, \ldots, \hat{\beta}_k^0)'$ ($\hat{\mathbf{p}}^0 := \mathbf{p}(\hat{\boldsymbol{\beta}}^0)$) is the MLE of $\boldsymbol{\beta}$ ($\mathbf{p}$) under H_0, and the score test statistic

$$[\mathbf{S}(\hat{\boldsymbol{\beta}}^0)]'[\mathbf{X}'\mathbf{D}(\hat{\boldsymbol{\beta}}^0)\mathbf{X}]^{-1}\mathbf{S}(\hat{\boldsymbol{\beta}}^0)$$

where $\mathbf{S}(\boldsymbol{\beta}) := \nabla \ln L(\boldsymbol{\beta}) = \mathbf{X}'(\mathbf{Y} - \hat{\mathbf{Y}})$ is the $k \times 1$ score vector. The latter only requires calculation of the restricted MLE $\hat{\boldsymbol{\beta}}^0$. Both the likelihood ratio and score tests have asymptotic null χ_1^2 distribution. All three tests can be generalized to test that a subset of $\boldsymbol{\beta}$ variables is zero or some other given vector (Problem 5.20).

Brand, Pinnock, and Jackson (1973) and Hauck (1983) construct large sample simultaneous confidence bands for the underlying success probability function $p(\mathbf{x}) = [1 + \exp\{-\mathbf{x}'\boldsymbol{\beta}\}]^{-1}$ over $\mathbf{x} \in \mathbb{R}^k$. Brand et al. consider only the simple linear logistic model $\beta_0 + x\beta_1$ while Hauck considers the general case. In addition, Brand et al. determine interval estimates of the inverse logistic curve; i.e., given a probability $p_0 \in (0, 1)$ they give an interval estimate for $\{x \in \mathbb{R}^1 \colon \operatorname{logit}(p_0) = \beta_0 + x\beta_1\}$. Carter et al. (1986) consider the problem of determining an asymptotic confidence region for the inverse logistic response surface for more general multivariate logistic models.

Hauck's proposal for constructing asymptotic joint confidence bands for $p(\mathbf{x})$ is to first observe that equation (5.3.12) implies that $(\hat{\boldsymbol{\beta}} - \boldsymbol{\beta})'[\mathbf{V}(\hat{\boldsymbol{\beta}})]^{-1}(\hat{\boldsymbol{\beta}} - \boldsymbol{\beta})$ is approximately distributed χ_k^2. This in turn implies that asymptotically

$$P[\mathbf{x}'\boldsymbol{\beta} \in \mathbf{x}'\hat{\boldsymbol{\beta}} \pm \{\mathbf{x}'\mathbf{V}(\hat{\boldsymbol{\beta}})\mathbf{x}\}^{1/2} \times \{\chi^2_{\alpha,k}\}^{1/2} \text{ for all } \mathbf{x} \in \mathbb{R}^k] \geq 1 - \alpha. \quad (5.3.13)$$

Equation (5.3.13) can be used to form simultaneous confidence intervals for the logistic probabilities $p(\mathbf{x})$ over $\mathbf{x} \in \mathbb{R}^k$ since the logistic distribution function $F(w) = [1 + \exp(-w)]^{-1}$ is strictly increasing in w. In practice, equation (5.3.13) is often used to form asymptotically conservative intervals for $\mathbf{x}'\boldsymbol{\beta}$ over $\mathbf{x}$ in a finite set. For example, it can be used to form conservative confidence intervals for

$$\{p(a) = [1 + \exp(-[a\mathbf{x}_0 + \mathbf{b}]'\boldsymbol{\beta})]^{-1} \colon a \in \mathbb{R}^1\},$$

where $\mathbf{x}_0$ and $\mathbf{b}$ are fixed in $\mathbb{R}^k$. In words, these bands are for the success probability along the ray $a\mathbf{x}_0 + \mathbf{b}$. The conservatism stems from the fact that (5.3.13) holds for all $\mathbf{x} \in \mathbb{R}^k$ whereas the confidence band considers only one particular direction. If one is interested in joint confidence intervals for $p(\mathbf{x})$ at a finite set of g covariate values, then an alternative conservative method is to use Bonferroni intervals obtained by replacing $\{\chi^2_{\alpha,k}\}^{1/2}$ by $z_{\alpha/2g}$. If confidence intervals are desired for covariates in the rectangular region

$$\{\mathbf{x} \in \mathbb{R}^k \colon a_i \leq x_i \leq b_i, i = 1(1)k\},$$

Piegorsch and Casella (1988) consider adaptations of the Casella and Strawderman (1980) normal theory rectangular regions and also of the Gafarian (1964) fixed-width bands.

The basic assumption underlying the large-sample tests and confidence intervals described above is that the loglikelihood function is well approximated by a second order Taylor expansion about $\hat{\boldsymbol{\beta}}$, i.e.,

$$\ln(L(\boldsymbol{\beta})) \simeq \ln(L(\hat{\boldsymbol{\beta}})) - \frac{1}{2}(\hat{\boldsymbol{\beta}} - \boldsymbol{\beta})'[\mathbf{V}(\hat{\boldsymbol{\beta}})]^{-1}(\hat{\boldsymbol{\beta}} - \boldsymbol{\beta}).$$

Jennnings (1986a) provides several examples where this quadratic approximation is inadequate and proposes a measure of its accuracy.

Tests of Fit to the Logistic Model

Assessing fit to a given logistic model can be approached in two ways: by formal tests of fit and by informal graphical diagnostics. A discussion of analytic tests of fit will be given next while a more lengthly description of graphical techniques is given in Section 5.4.

Two classes of tests of H_0: ((5.3.1) holds) have been considered in the literature. One class consists of "omnibus" tests which use the global alternative H_A: (not H_0); the second class consists of alternative-specific tests obtained by embedding H_0 in a parametric family. In the latter case if $\boldsymbol{\theta}$ is the parameter indexing the family and $\boldsymbol{\theta} = \boldsymbol{\theta}_0$ corresponds to the logistic model, then a test of $H_0^*: \boldsymbol{\theta} = \boldsymbol{\theta}_0$ versus $H_A^*: \boldsymbol{\theta} \neq \boldsymbol{\theta}_0$ is performed.

Two familiar omnibus test statistics are Pearson's X^2 statistic

$$X^2 = \sum_{i=1}^{T} [Y_i - m_i \hat{p}_i]^2 / m_i \hat{p}_i (1 - \hat{p}_i)$$

and the likelihood ratio statistic

$$G^2 = 2 \sum_{i=1}^{T} \left(Y_i \ln \left(\frac{Y_i}{m_i \hat{p}_i} \right) + (m_i - Y_i) \ln \left(\frac{m_i - Y_i}{m_i (1 - \hat{p}_i)} \right) \right),$$

both of which are approximately distributed as χ^2_{T-k} under H_0 and the large strata standard asymptotic model (k and T fixed, $N \to \infty$). Intuitively the degrees of freedom can be derived by noting that the loglinear model for the enlarged data has $T + k$ total parameters: T nuisance parameters and k covariate parameters. Testing fit of $\text{logit}(p(\mathbf{x})) = \mathbf{x}'\boldsymbol{\beta}$ against the global alternative is equivalent to comparing a $(T + k)$-parameter null model with the $2T$-parameter saturated alternative. The difference in the parameter dimensions is $T - k = 2T - (T + k)$. Both X^2 and G^2 are "fixed cell" tests in that the observed and expected numbers of successes are compared at T (fixed) "cells" corresponding to the distinct covariate vectors. Neither X^2 nor G^2 is appropriate in the many strata standard asymptotic model (k fixed, N and $T \to \infty$), because under this model there is no χ^2_{T-k} limiting distribution for either statistic (the degrees of freedom would be changing with the sample size).

There are several omnibus tests which are specifically designed to handle the many strata case. Tsiatis (1980) proposed a (fixed cell) chi-squared omnibus test of fit that is appropriate under either type of standard asymptotics. Partition the covariate space $\mathbb{R}^k$ into a fixed number of regions $R^1, \ldots, R^Q$, say, and consider the model

$$\text{logit}\{p(\mathbf{x}, \boldsymbol{\beta})\} = \mathbf{x}'\boldsymbol{\beta} + \sum_{j=1}^{Q} \gamma_j I[\mathbf{x} \in R^j]. \tag{5.3.14}$$

The variable $I[\mathbf{x} \in R^j]$ is the 0/1 indicator variable of the event $[\mathbf{x} \in R^j]$. Observe that H_0 holds if and only if $\gamma_1 = \ldots = \gamma_Q = 0$; Tsiatis proposes using the score test of H_0 based on the model (5.3.14) which has a chi-squared null distribution with degrees of freedom equal to the rank of a certain matrix under either standard asymptotic model (Problem 5.21). As with all fixed cell chi-squared tests, the cells $R^1, \ldots, R^Q$, although arbitrary, must be chosen independently of the data.

A second approach to construct valid many-strata omnibus tests uses random cell chi-squared tests. One such class of tests uses the data to define cells in the covariate space by dividing $[0, 1]$ into a fixed number of disjoint intervals and grouping those observations whose predicted probabilities fall in the same subinterval. For example, if $0 < .1 < .2 < \ldots < 1$ partitions $[0, 1]$, then observations with estimated probabilities falling in the same decile are grouped. Hosmer and Lemeshow (1980) propose tests of this form but use an incorrect asymptotic covariance matrix; Kwei (1983) gives the correct covariance. Andrews (1988a, 1988b) proposes a very general class of random cell chi-square tests and gives applications to categorical response models.

A third approach is taken by LeCessie and van Houwelingen (1989) who consider goodness-of-fit based on smoothing the standardized residuals defined in (5.4.11).

Tests of fit against restricted alternatives based on embedding the logistic link in a larger parameter family have been proposed by Prentice (1976), Pregibon (1980,1982a), Aranda-Ordaz (1981), Guerro and Johnson (1982), McCullagh and Nelder (1983), and Stukel (1988) (Problem 5.21). These tests are "directed" (toward the parametric family) so that their power is presumably superior to omnibus tests when the alternative is a member of the postulated family. In addition these tests can be inverted to form confidence regions for the parameter $\boldsymbol{\theta}$ which indexes the family and thus give a clearer picture of the evidence for/against H_0. On the other hand, likelihood ratio tests based on this approach can be computationally intensive as they require calculation of the MLE of $(\boldsymbol{\theta}, \boldsymbol{\beta})$. Of course, if the alternative is not of the form postulated by the family, link-based tests may have low power relative to omnibus tests.

Case Studies

Our discussion of likelihood methods is completed by applying some of the techniques sketched above to the data in Examples 5.1.1 and 5.3.1. In Example 5.1.1 the m_i are all large ($m_i = 240$ for all i) and hence both WLS and ML can be used; only ML is appropriate in Example 5.3.1.

Example 5.1.1 (continued). Table 5.3.2 lists the WLS and ML fits of Hoblyn and Palmer's data to the models

Model 1: $\text{logit}(p) = \beta_0 + \beta_1 I[\text{Length} = \text{long}] + \beta_2 I[\text{Plant} = \text{at once}]$

Model 2: $\text{logit}(p) = \beta_0 + \beta_1 I[\text{Length} = \text{long}]$

Model 3: $\text{logit}(p) = \beta_0 + \beta_1 I[\text{Plant} = \text{at once}]$.

Estimated coefficients, their estimated standard errors (ese), the ratio of the coefficients to their ese's (z-score), and goodness-of-fit (GOF) statistics are listed for each model. The G^2 and X^2 tests are appropriate for both ML and WLS as all m_i are large. Well-fitting models have nonsignificant X^2 and G^2 statistics (large P-values).

The agreement between the WLS and ML fits is very good. The precision estimates are within .01 for all coefficients in all models. The WLS method gives coefficient estimates slightly smaller in absolute value than ML which yields smaller z-scores for WLS coefficients. This trend becomes more evident as the fit of the model worsens; that is, the largest discrepancies are in the fits to Model 2, the worst fitting model, and smallest discrepancies are in the fits to Model 1, the best fitting model.

The likelihood ratio test for individual coefficients compares differences in G^2 values for the model with and without that variable to a χ^2_1 critical point. For example, the LR statistic of

$$H_0\text{: the coefficient of } I[\text{Length} = \text{long}] \text{ is zero}$$

in Model 1 is $G^2(\text{Model 3}) - G^2(\text{Model 1}) = 53.43 - 2.27 = 51.16$, which compares favorably with the squared z-score for this coefficient in Model 1 of $49.0 = (7.0)^2$. Clearly there is strong evidence that survival is related to length of cutting.

Model 2 (survival independent of the time of planting) and Model 3 (survival independent of the length of cutting) fit poorly using either WLS or MLE. In contrast, Model 1 fits well by both methods implying that root stock survival depends on both the time of planting and the length of cutting and that there is no interaction between the two factors.

The coefficient β_1 of Model 1 is the log odds of survival of long to short cuttings when planting either at once or in the spring. A point estimate of the odds of survival of long to short cuttings based on the ML (WLS) fit is $\exp\{1.02\} = 2.77$ ($\exp\{1.10\} = 2.75$). An approximate 95% confidence interval for this odds ratio calculated from the ML (WLS) fit is (2.07, 3.72)

Table 5.3.2. WLS and ML Fits for Three Models Based on Example 5.1.1 Data

(a) *Model* 1: Constant Odds of Root Stock Survival for Different Cutting Lengths and Time of Planting (across values of the other variable)

	WLS			ML		
Coefficient	$\hat{\beta}$	ese($\hat{\beta}$)	z-score	$\hat{\beta}$	ese($\hat{\beta}$)	z-score
Constant	−1.71	.15	−11.58	−1.73	.14	−12.11
I[Length = long]	1.01	.15	6.86	1.02	.15	7.00
I[Plant = at once]	1.41	.15	9.57	1.43	.15	9.75
GOF statistic	$X^2 = 2.27$,		$G^2 = 2.31$	$X^2 = 2.23$,		$G^2 = 2.27$
P-value	.13		.13	.13		.13

(b) *Model* 2: Probability of Root Stock Survival Independent of Time of Planting

	WLS			ML		
Coefficient	$\hat{\beta}$	ese($\hat{\beta}$)	z-score	$\hat{\beta}$	ese($\hat{\beta}$)	z-score
Constant	−.74	.11	−6.89	−.91	.10	−9.00
I[Length = long]	.74	.15	5.15	.91	.14	6.67
GOF statistic	$X^2 = 101.0$		$G^2 = 108.0$	$X^2 = 105.1$,		$G^2 = 101.9$
P-value	$< .0001$		$< .0001$	$< .0001$		$< .0001$

(c) *Model* 3: Probability of Root Stock Survival Independent of Cutting Length

	WLS			ML		
Coefficient	$\hat{\beta}$	ese($\hat{\beta}$)	z-score	$\hat{\beta}$	ese($\hat{\beta}$)	z-score
Constant	−1.04	.11	−9.39	−1.15	.11	−10.80
I[Plant = at once]	1.22	.15	8.43	1.35	.14	9.56
GOF statistic	$X^2 = 51.7$,		$G^2 = 54.6$	$X^2 = 52.3$,		$G^2 = 53.43$
P-value	$< .001$		$< .001$	$< .001$		$< .001$

((2.05, 3.65)). A similar interpretation holds for β_2. Conservative simultaneous 95% confidence intervals for the survival probabilities under the four experimental conditions can be calculated from the ML fit and (5.3.13). However, Bonferroni intervals based on (5.3.13) with $z_{.05/8} = 2.50$ used in place of $\{\chi^2_{.05,3}\}^{1/2} = 2.79$ are shorter; these intervals are listed in Table 5.3.3.

Table 5.3.3. Approximate 95% Bonferroni Joint Confidence Intervals for the Probability of Survival

Treatment	95% Confidence Interval for p
long, at once	(.60, .73)
long, spring	(.27, .40)
short, at once	(.36, .50)
short, spring	(.11, .20)

Comparing the observed best treatment condition to the worst, the long shoots planted at once have odds of survival approximately 11 times larger than short shoots planted in the spring ($2.77 \times 4.18 = 11.58$ by ML; $2.75 \times 4.10 = 11.28$ by WLS).

Example 5.3.1 (continued). Table 5.3.4 gives ML fits for the models

Model 1: $\text{logit}(p) = \beta_0 + \beta_1 \ln(LI)$

Model 2: $\text{logit}(p) = \beta_0 + \beta_1 \ln(LI) + \beta_2 \text{TEMP}$

Model 3: $\text{logit}(p) = \beta_0 + \beta_1 \ln(LI) + \beta_2 I[\text{TEMP} \leq 1.002]$

to the data in Table 5.3.1 where p is the probability of being in remission. The Tsiatis test is computed for the four regions formed by computing the midpoints of the range of the $\ln(LI)$ and TEMP values (−.137 and 1.009, respectively). None of the three tests shows significant lack of fit. Of course, the data set is small and the power to detect any but gross model deviations is presumably low.

Comparing Model 2 with Model 1, the addition of TEMP as a covariate is seen to greatly increase the magnitudes of the estimated constant and its standard error. One reason for this (and for the large magnitude of the TEMP coefficient) is that the temperature range is very small (.98 to 1.038).

Model 3 discretizes temperature into high and low values. The cut-off point (1.002) is empirically chosen to maximize the chi-squared statistic for independence in the 2×2 table with margins; temperature (high, low)

Table 5.3.4. ML Fits for Models 1, 2, and 3 Based on Example 5.3.1 Data

(a) *Model* 1: Constant and $\ln(LI)$

Coefficient	$\hat{\beta}$	ese($\hat{\beta}$)	z-score
Constant	−.70	.51	−1.36
$\ln(LI)$	3.60	1.43	2.51
GOF statistic		$A^2 = 4.46$	
P-value		.35	

(b) *Model* 2: Constant, $\ln(LI)$, and TEMP

Coefficient	$\hat{\beta}$	ese($\hat{\beta}$)	z-score
Constant	41.43	48.31	.86
$\ln(LI)$	4.02	1.67	2.41
TEMP	−42.46	48.72	−.87
GOF statistic		$A^2 = 2.63$	
P-value		.62	

(c) *Model* 3: Constant, $\ln(LI)$, and $I[\text{TEMP} \leq 100.2]$

Coefficient	$\hat{\beta}$	ese($\hat{\beta}$)	z-score
Constant	−3.14	1.79	−1.76
$\ln(LI)$	4.74	21.03	2.33
$I[\text{TEMP} \leq 1.002]$	2.77	1.86	1.50
GOF statistic		$A^2 = 1.71$	
P-value		.79	

and remission status (yes, no). This is sometimes a convenient method for categorizing a quantitative covariate. Halperin (1982) and Miller and Siegmund (1982) discuss the distribution of maximally selected chi-square statistics. It is important to remember that the TEMP value 1.002 is based on the sample rather than on any physical phenomenon, and thus exaggerates the importance of the high versus low temperature distinction.

The fitted Model 3,

$$\text{logit}(p) = -3.41 + 4.79 \ln(LI) + 2.77 I[\text{TEMP} \leq 1.002],$$

says that the probability of remission *increases* when LI increases or when TEMP is less than 1.002. For any fixed value of LI, the log odds of remission of low temperature to high temperature patients is 2.77. An analogous interpretation holds for LI although one must be careful to adjust for the fact that the model is in terms of the natural logarithm of LI.

Figure 5.3.2 adds the estimated $p = .1, .5$, and $.9$ contours of the ML fit to a coded scatterplot of the data; for example, the line labeled ".5" shows covariate pairs with estimated remission probability equal to 1/2. Several aspects of the fit deserve comment. First, there is only one observed remission among the high temperature group (case 16). Second, two high temperature observations have identical LI covariates but *different* responses (cases 8 and 16). Lastly, there is one relapsed patient who has high predicted probability of remission (case 24 with $\hat{p}_{24} = .87$). These points will be discussed further when residuals and influence for logistic regression are discussed in the next section.

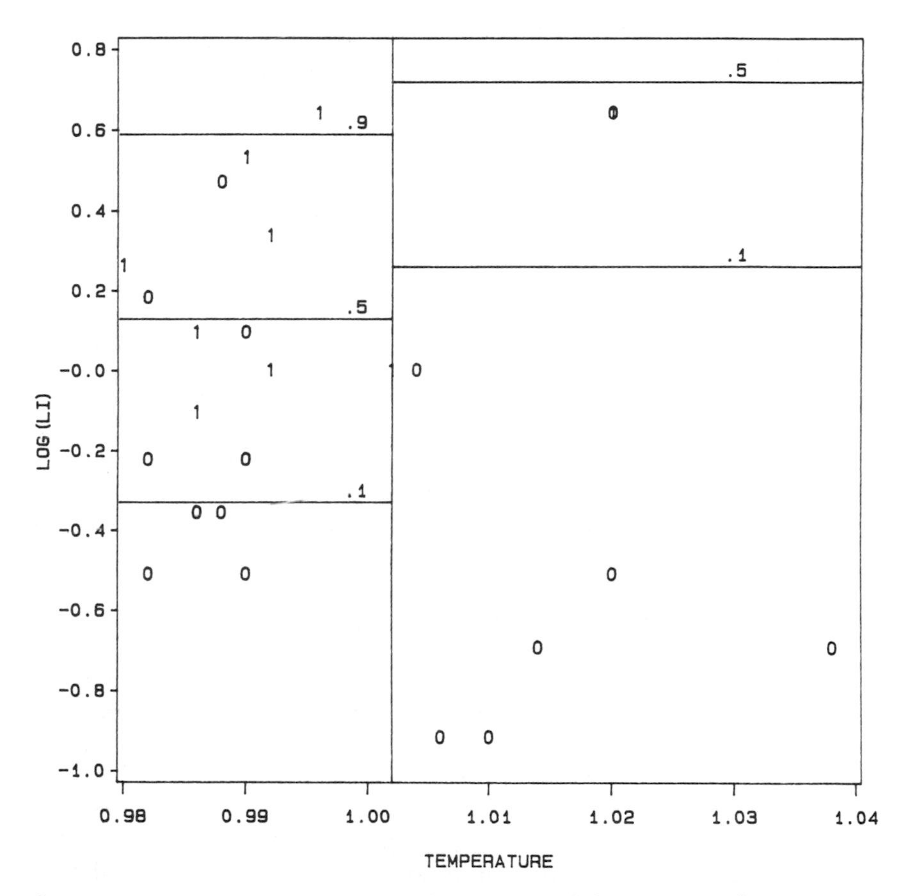

Figure 5.3.2. Coded scatterplot of the MLE of the .1, .5, and .9 contours of the probability of remission for Example 5.3.1 data based on Model 3 of Table 5.3.4.

Figure 5.3.3 shows conservative 90% confidence bands for the probability of remission as a function of $\ln(LI)$ and TEMP calculated from (5.3.13) (one set of bands for the low and one for the high temperature groups). Observe that the high temperature bands are wider than the low temperature bands (the former are based on only 8 patients).

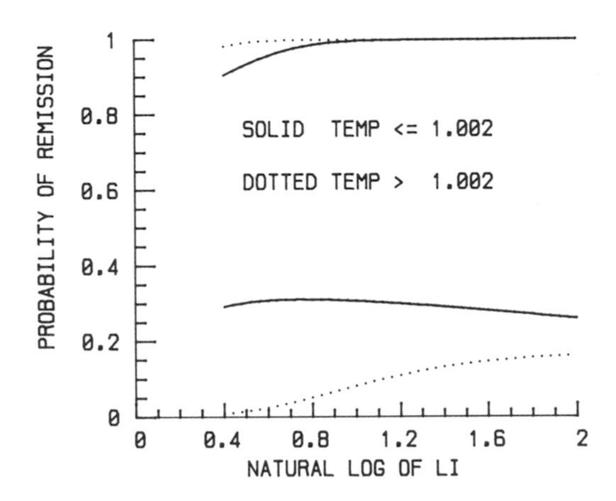

Figure 5.3.3. Ninety percent confidence bands for the probability of remission for Example 5.3.1 data based on Model 3 of Table 5.3.4.

5.4 Logistic Regression: Recent Developments

This section first considers point and interval estimation of logistic regression parameters and success probabilities based on Bayes and related methods. The second part of this section discusses local assessments of fit using influence and residual analyses.

A. Alternative Estimation Methods

The small sample Monte Carlo studies of $\hat{\boldsymbol{\beta}}$ and $\hat{\mathbf{p}}$ summarized in Section 5.3 show the MLE has worst MSE when the true $\boldsymbol{\beta}$ is near the origin ($||\boldsymbol{\beta}||$ is small) or equivalently when the true $\mathbf{p}$ is "central." This observation motivates the study of alternative estimators which have better MSE properties in this case (and not much worse for extreme $\boldsymbol{\beta}$). From the discussion in Section 2.1, one method for generating such estimators is Bayesian analysis with respect to priors putting mass on $\boldsymbol{\beta}$ near the origin. Unlike the binomial and multinomial problems there is no conjugate prior family for the logistic regression model. (Rubin and Schenker (1986) consider a prior which is both analytically feasible and interpretable; most priors considered in the literature lack these features.) Without the convenience of conjugate priors, normal priors have been investigated in Zellner and Rossi (1984),

Duffy (1986 and 1988), and Duffy and Santner (1987a and 1989). Normal priors not only have intuitive appeal, but also lead to a simple class of restricted MLEs. The derivation of these alternative estimators and some of their properties are given below.

Suppose $\boldsymbol{\beta}$ has the $N_k(\boldsymbol{\mu}, \sigma^2\mathbf{I}_k)$ prior where $\boldsymbol{\mu} \in \mathbb{R}^k$ and $\sigma^2 > 0$ are given. If, for example $\boldsymbol{\mu} = \mathbf{0}_k$, then the Bayes estimator should shrink the MLE toward the origin with the prior variance σ^2 controlling the amount of shrinkage. The assumption of a common variance for the β_j means that care must be taken in the choice of measurement scales for the covariates. A similar issue arises in normal theory ridge regression where centering and scaling of nonconstant covariates is ordinarily performed to create dimensionless covariates before applying ridge methods. Centering and scaling of nonconstant columns makes the resulting inferences invariant with respect to the original choice of scale. Thus centering and scaling is likewise recommended in the logistic case.

Let $\ln L_p(\boldsymbol{\beta})$ denote the log posterior likelihood of the data. Then, aside from constants,

$$\ln L_p(\boldsymbol{\beta}) = \mathbf{Y}'\mathbf{X}\boldsymbol{\beta} + \mathbf{m}'\ln(\mathbf{1}_T + \exp[\mathbf{X}\boldsymbol{\beta}]) - ||\boldsymbol{\beta} - \boldsymbol{\mu}||^2/2\sigma^2. \tag{5.4.1}$$

Calculation of the Bayes estimator of $\mathbf{p}$ with respect to even the simple case of SEL requires the minimization of the k-dimensional integral of the loss times $L_p(\boldsymbol{\beta})$. Further, since the parameter space defined by the logistic model is not convex in $\mathbf{p}$, the posterior expected mean of $\mathbf{p}$ is *not* the Bayes estimator of $\mathbf{p}$ with respect to SEL. Zellner and Rossi (1984) review methods of computing high-dimensional integrals but conclude that loss specific Bayesian estimators are currently too computationally intensive for ordinary use. (However the work of Stewart (1987), van Dijk (1987), and the references therein show that progress is being made in this regard.)

This section focuses attention on the mode of the posterior distribution—a Bayesian MLE of $\boldsymbol{\beta}$ which does not require specification of a loss function. Denoted $\hat{\boldsymbol{\beta}}^B$, the Bayesian MLE is defined by

$$\ln L_p(\hat{\boldsymbol{\beta}}^B) = \sup_{\boldsymbol{\beta} \in \mathbb{R}^k} \ln L_p(\boldsymbol{\beta}) \tag{5.4.2}$$

with $\hat{\mathbf{p}}^B := \mathbf{p}(\hat{\boldsymbol{\beta}}^B)$ the corresponding estimator of $\mathbf{p}$.

Since $\ln L_p(\boldsymbol{\beta})$ is strictly concave in $\boldsymbol{\beta}$ and $\ln L_p(\boldsymbol{\beta}) \to -\infty$ as $||\boldsymbol{\beta}|| \to \infty$, both $\hat{\boldsymbol{\beta}}^B$ and $\hat{\mathbf{p}}^B$ exist uniquely for all data $(\mathbf{Y},\mathbf{X})$. Differentiating (5.4.1) shows that $\hat{\boldsymbol{\beta}}^B$ solves

$$\mathbf{X}'(\mathbf{Y} - \hat{\mathbf{Y}}^B) - (\hat{\boldsymbol{\beta}}^B - \boldsymbol{\mu})/\sigma^2 = \mathbf{0}_k \tag{5.4.3}$$

where $\hat{Y}_i^B := m_i\hat{p}_i^B$. Thus $\hat{\boldsymbol{\beta}}^B$ requires the same order of computation as $\hat{\boldsymbol{\beta}}$.

The estimator $\hat{\boldsymbol{\beta}}^B$ has at least two attractive alternative motivations. First, it can be regarded as a maximum penalized likelihood estimator with penalty term $||\boldsymbol{\beta} - \boldsymbol{\mu}||^2/2\sigma^2$. Second, it is a restricted MLE based on the logistic likelihood (5.3.4). The characterization is analogous to that of the ridge estimator in normal theory regression analysis demonstrated by Hoerl and Kennard (1970) (Problem 5.27).

Several small sample properties of $\hat{\boldsymbol{\beta}}^B$ and $\hat{\mathbf{p}}^B := \mathbf{p}(\hat{\boldsymbol{\beta}}^B)$ can be derived from the characterizing equation (5.4.3). First, if successes and failures are interchanged so that $\mathbf{Y} \to \mathbf{m} - \mathbf{Y}$ and if $\boldsymbol{\mu} \to -\boldsymbol{\mu}$, then $\hat{\boldsymbol{\beta}}^B \to -\hat{\boldsymbol{\beta}}^B$ (hence $\hat{\mathbf{p}}^B \to 1 - \hat{\mathbf{p}}^B$). Second, an immediate consequence of Problem 5.27 is that $\hat{\boldsymbol{\beta}}^B$ does indeed pull the estimate of $\boldsymbol{\beta}$ toward $\boldsymbol{\mu}$ in the sense that $||\hat{\boldsymbol{\beta}}^B - \boldsymbol{\mu}|| \leq ||\hat{\boldsymbol{\beta}} - \boldsymbol{\mu}||$ whenever $\hat{\boldsymbol{\beta}}$ exists. Lastly, a more detailed analysis of the effect of σ on $\hat{\boldsymbol{\beta}}^B$ can be performed. Let $\hat{\boldsymbol{\beta}}^B(\sigma)$ explicitly denote the dependence of $\hat{\boldsymbol{\beta}}^B$ on σ and $\hat{\mathbf{p}}^B(\sigma) = \mathbf{p}(\hat{\boldsymbol{\beta}}^B(\sigma))$. Consider the path that $\hat{\boldsymbol{\beta}}^B(\sigma)$ traces as σ increases from 0 to infinity. Applying the Implicit Function Theorem to the mapping $\sigma \to \hat{\boldsymbol{\beta}}^B(\sigma)$ shows that $\hat{\boldsymbol{\beta}}^B(\sigma)$ is a continuous function of σ. Thus $\hat{\boldsymbol{\beta}}^B(\sigma) \to \boldsymbol{\mu}$ and $\hat{\mathbf{p}}^B(\sigma) \to \mathbf{p}(\boldsymbol{\mu})$ as $\sigma \to 0$ since $\ln L_p(\boldsymbol{\beta}) \to -\infty$ as $\sigma \to 0$ for all $\boldsymbol{\beta} \neq \boldsymbol{\mu}$. Similarly, $\hat{\boldsymbol{\beta}}^B(\sigma) \to \hat{\boldsymbol{\beta}}$ and $\hat{\mathbf{p}}^B(\sigma) \to \hat{\mathbf{p}}$ as $\sigma \to \infty$ when these exist since $L_p(\boldsymbol{\beta}) \to L(\boldsymbol{\beta})$ as $\sigma \to \infty$ for all $\boldsymbol{\beta} \in \mathbb{R}^k$. Thus $\hat{\boldsymbol{\beta}}^B$ ($\hat{\mathbf{p}}^B$) can be viewed as pulling $\hat{\boldsymbol{\beta}}$ ($\hat{\mathbf{p}}$) toward $\boldsymbol{\mu}$ ($\mathbf{p}(\boldsymbol{\mu})$) to a greater or lesser extent depending on the magnitude of σ. The following result, established in Duffy and Santner (1989), makes this precise.

Proposition 5.4.1. *The norm* $||\hat{\boldsymbol{\beta}}^B(\sigma) - \boldsymbol{\mu}||$ *increases in* σ.

Duffy (1986, 1987) studies the large sample properties of $\hat{\boldsymbol{\beta}}^B$ for fixed $\boldsymbol{\mu}$ and σ. She shows that $\hat{\boldsymbol{\beta}}^B$ is consistent, asymptotically normal, and efficient under the same conditions (5.3.10) and (5.3.11) as the MLE. In particular, she proves that

$$\boldsymbol{\Sigma}^{-1/2}(\hat{\boldsymbol{\beta}}^B - \boldsymbol{\beta}) \xrightarrow{\mathcal{L}} N_k[\mathbf{0}_k, \mathbf{I}_k] \tag{5.4.4}$$

where

$$\boldsymbol{\Sigma} = \boldsymbol{\Sigma}(\boldsymbol{\beta}, \sigma^2) = \left[\mathbf{X}'\mathbf{D}(\boldsymbol{\beta})\mathbf{X} + \frac{(\boldsymbol{\beta} - \boldsymbol{\mu})(\boldsymbol{\beta} - \boldsymbol{\mu})'}{\sigma^4}\right]^{-1} \tag{5.4.5}$$

and $\mathbf{D}(\boldsymbol{\beta}) = \text{Diag}(\ldots, m_i p_i(\boldsymbol{\beta})[1 - p_i(\boldsymbol{\beta})], \ldots)$.

Two choices for the prior mean $\boldsymbol{\mu}$ are of particular interest. The first is $\boldsymbol{\mu} = \mathbf{0}_k$, which pulls $\hat{\boldsymbol{\beta}}^B$ toward the origin (and each $\hat{p}_i^B$ toward 1/2). The second is $\boldsymbol{\mu} = \boldsymbol{\mu}^* := (\text{logit}[\overline{y}], 0, \ldots, 0)'$ where $\overline{y} := N^{-1} \sum_{i=1}^T y_i$ is the observed proportion of successes. The latter is an adaptive (data-selected) prior mean which is intuitively reasonable when the first column of $\mathbf{X}$ is $\mathbf{1}_T$ since the constant coefficient is pulled toward logit$[\overline{y}]$. Of course when $\boldsymbol{\mu}$ is data-selected the asymptotic normality result (5.4.4) no longer holds.

Clearly the choice of σ is crucial because it governs the degree of shrinkage of $\hat{\beta}^B$. However in many practical problems there will be little information available to guide the selection of σ. Thus Duffy and Santner (1987a and 1989) consider a data-selected choice of σ. Intuitively, the goal is to choose a small value of σ when the data indicate $\boldsymbol{\beta}$ is near $\boldsymbol{\mu}$.

Empiricial Bayes analysis suggests estimating σ from the marginal probability of $\mathbf{Y}$ given σ^2

$$m(\mathbf{y} \mid \sigma^2) = \int_{\mathbb{R}^k} \left[\prod_{i=1}^{T} \frac{\exp(y_i \mathbf{x}_i' \boldsymbol{\beta})}{\{1 + \exp(\mathbf{x}_i' \boldsymbol{\beta})\}^{m_i}} \right]$$

$$\times \left[\prod_{j=1}^{k} \frac{1}{(2\pi\sigma^2)^{1/2}} \exp\{-(\beta_j - \mu_j)^2 / 2\sigma^2\} \right] d\boldsymbol{\beta}. \qquad (5.4.6)$$

Unfortunately directly maximizing $m(\mathbf{y} \mid \sigma^2)$ over $\sigma^2 \in (0, \infty)$ requires k-dimensional integration and is computationally intractable. Instead Duffy and Santner (1987a) cast the maximization problem as an "incomplete data" problem and apply the EM algorithm of Dempster, Laird, and Rubin (1977) which is an iterative maximization technique. The exact EM algorithm also requires k-fold (numerical) integration in the so-called E-step. Following Leonard (1972, 1975) and Laird (1978a), the conditional distribution of $\boldsymbol{\beta}$ given $\mathbf{y}$ and σ^2 can be approximated by a multivariate normal distribution to avoid integration. The mean of the approximating normal is taken to be $\hat{\boldsymbol{\beta}}^B(\sigma)$; i.e., the mode of the conditional distribution of $\boldsymbol{\beta}$ given $\mathbf{y}$ and σ^2. The covariance is taken to be $\Sigma(\hat{\boldsymbol{\beta}}^B(\sigma), \sigma^2)$. The vector $\hat{\boldsymbol{\beta}}^B(\sigma)$ is an obvious choicc for the mean of the approximating distribution and the matrix $\boldsymbol{\Sigma}(\cdot, \cdot)$ is the consistent estimator (5.4.5).

With this approximation, the EM algorithm has pth iteration with current guess σ_p^2

E-Step: Estimate $||\boldsymbol{\beta} - \boldsymbol{\mu}||^2$ by

$$t_p = ||\hat{\boldsymbol{\beta}}^B(\sigma_p) - \boldsymbol{\mu}||^2 + \operatorname{Tr}\left\{\Sigma(\hat{\boldsymbol{\beta}}^B(\sigma_p), \sigma_p^2)\right\}. \qquad (5.4.7)$$

M-Step: Set $\sigma_{p+1}^2 = t_p / k$.

If the algorithm (5.4.7) converges to $\hat{\sigma}$, then the empiric mode estimator is defined as $\hat{\boldsymbol{\beta}}^E := \hat{\boldsymbol{\beta}}^B(\hat{\sigma})$ and $\hat{\mathbf{p}}^E := \mathbf{p}(\hat{\boldsymbol{\beta}}^E)$. When $\boldsymbol{\mu} = \boldsymbol{\mu}^*$, (5.4.7) essentially ignores the randomness in $\boldsymbol{\mu}$ since the normal approximation is based on (5.4.4).

The adequacy of the normal approximation to the (posterior) distribution of $\boldsymbol{\beta}$ given $\mathbf{Y}$ and σ^2, has been examined by Zellner and Rossi (1984) and is related to the adequacy of the normal approximation to the loglikelihood function $\ln(L(\boldsymbol{\beta}))$. These approximations are generally poor in small and moderate samples (Jennings (1986a)).

Several small sample properties of $\hat{\boldsymbol{\beta}}^E$ and $\hat{\mathbf{p}}^E$ follow from those of $\hat{\boldsymbol{\beta}}^B(\sigma)$. If successes and failures are interchanged so that $\mathbf{Y} \to \mathbf{m} - \mathbf{Y}$ and $\boldsymbol{\mu} \to -\boldsymbol{\mu}$, then $\hat{\boldsymbol{\beta}}^E \to -\hat{\boldsymbol{\beta}}^E$ (hence $\hat{\mathbf{p}}^E \to 1 - \hat{\mathbf{p}}^E$). If centering and scaling is first performed on the covariates, then $\hat{\boldsymbol{\beta}}^E$ is invariant under rescaling of any covariate. Also $\hat{\boldsymbol{\beta}}^E$ pulls the estimate of $\boldsymbol{\beta}$ toward $\boldsymbol{\mu}$ in that $||\hat{\boldsymbol{\beta}}^E - \boldsymbol{\mu}|| \leq ||\hat{\boldsymbol{\beta}} - \boldsymbol{\mu}||$ whenever $\hat{\boldsymbol{\beta}}$ exists.

Duffy (1986) and Duffy and Santner (1987b) study the small-sample MSE of $\hat{\boldsymbol{\beta}}^B$ ($\sigma^2 = 1$) and $\hat{\boldsymbol{\beta}}^E$ with $\boldsymbol{\mu} = \mathbf{0}_k$ and $\boldsymbol{\mu} = \boldsymbol{\mu}^*$. They also discuss their experience with the implementation of the algorithm (5.4.7). In short, although this class of alternative estimators appears promising (especially $\hat{\boldsymbol{\beta}}^B$ with $\sigma^2 = 1$ and $\boldsymbol{\mu} = \boldsymbol{\mu}^*$). Further research is needed to better understand its behavior.

Several other alternatives to $\hat{\boldsymbol{\beta}}$ have been proposed in the literature. Anderson and Richardson (1979) and Schaefer (1983) discuss a bias-correction to the MLE arising from a two-term Taylor expansion of the score function evaluated at $\hat{\boldsymbol{\beta}}$. Suppose that $m_i = 1$ for $1 \leq i \leq T$. Then their analysis yields the estimator

$$\hat{\boldsymbol{\beta}}^C := \hat{\boldsymbol{\beta}} + .5(\mathbf{X}'\mathbf{D}(\hat{\boldsymbol{\beta}})\mathbf{X})^{-1}\mathbf{X}'\mathbf{D}(\hat{\boldsymbol{\beta}})\mathbf{V} \tag{5.4.8}$$

where $\mathbf{V}' = (V_1, \ldots, V_T)$ and $V_i := (1 - 2\hat{p}_i)\mathbf{x}_i'(\mathbf{X}\mathbf{D}(\hat{\boldsymbol{\beta}})\mathbf{X})^{-1}\mathbf{x}_i$. While this estimator has low MSE compared to $\hat{\boldsymbol{\beta}}$ in Schaefer's simulation, there are several nonstandard features in the study design. In particular, the design matrix $\mathbf{X}$ and the true $\boldsymbol{\beta}$ change from replication to replication. Thus he report Bayes risks for a model with stochastic covariates rather than the MSE.

Schaefer, Roi, and Wolfe (1984) and Schaefer (1986) study a "ridge" logistic estimator defined by

$$\hat{\boldsymbol{\beta}}^R = (\mathbf{X}'\mathbf{D}(\hat{\boldsymbol{\beta}})\mathbf{X} + k\mathbf{I}_k/||\boldsymbol{\beta}||^2)^{-1}\mathbf{X}'\mathbf{D}(\hat{\boldsymbol{\beta}})\mathbf{X}\hat{\boldsymbol{\beta}}. \tag{5.4.9}$$

This estimator algebraically mimics the normal theory ridge estimator and was proposed primarily for situations in which $\mathbf{X}$ is ill-conditioned. Schaefer also studies two other modifications of $\hat{\boldsymbol{\beta}}$ for collinear $\mathbf{X}$. Again his numerical studies, which favor the modified estimators, are of the Bayes risk with stochastic covariates.

Rubin and Schenker (1986) develop a promising empirical Bayes estimator with respect to a "pseudo-conjugate prior." Their prior distribution depends on $\mathbf{X}$ and the data and corresponds to a prior sample consisting of $r\overline{y}m_i$ (fractional) successes and $r(1 - \overline{y})m_i$ (fractional) failures at each design point $\mathbf{x}_i$ where $r = k/N$ is the ratio of number of parameters to the total number of observations. Rubin and Schenker (1986) consider estimating $\boldsymbol{\beta}$ by the posterior mode $\hat{\boldsymbol{\beta}}^P$ which solves

$$\mathbf{X}'(\mathbf{Y} + r\overline{y}\mathbf{m} - (1 + r)\hat{\mathbf{Y}}^P) = \mathbf{0}_k \tag{5.4.10}$$

where $\hat{\mathbf{Y}}^P = (\ldots, \hat{Y}_i^P := m_i \exp(\mathbf{x}_i'\hat{\boldsymbol{\beta}}^P)/\{1+\exp(\mathbf{x}_i'\hat{\boldsymbol{\beta}}^P)\}, \ldots)'$ is the $T \times 1$ vector of estimated mean successes.

Before illustrating some of the methods on the data in Example 5.3.1, a different approach to alternative estimators will be noted. Pregibon (1982b), Stefanski, Carroll, and Ruppert (1986) and Copas (1988) consider robust or resistant modifications of $\hat{\boldsymbol{\beta}}$; these modifications are designed for stability when either any single observation is drastically modified or when all observations are changed slightly.

Example 5.3.1 (continued). In Table 5.4.1 below columns 2–5 show $\hat{\boldsymbol{\beta}}^B(\sigma = 1)$ and $\hat{\boldsymbol{\beta}}^E$ with $\boldsymbol{\mu} = \mathbf{0}_k$ and $\boldsymbol{\mu} = \boldsymbol{\mu}^*$ for Model 1 ($\text{logit}(p) = \beta_0 + \beta_1 \ln(LI)$) based on the data in Example 5.3.1. Columns 6–8 show $\hat{\boldsymbol{\beta}}^C$ (from (5.4.8)), $\hat{\boldsymbol{\beta}}^R$ (5.4.9), and $\hat{\boldsymbol{\beta}}^P$ (5.4.10) for these same data. To facilitate comparisons, the MLE $\hat{\boldsymbol{\beta}}$ is given in the last column. With the exception of $\hat{\boldsymbol{\beta}}^C$, all the alternative estimators have coefficients componentwise smaller in magnitude than $\hat{\boldsymbol{\beta}}$. Anderson, Richardson, and Schaefer's estimate $\hat{\boldsymbol{\beta}}^C$ differs comparatively little from $\hat{\boldsymbol{\beta}}$, whereas $\hat{\boldsymbol{\beta}}^P$, $\hat{\boldsymbol{\beta}}^B$ and $\hat{\boldsymbol{\beta}}^E$ exhibit stronger pull. Geometrically, $\hat{\boldsymbol{\beta}}$ and $\hat{\boldsymbol{\beta}}^C$ produce the steepest curves of fitted probability, with $\hat{\boldsymbol{\beta}}^P$ slightly flatter and $\hat{\boldsymbol{\beta}}^R$, $\hat{\boldsymbol{\beta}}^B$ and $\hat{\boldsymbol{\beta}}^E$ flatter still. In this example, $\text{logit}(\overline{y}) = -.69$ so $\boldsymbol{\mu}^* = (-.69, 0)'$. The difference between using $\mathbf{0}_2$ and $\boldsymbol{\mu}^*$ as the prior mean for $\hat{\boldsymbol{\beta}}^B$ and $\hat{\boldsymbol{\beta}}^E$ can be clearly seen in the estimated constant coefficients. Geometrically, when $\boldsymbol{\mu} = \mathbf{0}_2$, the curves of fitted probability for $\hat{\boldsymbol{\beta}}^B$ and $\hat{\boldsymbol{\beta}}^E$ are shifted slightly to the left (toward higher values). Duffy (1988) contains more detailed comparisons of the various estimators on this and other data sets.

Table 5.4.1. Estimators $\hat{\boldsymbol{\beta}}^B$, $\hat{\boldsymbol{\beta}}^E$, $\hat{\boldsymbol{\beta}}^C$, $\hat{\boldsymbol{\beta}}^R$, and $\hat{\boldsymbol{\beta}}^P$ for Model 1 of Example 5.3.1

Coefficient	$\hat{\boldsymbol{\beta}}^B$ $\sigma = 1$ $\boldsymbol{\mu} = 0$	$\hat{\boldsymbol{\beta}}^B$ $\sigma = 1$ $\boldsymbol{\mu} = \boldsymbol{\mu}^*$	$\hat{\boldsymbol{\beta}}^E$ $\boldsymbol{\mu} = 0$	$\hat{\boldsymbol{\beta}}^E$ $\boldsymbol{\mu} = \boldsymbol{\mu}^*$	$\hat{\boldsymbol{\beta}}^C$	$\hat{\boldsymbol{\beta}}^R$	$\hat{\boldsymbol{\beta}}^P$	$\hat{\boldsymbol{\beta}}$
Constant	−.48	−.60	−.52	−.60	−.69	−.62	−.67	−.70
$\ln(LI)$	2.53	2.60	2.69	2.56	3.65	2.75	3.17	3.60
			$\hat{\sigma} = 1.15$	$\hat{\sigma} = .97$				

B. Graphical Assessment of Fit to the Logistic Model

A wide variety of diagnostic tools have been proposed for assessing different aspects of the maximum likelihood fit to a logistic regression model (Pregibon, 1981; Cook and Weisberg, 1982; McCullagh and Nelder, 1983; Landwehr, Pregibon, and Shoemaker, 1984; Williams, 1984 and 1987; Copas, 1988; A.H. Lee, 1987 and 1988). Many of these tools rely on the iteratively reweighted least squares interpretation of the MLE to provide analogs of diagnostic tools that have been developed for linear regression models. Following a brief summary of logistic regression residuals, three such diagnostics are defined and illustrated: leverages, partial residual plots, and Cook's distance.

Residuals

Proceeding in analogy to the discussion in Section 4.4, raw residuals are defined as,

$$e_i^r = Y_i - m_i\hat{p}_i,$$

Pearson residuals as,

$$e_i^P = (Y_i - m_i\hat{p}_i)/(m_i\hat{p}_i[1-\hat{p}_i])^{1/2}, \tag{5.4.11}$$

and deviance residuals as,

$$e_i^d = \pm(|2Y_i \ln\{Y_i/m_i\hat{p}_i\} + 2[m_i - Y_i]\ln\{(m_i - Y_i)/m_i(1-\hat{p}_i)\}|)^{1/2},$$

where the "+" sign is used if $Y_i > m_i\hat{p}_i$ and "−" is used otherwise. Arguing from the IRLS interpretation of the maximum likelihood calculation, the adjusted residuals are

$$e_i^a = \frac{e_i^P}{[1 - m_i\hat{p}_i(1-\hat{p}_i)\{\mathbf{X}(\mathbf{X}'\hat{\mathbf{D}}\mathbf{X})^{-1}\mathbf{X}'\}_{ii}]^{1/2}} = \frac{e_i^P}{(1-h_i)^{1/2}} \tag{5.4.12}$$

where $\hat{\mathbf{D}}$ is the $T \times T$ diagonal matrix with ith diagonal element $m_i\hat{p}_i(1-\hat{p}_i)$ and

$$h_i = m_i\hat{p}_i(1-\hat{p}_i)\{\mathbf{X}(\mathbf{X}'\hat{\mathbf{D}}\mathbf{X})^{-1}\mathbf{X}'\}_{ii}. \tag{5.4.13}$$

Problem 5.8 shows the relationship between the adjusted residuals (5.4.12) and those defined for the corresponding loglinear model based on the enlarged data. The adjusted residuals $\{e_i^a\}_{i=1}^T$ have asymptotic $N(0,1)$ distributions under the large strata model (only).

Williams (1984, 1987), Jennings (1986b), and Copas (1988) contain further discussion of residuals including proposed models for the distribution of outliers. Williams (1984, 1987), for example, considers the mean shift outlier model:

$$\text{logit}(\mathbf{Y}) = \mathbf{X}\boldsymbol{\beta} + \mathbf{u}_i\gamma_i$$

where $\mathbf{u}_i$ is the $T \times 1$ unit vector with ith component 1. Based on this model, Y_i is *not* an outlier if and only if $\gamma_i = 0$. Williams derives the following approximation to the likelihood ratio statistic for H_0: $\gamma_i = 0$:

$$(e_i^o)^2 = (e_i^d)^2 + h_i(e_i^P)^2.$$

Based on this approximation he proposes outlier residuals

$$e_i^o = \pm\sqrt{(e_i^d)^2 + h_i(e_i^P)^2}$$

with the "+" sign if $Y_i > m_i\hat{p}_i$ and the "−" sign otherwise. The outlier residuals can also be written as

$$e_i^o = \pm\sqrt{(1-h_i)(e_i^{ad})^2 + h_i(e_i^P)^2} \tag{5.4.14}$$

where $e_i^{ad} = e_i^d/(1-h_i)^{1/2}$ are adjusted deviance residuals. Equation (5.4.14) shows that the $(e_i^o)^2$ are convex combinations of the $(c_i^P)^2$ and the $(e_i^{ad})^2$; in practice, the e_i^o are often very nearly equal to the e_i^d. Williams (1984) and Duffy (1989) compare these and other residuals. Based on their comparisons, e_i^a or e_i^o are recommended for general use.

Leverages

The first diagnostic developed is that of *leverage.* Recall that in the standard linear regression model $\mathbf{Y}^* = \mathbf{X}^*\boldsymbol{\beta}^* + \boldsymbol{\epsilon}^*$, the least squares estimate of $\boldsymbol{\beta}^*$ is $\hat{\boldsymbol{\beta}}^* = ((\mathbf{X}^*)'\mathbf{X}^*)^{-1}(\mathbf{X}^*)'\mathbf{Y}^*$, and the fitted means are $\hat{\mathbf{Y}}^* = \mathbf{X}^*\hat{\boldsymbol{\beta}}^* = \mathbf{H}^*\mathbf{Y}^*$ with $\mathbf{H}^* = \mathbf{X}^*((\mathbf{X}^*)'\mathbf{X}^*)^{-1}(\mathbf{X}^*)'$. The matrix $\mathbf{H}^*$ is called the "hat matrix" since it produces "$\mathbf{Y}^*$ hat" when applied to $\mathbf{Y}^*$. The diagonal elements of $\mathbf{H}^*$ are called leverages; large leverages identify points which have a strong influence on the fitted model. If $\hat{\mathbf{e}}^* = \mathbf{Y}^* - \hat{\mathbf{Y}}^*$ denotes the raw residuals then

$$\begin{aligned}
&\hat{\mathbf{Y}}^* = \mathbf{H}^*\mathbf{Y}^* \\
\iff\ &\hat{\mathbf{e}}^* = \mathbf{Y}^* - \hat{\mathbf{Y}}^* = (\mathbf{I} - \mathbf{H}^*)\mathbf{Y}^* \\
\iff\ &\hat{\mathbf{e}}^* = (\mathbf{I} - \mathbf{H}^*)(\mathbf{Y}^* - \hat{\mathbf{Y}}^*) \\
\iff\ &\hat{\mathbf{e}}^* = (\mathbf{I} - \mathbf{H}^*)\hat{\mathbf{e}}^*.
\end{aligned} \tag{5.4.15}$$

For logistic regression the matrix

$$\mathbf{H} = \hat{\mathbf{D}}^{1/2}\mathbf{X}(\mathbf{X}'\hat{\mathbf{D}}\mathbf{X})^{-1}\mathbf{X}'\hat{\mathbf{D}}^{1/2}$$

satisfies the equation

$$\mathbf{e}^P = (\mathbf{I}_T - \mathbf{H})\mathbf{e}^P \tag{5.4.16}$$

which is the analog of (5.4.15). To see this let $\hat{\mathbf{Y}}' = (\ldots, m_i\hat{p}_i, \ldots)$. Then

$$\begin{aligned}
\hat{\mathbf{D}}^{-1/2}\mathbf{e}^P &= \hat{\mathbf{D}}^{-1}(\mathbf{Y} - \hat{\mathbf{Y}}) \\
&= \hat{\mathbf{D}}^{-1}(\mathbf{Y} - \hat{\mathbf{Y}}) - \mathbf{X}(\mathbf{X}'\hat{\mathbf{D}}\mathbf{X})^{-1}\mathbf{X}'(\mathbf{Y} - \hat{\mathbf{Y}}), \quad \text{by (5.3.6)} \\
&= \hat{\mathbf{D}}^{-1/2}\{\mathbf{I}_T - \hat{\mathbf{D}}^{1/2}\mathbf{X}(\mathbf{X}'\hat{\mathbf{D}}\mathbf{X})^{-1}\mathbf{X}'\hat{\mathbf{D}}^{1/2}\}\hat{\mathbf{D}}^{-1/2}(\mathbf{Y} - \hat{\mathbf{Y}}) \\
&= \hat{\mathbf{D}}^{-1/2}\{\mathbf{I}_T - \mathbf{H}\}\mathbf{e}^P
\end{aligned}$$

which is equivalent to (5.4.16). This analogy suggests that an index plot of the diagonal elements of **H** may be useful for identifying those points which contributed heavily to the fitted model. In fact these diagonal elements are exactly the h_i as defined in (5.4.13). Pregibon (1981) contains examples of the use and interpretation of the h_i. Since $\sum_{i=1}^T h_i = k$, individual h_i larger than $2k/T$ are often earmarked for further investigation.

Partial Residual Plots

The second diagnostic is the partial residual plot which is useful for studying both the effects of covariates not in the model and the functional form of covariates in the model. In the standard linear model the partial residual vector for the jth covariate is

$$\mathbf{r}_j^* = \hat{\mathbf{e}}^* + \boldsymbol{\xi}_j^* \hat{\beta}_j^* \tag{5.4.17}$$

where $\boldsymbol{\xi}_j^*$ is the jth column of $\mathbf{X}^*$. Non-linearity in the scatterplot of $\mathbf{r}_j^*$ versus $\boldsymbol{\xi}_j^*$ is evidence that the effect of jth covariate is different than postulated. To study variables not in the model, fit the extended model

$$\mathbf{Y}^* = (\mathbf{X}^*)'\boldsymbol{\beta}^* + \mathbf{Z}^*\gamma^* \tag{5.4.18}$$

where $\mathbf{Z}^*$ is the *vector* of values for the potential covariate of interest. If $\boldsymbol{\epsilon}$ is the vector of raw residuals and $\hat{\gamma}^*$ is the estimate of γ^* based on model (5.4.18), then the partial residual vector is

$$\mathbf{r}_{\mathbf{Z}}^* = \boldsymbol{\epsilon} + \mathbf{Z}^*\hat{\gamma}^*. \tag{5.4.19}$$

Under the extended model (5.4.18), the plot of $\mathbf{r}_{\mathbf{Z}}^*$ versus $\mathbf{Z}^*$ will be linear with slope γ^*.

Landwehr, Pregibon, and Shoemaker (1984) propose the following analog of the partial residual vector (5.4.17) for logistic regression

$$\mathbf{r}_j := \frac{\mathbf{Y} - \hat{\mathbf{Y}}}{\hat{\mathbf{Y}}(\mathbf{1}_T - \hat{\mathbf{p}})} + \boldsymbol{\xi}_j \hat{\beta}_j$$

where $\boldsymbol{\xi}_j$ is the jth column of $\mathbf{X}$. Plots of $\mathbf{r}_j$ versus $\boldsymbol{\xi}_j$ can be used in the same manner as in the linear regression case. Wang (1985) and A.H. Lee (1988) consider extensions of (5.4.19) for logistic regression.

Cook's Distance

The final diagnostic considered is Cook's distance (Cook, 1977, 1979) which measures the effect of deleting a single observation on the estimated parameter vector. For the standard linear model with h_i^* the ith diagonal element of $\mathbf{H}^*$,

$$c_i^* = \frac{(\hat{e}_i^*)^2 h_i^*}{s^2(1-h_i^*)^2}$$

measures the likelihood displacement of $\hat{\boldsymbol{\beta}}^*$ when the ith case is deleted. Here s^2 is the usual unbiased estimate of error variance.

An analog of Cook's distance can be obtained by deleting a single point $(Y_i,\ m_i, \mathbf{x}_i)$ from the data and assessing the effect on the approximate $100(1-\alpha)\%$ likelihood confidence region for $\boldsymbol{\beta}$. The region is given by

$$\{\boldsymbol{\beta}: -2\ln\{L(\boldsymbol{\beta})/L(\hat{\boldsymbol{\beta}})\} \leq \chi^2_{\alpha,k}\}. \tag{5.4.20}$$

Thus if $\hat{\boldsymbol{\beta}}_{-i}$ is the MLE of $\boldsymbol{\beta}$ when the ith point is deleted, then comparison of $c_i := -2\ln\{L(\hat{\boldsymbol{\beta}}_{-i})/L(\hat{\boldsymbol{\beta}})\}$ with the percentage point of the χ_k^2 distribution roughly determines the contour of the confidence region (5.4.20), to which the maximum likelihood estimator is displaced due to deleting the ith observation. An index plot of $\{c_i : i = 1(1)T\}$ can be examined for large values to determine influential points.

Unfortunately, c_i is not readily computable from the output of fitting a logistic model. Pregibon (1981) proposed approximating c_i by replacing the likelihood confidence region (5.4.20) by

$$\{\boldsymbol{\beta}: (\boldsymbol{\beta}-\hat{\boldsymbol{\beta}})'(\mathbf{X}'\hat{\mathbf{D}}\mathbf{X})^{-1}(\boldsymbol{\beta}-\hat{\boldsymbol{\beta}}) \leq \chi^2_{\alpha,k}\}$$

and the delete-one MLE $\hat{\boldsymbol{\beta}}_{-i}$ by

$$\hat{\boldsymbol{\beta}}^a_{-i} = \hat{\boldsymbol{\beta}} - \frac{(\mathbf{X}'\hat{\mathbf{D}}\mathbf{X})^{-1}\mathbf{e}_i^P(Y_i - m_i\hat{p}_i)}{1-h_i}.$$

The latter approximation comes from performing one step in the Newton–Raphson algorithm for computing $\hat{\boldsymbol{\beta}}_{-i}$ starting with value $\hat{\boldsymbol{\beta}}$. These approximations yield the formula

$$c_i^a = \frac{(e_i^a)^2 h_i}{(1-h_i)^2}. \tag{5.4.21}$$

In practice, the approximate values c_i^a tend to underestimate the exact c_i's.

The influence values c_i and c_i^a assess the effect of deleting the ith observation on the estimate of $\boldsymbol{\beta}$. Pregibon (1981), Johnson (1985), A.H. Lee (1987, 1988), and Williams (1987) derive other influence values obtained by measuring the changes in various aspects of fit. For example, Johnson (1985) measures the change in the estimate of $\mathbf{p}$ and in the classification of future observations and Williams (1987) measures the change in likelihood ratio test statistics.

Example 5.3.1 (continued). Figure 5.4.1 is a coded index plot of $\{e_i^P\}_{i=1}^T$ (code $= S$), $\{e_i^d\}_{i=1}^T$ (code $= D$), and $\{e_i^a\}_{i=1}^T$ (code $= A$) based on fitting the model

$$\text{logit}(p_i) = \beta_0 + \beta_1 \ln(LI) + \beta_2 I[\text{TEMP} \leq 1.002]. \tag{5.4.22}$$

While the overall pattern is similar for S, D, and A, it should be noted that the points with largest magnitudes are the adjusted residuals for cases (24, 26, and 16). Figure 5.4.2 is a probability plot of the $\{e_i^a\}$. The major feature of Figure 5.4.2 is the potential outlier in the lower left corner; as illustrated in Example 4.1.1, simulated bands could be added to the plot to aid in this assessment. The adjusted residual for the potential outlier corresponds to case 24 which, as remarked earlier, has fitted probability near .9 but the patient is relapsed ($Y_{24} = 0$)

Figures 5.4.3 and 5.4.4 are index plots of h_i and c_i^a, respectively. The leverage plot shows that cases 8 and 16 have $h_i = .46$ which is more than four times the average value of $k/N = 3/27 = .11$. Recall that cases 8 and 16 have identical covariate vectors and opposite responses. They are extreme in the design space being the only patients with high temperature and high LI values. Figure 5.4.4 highlights cases 16, 8, and 24 as having large c_i^a and being highly influential in determining the estimated $\boldsymbol{\beta}$.

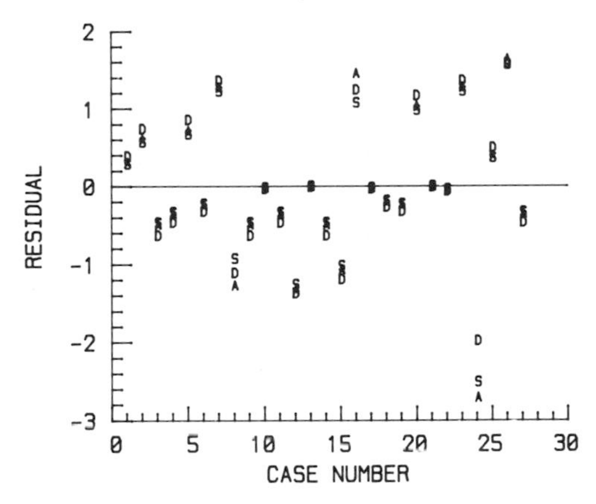

Figure 5.4.1. Coded index plot of Pearson residuals, deviance residuals, and adjusted residuals for Example 5.3.1 data based on Model 5.4.22.

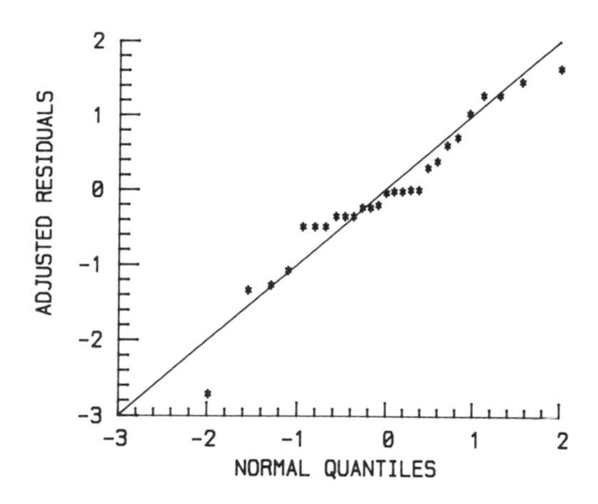

Figure 5.4.2. Normal probability plot of adjusted residuals for Example 5.3.1 data based on Model 5.4.22.

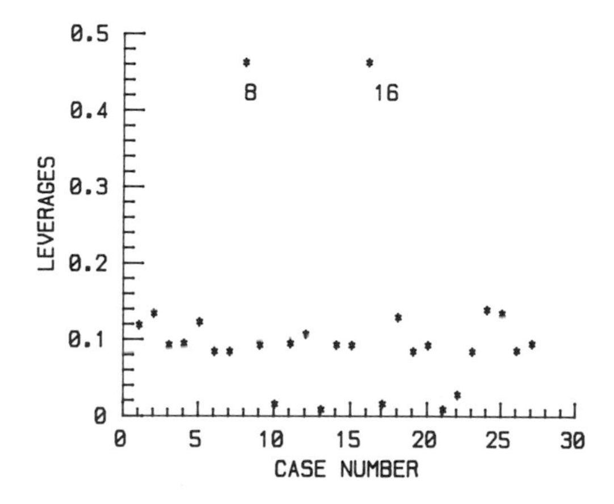

Figure 5.4.3. Index plot of leverages for Example 5.3.1 data based on Model 5.4.22.

For contrasts' sake Figure 5.4.5 is an index plot of the exact Cook's distances c_i. These are easily computed because of the small size of the data set. This plot is qualitatively different than Figure 5.4.4 for c_i^a. It clearly singles out case 16 as highly influential with $c_{16} = 36.49$ (whereas $c_{16}^a = 1.75$). Cases 8 and 24, on the other hand, appear qualitatively much less influential in Figure 5.4.5 than in Figure 5.4.4 because of the scaling of the second plot; however, for example, $c_8 = 1.52$ while $c_8^a = 1.43$.

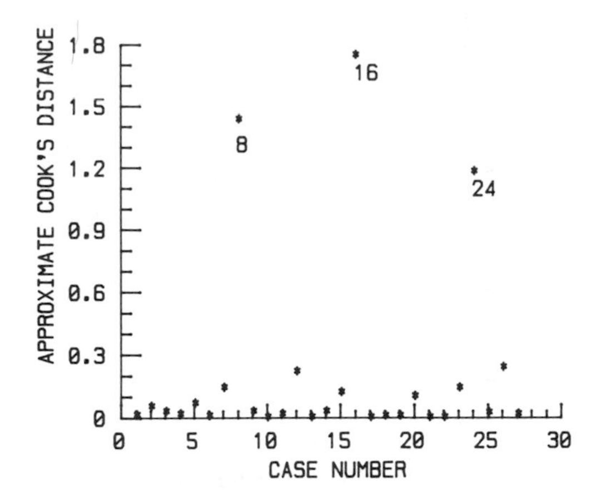

Figure 5.4.4. Index plot of approximate Cook's distance (5.4.21) for Example 5.3.1 data based on Model 5.4.22.

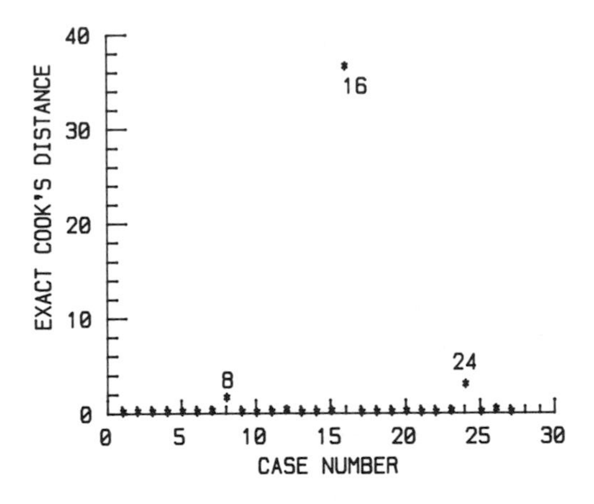

Figure 5.4.5. Index plot of exact Cook's distance for Example 5.3.1 data based on Model 5.4.22.

In sum the five diagnostic plots presented suggest two things.

(i) Because only two points have large values of LI *and* temperature, these points are highly influential. Any inferences made about patients with high temperature and high LI will be very tentative.

(ii) Observation 24 may be an outlier.

The comments above suggest additional investigation of cases 8, 16, and 24 is appropriate. To illustrate the impact of removing one of these points supposed it is determined that case 24 is abherent in some way and should not be used in the analysis. Table 5.4.2 shows the ML fit to the data with this case deleted. The most striking differences between this fit and the fit based on the full data are (i) the increase in magnitude of the coefficients and their estimated standard errors, (ii) the increase in magnitude of the z-score for the temperature coefficient, and (iii) the dramatic drop in the fit statistic. Diagnostic plots for the fit without observation 24 show no unusually large adjusted residuals and all influence values are less than .62. The largest approximate Cook's distance is now that of case 12. The leverages, as might be expected, hardly change.

Table 5.4.2. ML Fit of Model with Constant, $\ln(LI)$, and $I[\text{TEMP} \leq 100.2]$ Omitting Case 24

Coefficient	$\hat{\beta}$	ese$(\hat{\beta})$	z-score
Constant	−5.01	2.60	−1.93
$\ln(LI)$	7.79	3.45	2.26
$I[\text{TEMP} \leq 100.2]$	5.05	2.77	1.82
GOF statistic		$A^2 = 1.09$	
P-value		.90	

While it is interesting to observe the effects of deleting cases and important to know which observations are most strongly influencing the fit, extreme caution must be exercised in basing conclusions on reduced data sets. In this context, if a careful review of patient 24 yields no indication of error, deletion seems inadvisable. An alternative to completely deleting the case is to downweight it; see the earlier references on robust estimators.

The final method illustrated here is the partial residual plot. Suppose it is desired to investigate the form in which temperature affects p_i. Consider fitting the model

$$\text{logit}(p_i) = \beta_0 + \beta_1 \ln(LI) + \beta_2 \text{TEMP}. \tag{5.4.23}$$

The partial residual plot for TEMP should provide an indication of the appropriateness of the linear term in TEMP. Figure 5.4.6 is a plot of

$$\mathbf{r}_2 := \frac{\mathbf{Y} - \hat{\mathbf{p}}}{\hat{\mathbf{Y}}(\mathbf{1}_T - \hat{\mathbf{p}})} - \hat{\beta}_2 \text{TEMP}$$

($\hat{\beta}_2 = -42.46$ from Table 5.3.4). The dashed line is a smooth using 20% of the points and the solid line is linear with slope $\hat{\beta}_2$. The scatter is clearly nonlinear and appears qualitatively different for the low and high temperature subgroups. The dotted vertical line is TEMP $= 100.2$. The partial residual plot gives some evidence that the threshold model developed using maximally selected chi-squared statistics is more appropriate than (5.4.23).

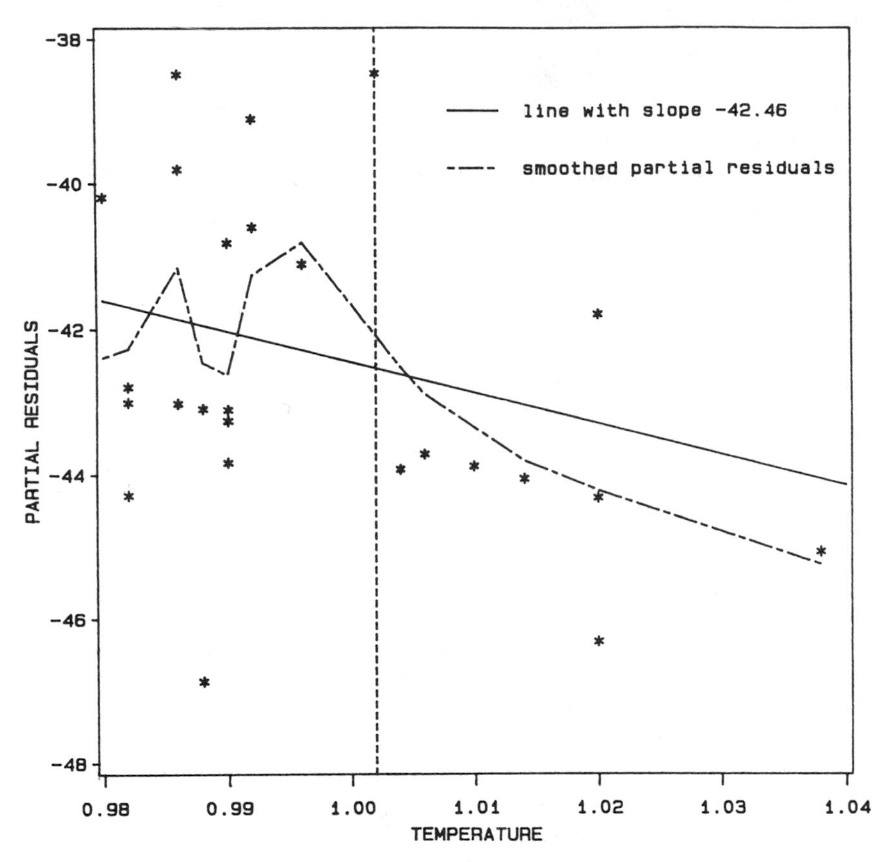

Figure 5.4.6. Partial residual plot for Example 5.3.1 data based on Model 5.4.23.

There are many other specialized diagnostics that have been proposed in the literature. A recent proposal by Fowlkes (1987) defines smoothed analogs of e_i^P and $\mathbf{r}_j$ which behave more like their normal theory counterparts. While the proposed diagnostics are very powerful, large sample sizes (300 or more) are desirable for the smoothing.

Lastly it should be noted that in some applications, the covariates are measured with error. An example is an epidemiological investigation of the factors associated with the occurrence of myocardial infarction; covariates such as the subject's smoking patterns over the past 25 years would rely on their recall. Carrol et al. (1984) and Stefanski and Carroll (1985) have considered such errors-in-variables problems for logistic regression models.

5.5 Two by Two by S Tables: Matched Analyses

A. Introduction

This section introduces some statistical methods that are useful for analyzing binary response data subject to block and treatment effects. The example considered throughout is that of S, 2×2 tables where the sth table ("strata", "block") consists of Y_{s1} successes out of m_{s1} trials from a binomial population Π_{s1} with success probability p_{s1}, and Y_{s2} successes out of m_{s2} trials from a second binomial population Π_{s2} with success probability p_{s2}, $1 \leq s \leq S$. It is assumed that $\{Y_{sj}\colon s = 1(1)S,\ j = 1,2\}$ are mutually independent. The data for strata s is presented in generic form in Table 5.5.1. In epidemiological and other applications, there are often *many* strata each with small numbers of trials. The stochastic models used to analyze such data have many nuisance parameters and special methods are required to analyze them because the techniques of Sections 5.2–5.4 are inefficient in this situation.

Table 5.5.1. Notation for Strata s Data from a $2 \times 2 \times S$ Table

	Π_{s1}	Π_{s2}	Total
Success	Y_{s1}	Y_{s2}	Y_{s+}
Failure	$m_{s1} - Y_{s1}$	$m_{s2} - Y_{s2}$	$m_{s+} - Y_{s+}$
Total	m_{s1}	m_{s2}	m_{s+}

Let $\psi_s := \dfrac{p_{s1}(1-p_{s2})}{p_{s2}(1-p_{s1})}$ denote the odds ratio for strata s, $1 \leq s \leq S$. The parameters for the model can equivalently be considered as $\{(p_{s1}, p_{s2})\colon 1 \leq s \leq S\}$ or $\{(\psi_s, p_{s1})\colon 1 \leq s \leq S\}$. A straightforward extension of equation (5.2.1) for a single 2×2 table shows that the conditional probability mass function of $(Y_{11}, Y_{21}, \ldots, Y_{S1})$ given $(Y_{1+} = t_1,\ Y_{2+} = t_2, \ldots, Y_{S+} = t_S)$ is

$$\prod_{s=1}^{S} \frac{\binom{m_{s1}}{y_{s1}}\binom{m_{s2}}{t_s - y_{s1}}\psi_s^{y_{s1}}}{\sum_{j=L_s}^{U_s}\binom{m_{s1}}{j}\binom{m_{s2}}{t_s - j}\psi_s^{j}} = \prod_{s=1}^{S} f(y_{s1} \,|\, t_s; \psi_s) \tag{5.5.1}$$

for $L_s \leq y_{s1} \leq U_s$, $1 \leq s \leq S$, where $L_s := \max\{0, t_s - m_{s2}\}$ and $U_s := \min\{m_{s1}, t_s\}$. Here $f(\cdot)$ is defined in equation (5.2.1).

Most of the statistical procedures for analyzing $2 \times 2 \times S$ tables assume either homogeneity of odds ratios; i.e., $\psi_1 = \ldots = \psi_S$, or arbitrary odds ratios $\psi_1, \ldots, \psi_S$. An intermediate model, used in relative risk regression, assumes that the $\{\psi_s\}$ depend on a set of (regression) parameters. Relative risk regression will not be discussed here; see Breslow (1976), Breslow and Day (1980), and the references therein.

As in the case of logistic regression, large sample analysis of $2 \times 2 \times S$ tables is studied under one of two models—the "large strata" model and the "sparse strata" model. In the large strata model all the strata sample sizes $\{m_{sj}: s = 1(1)S, j = 1,2\}$ grow, whereas in the sparse strata model S is large but the $\{m_{sj}\}$ are small. More specifically, large strata asymptotics assumes (i) S is fixed, (ii) $0 < p_{sj} < 1$ for all $s = 1(1)S$ and $j = 1,2$, and (iii) $\min\{m_{sj}\} \to \infty$ so that $m_{sj}/m_{++} \to \gamma_{sj} \in (0,1)$. Sparse strata asymptotics considers a sequence of problems with increasing numbers of strata. If the Sth problem in the sequence has S strata with sample sizes $\{m_{sj}^S: 1 \le s \le S, j = 1,2\}$ and success probabilities $\{p_{sj}^S: 1 \le s \le S, j = 1,2\}$ for $1 \le S < \infty$, then sparse strata asymptotics assumes that (i) $\{m_{sj}^S: 1 \le s \le S, j = 1,2, 1 \le S < \infty\}$ are uniformly bounded and (ii) for every j and s, $\{p_{sj}^S: s \le S < +\infty\}$ are bounded away from zero and one.

The usefulness of the sparse strata model is motivated by epidemiological investigations such as matched pairs studies. In a matched pairs study, S *pairs* of (similar) individuals are observed in which one member of each pair receives Treatment 1 while the other receives Treatment 2. Thus $m_{sj} = 1$ for all pairs (strata) s and both treatment groups ($j = 1,2$). If p_{sj} is the probability of a successful response for individuals on treatment j in the sth pair, then the usual model is

$$\text{logit}(p_{sj}) = \begin{cases} \alpha_s + \beta, & 1 \le s \le S, \quad j = 1 \\ \alpha_s, & 1 \le s \le S, \quad j = 2, \end{cases}$$

so that e^{β} is the *common* odds ratio for Treatment 1 to Treatment 2 and the $\{\alpha_s\}$ are (nuisance) strata effects.

Three problems are discussed in this section. First, two large sample tests of the homogeneity hypothesis $H_=: \psi_1 = \ldots = \psi_S$ are described. The alternative for both is the global hypothesis $H_{\neq}$: (not $H_=$). Second, large and small sample tests of zero partial association (zpa) are considered. The zpa hypothesis, $H_Z: \psi_1 = \ldots = \psi_S = 1$ means that the treatment and response categories are conditionally independent given the strata category. As mentioned in Section 4.5 (see (4.5.2) and (4.5.3)), two possible alternatives are the global hypothesis $H_{\neq}$, and the restricted hypothesis $H_R: \psi_1 = \ldots = \psi_S \neq 1$. Lastly, assuming homogeneity holds, five methods of estimating the common odds ratio are discussed.

B. Tests for Homogeneity of Odds Ratios

The likelihood ratio test is appropriate for testing $H_=$ versus $H_{\neq}$ in the large strata setting. The likelihood function is

$$L = \prod_{s=1}^{S} \binom{m_{s1}}{y_{s1}} \binom{m_{s2}}{y_{s2}} p_{s1}^{y_{s1}} (1 - p_{s1})^{m_{s1} - y_{s1}} p_{s2}^{y_{s2}} (1 - p_{s2})^{m_{s2} - y_{s2}}. \quad (5.5.2)$$

The denominator of the LR statistic is $L(\cdot)$ evaluated at its unrestricted maximum, $\hat{p}_{sj} = Y_{sj}/m_{sj}$, $1 \le s \le S$ and $j = 1, 2$. There are two methods of calculating the numerator of the LR statistic. First, $L(\cdot)$ can be expressed in terms of the parameters $\{(\psi_s, p_{s1})\colon 1 \le s \le S\}$ and numerically maximized subject to the restriction $\psi_1 = \ldots = \psi_S$. Alternately, one can observe that $\psi_1 = \ldots = \psi_S$ holds if and only if $\boldsymbol{\lambda}^{123} = \mathbf{0}$ in the $2 \times 2 \times S$ table of successes and failures with data $\{Z_{ijk}\colon 1 \le s \le S, j = 1, 2, k = 1, 2\}$ defined by $Z_{s11} = Y_{s1}$, $Z_{s12} = Y_{s2}$, $Z_{s21} = m_{s1} - Y_{s1}$, and $Z_{s22} = m_{s2} - Y_{s2}$ (see Section 4.3 and Problem 5.33). The Iterative Proportional Fitting (IPF) algorithm (Section 4.4) can then be used to find the MLE of $\{p_{sj}\}$ under $\boldsymbol{\lambda}^{123} = \mathbf{0}$, and hence to calculate the numerator of the LR statistic. The approximate size α test rejects $H_=$ when

$$-2 \ln \left(L(\tilde{\mathbf{p}}) / L(\hat{\mathbf{p}}) \right) \ge \chi^2_{\alpha, S-1} \tag{5.5.3}$$

where $\tilde{\mathbf{p}}$ is the MLE of $\mathbf{p} = \{p_{sj}\}$ under $H_=$.

An alternative test also valid for the large strata model which does not require the iterative computation of the LRT was proposed by Woolf (1955). (See also Grizzle, Starmer, and Koch, 1969.) Let $\hat{\psi}_s$ denote the estimated odds ratio in the sth strata; i.e.,

$$\hat{\psi}_s = \frac{Y_{s1}(m_{s2} - Y_{s2})}{Y_{s2}(m_{s1} - Y_{s1})}. \tag{5.5.4}$$

Many authors modify (5.5.4) by adding .5 to each Y_{sj} and $m_{sj} - Y_{sj}$. Recall from Section 5.2 (see equations (5.2.17) and (5.2.18)) that under large strata asymptotics the empiric logits $\ln(\hat{\psi}_s)$ are approximately distributed as

$$N\left[\ln(\psi_s), \hat{v}_s := \left\{\frac{1}{Y_{s1} + .5} + \frac{1}{Y_{s2} + .5} + \frac{1}{m_{s1} - Y_{s1} + .5} + \frac{1}{m_{s2} - Y_{s2} + .5}\right\}\right] \tag{5.5.5}$$

for $1 \le s \le S$. The hypothesis $H_=$ is equivalent to assuming that the $\ln(\hat{\psi}_1), \ldots, \ln(\hat{\psi}_S)$ have a common mean which is a linear hypothesis in the logits. Since $\{\ln(\hat{\psi}_s)\colon 1 \le s \le S\}$ are mutually independent, standard linear model theory rejects $H_=$ if and only if

$$\sum_{s=1}^{S} \left(\frac{(\ln \hat{\psi}_s)^2}{\hat{v}_s} \right) - \left(\frac{\{\sum_{s=1}^{S} \ln(\hat{\psi}_s)/\hat{v}_s\}^2}{\sum_{s=1}^{S} (\hat{v}_s)^{-1}} \right) \ge \chi^2_{\alpha, S-1}. \tag{5.5.6}$$

Tests of $H_=$ versus $H_{\neq}$ appropriate in the sparse strata framework are discussed by Liang and Self (1985), Hauck (1989), and Jones et al. (1989). The following example illustrates the two large strata tests.

Example 1.2.6 (continued). Example 1.2.6 concerned a tumorigenicity experiment in which mice are treated with the fungicide Avadex. There are $S = 4$ strata: (i) strain X male mice, (ii) strain X female mice, (iii) strain Y male mice, and (iv) strain Y female mice. For convenience the data are redisplayed as Table 5.5.2.

Table 5.5.2. Numbers of Mice Developing Tumor within Two Years in a Tumorigenicity Experiment

			Tumor	
Strain	Sex	Treatment	Y	N
X	M	Rx	4	12
		C	5	74
	F	Rx	2	14
		C	3	84
Y	M	Rx	4	14
		C	10	80
	F	Rx	1	14
		C	3	79

The estimated odds ratios in the 4 strata are $\hat{\psi}_1 = 4.93$, $\hat{\psi}_2 = 4.00$, $\hat{\psi}_3 = 2.29$, and $\hat{\psi}_4 = 1.88$. The likelihood ratio statistic (5.5.3) is .864 and Woolf's statistic (5.5.6) is .851; the respective P-values are .834 and .837. Clearly there is no evidence to suggest that the odds ratio of tumor incidence differ among the sex and strain combinations.

C. Tests of Zero Partial Association

The hypothesis of zpa is $H_Z\colon \psi_1 = \ldots = \psi_S = 1$. H_Z holds if and only if $p_{s1} = p_{s2}$ for $1 \le s \le S$; i.e., in each strata the probability of success is the same for the treated and the control groups (but the common strata probability can vary among the strata). Both small and large sample tests of H_Z will be discussed.

The literature on small sample tests for zpa focuses on the construction of UMPU tests. Birch (1964) assumes $\psi_1 = \ldots = \psi_S = e^\theta$ for some $\theta \in \mathbb{R}$ and considers a one-sided test of $\theta = 0$ (i.e., H_Z) versus $H_>\colon \theta > 0$. He shows that the UMPU test of H_Z versus $H_>$ is randomized with critical function

$$\phi_1^R = \begin{cases} 1 & \text{if } \sum_{s=1}^S Y_{s1} > c \\ \gamma & \text{if } \sum_{s=1}^S Y_{s1} = c \\ 0 & \text{if } \sum_{s=1}^S Y_{s1} < c, \end{cases}$$

where c and γ are functions of $Y_{1+}, \ldots, Y_{S+}$ chosen to satisfy $E_{\theta=0}[\phi_1^R \mid Y_{1+}, \ldots, Y_{S+}] = \alpha$. Thus the UMPU test is conditionally size α given the S-dimensional sufficient statistic $\mathbf{T} = (Y_{1+}, \ldots, Y_{S+})'$ (i.e., ϕ_1^R has Neyman structure with respect to $\mathbf{T}$); clearly, it is also unconditionally size α.

The numbers c and γ solve the equation

$$\sum_{\mathbf{y} \in \mathcal{U}(c)} \prod_{s=1}^{S} f(y_{s1} \mid Y_{s+}, \psi_s = 1) + \gamma \sum_{\mathbf{y} \in \mathcal{E}(c)} \prod_{s=1}^{S} f(y_{s1} \mid Y_{s+}, \psi_s = 1) = \alpha \tag{5.5.7}$$

where $\mathbf{y}' = (y_{11}, \ldots, y_{S1})$, $\mathcal{U}(c) = \{\mathbf{y}: L_s \leq y_{s1} \leq U_s \text{ and } \sum_{s=1}^{S} y_{s1} > c\}$ is an "upper tail" of outcomes, and $\mathcal{E}(c) = \{\mathbf{y}: L_s \leq y_{s1} \leq U_s \text{ and } \sum_{s=1}^{S} y_{s1} = c\}$ is a "slice" of outcomes.

Related quantities of interest can also be calculated. For example, the P-value of ϕ_1^R is given by

$$P-\text{value} = P\left[\sum_{s=1}^{S} Y_{s1} \geq y_{+1} \mid Y_{1+}, \ldots, Y_{S+}, \psi_1 = \ldots = \psi_S = 1\right]. \tag{5.5.8}$$

EGRET and StatXact contain programs to calculate (5.5.8). These programs also calculate conditional (hence, unconditional) tail intervals for the common odds ratio $\psi := \psi_1 = \ldots = \psi_S$ by solving equations (5.5.9) and (5.5.10) below for $\underline{\psi}$ and $\overline{\psi}$, respectively:

$$P[Y_{+1} \geq y_{+1} \mid Y_{1+} = y_{1+}, \ldots, Y_{S+} = y_{S+}, \psi_1 = \ldots = \psi_S = \underline{\psi}] \overset{set}{=} \alpha/2 \tag{5.5.9}$$

$$P[Y_{+1} \leq y_{+1} \mid Y_{1+} = y_{1+}, \ldots, Y_{S+} = y_{S+}, \psi_1 = \ldots = \psi_S = \overline{\psi}] \overset{set}{=} \alpha/2. \tag{5.5.10}$$

See Mehta, Patel, and Gray, 1985.

UMPU tests of H_Z have also been derived for alternative hypotheses other than $H_>$. Berger, Wittes, and Gold (1979) consider the setup in which a unit vector $\mathbf{v} \in \mathbb{R}^S$ is fixed and it is assumed that $(\ln(\psi_1), \ldots, \ln(\psi_S))' = \theta\mathbf{v}$ for some $\theta \in \mathbb{R}$. Under this model, they develop the UMPU test of H_Z: $\theta = 0$ versus $H_{\mathbf{v}}: \theta > 0$. As a special case, if $\mathbf{v}' = (1/\sqrt{S}, \ldots, 1/\sqrt{S})$ then the model implies that $\psi_1 = \ldots = \psi_S$ and $H_{\mathbf{v}}$ specializes to $H_>$. However, if $\mathbf{v} \neq (1/\sqrt{S}, \ldots, 1/\sqrt{S})$ then the alternative $H_{\mathbf{v}}$ represents a different line through the origin. Another generalization is testing zpa in general $R \times C \times S$ tables for which Birch (1965) discusses UMPU tests.

Two *large strata* tests of H_Z are described next. The numerator of the likelihood ratio statistic of H_Z versus the global alternative H_G: (not H_Z) is the likelihood function (5.5.2) evaluated at the MLEs of the probabilities $\{p_{sj}\}$ under H_Z which are trivially $\hat{p}_{s1} = \hat{p}_{s2} = Y_{s+}/m_{s+}$, $1 \leq s \leq S$. The denominator is the likelihood function evaluated at $\hat{p}_{sj} = Y_{sj}/m_{sj}$, and the critical point for minus two times the log likelihood ratio is $\chi^2_{\alpha,S}$. More often the alternative of interest is H_R: $\psi_1 = \ldots = \psi_S \neq 1$. In this case an iterative calculation is necessary to obtain the denominator of the likelihood ratio statistic. The IPF algorithm can be used since H_Z holds if and only if $\boldsymbol{\lambda}^{123} = \mathbf{0} = \boldsymbol{\lambda}^{12}$ in the $2 \times 2 \times S$ table (see Problem 5.33). With alternative H_R, the critical point for minus two times the likelihood ratio is $\chi^2_{\alpha,1}$.

A computationally simpler large strata test of H_Z versus H_R is the Cochran–Mantel–Haenszel test (Cochran (1954), Mantel and Haenszel (1959)). The CMH test rejects H_Z if and only if

$$X^2_{CMH} := \frac{(\sum_{s=1}^{S}\{Y_{s1} - E^c_s\})^2}{\sum_{s=1}^{S} \hat{v}^c_s} \geq \chi^2_{\alpha,1} \tag{5.5.11}$$

where

$$E^c_s = Y_{s+}\left(\frac{m_{s1}}{m_{s+}}\right) \quad \text{and} \quad \hat{v}^c_s = \frac{Y_{s+}(m_{s+} - Y_{s+})m_{s1}m_{s2}}{m^2_{s+}(m_{s+} - 1)}.$$

Intuitively, E^c_s is the expected number of successes on treatment 1 in strata s based on the conditional distribution of Y_{s1} given Y_{s+} when $\psi_s = 1$; i.e., based on the hypergeometric distribution. Similarly, $\hat{v}^c_s$ is the conditional variance of Y_{s1} based on the same hypergeometric distribution.

The CMH test relies on the conditional distribution (5.5.1) and thus does not require estimation of nuisance parameters. This test was first proposed by Cochran on intuitive grounds. Day and Byar (1979) demonstrated that it is the score test of H_Z versus H_R (Problem 5.34). Birch (1964, 1965) showed that with an appropriate small-sample critical point, the CMH test is the UMPU test of H_Z versus H_R. A one-tailed version of the CMH test with null hypothesis H_Z and alternative H_+: $\psi_1 = \ldots = \psi_S > 1$ is studied by Gart (1971). His proposal rejects H_Z if and only if

$$\frac{\sum_{s=1}^{S}(Y_{s1} - E^c_s)}{(\sum_{s=1}^{S} \hat{v}_s)^{1/2}} \geq z_\alpha$$

where E^c_s is defined by (5.5.5) and $\hat{v}_s$ is defined in (5.5.5). The power function of the two-sided CMH test has been studied by Bennett and Kaneshiro (1974), Li, Simon, and Gart (1979), and Wittes and Wallenstein (1987).

Example 1.2.6 (continued). The small sample one-sided UMPU test of H_Z:$\theta = 0$ versus $H_>$: $\theta > 0$ for the data in Table 5.5.2 has a P-value of .0072. A 95% tail interval for ψ is (1.24, 7.13). The two-tailed CMH test of H_Z: $\psi_1 = \ldots = \psi_S = 1$ versus H_R: $\psi_1 = \ldots = \psi_S \neq 1$ has a P-value of .0086. The data provide strong evidence that the common odds ratio is larger than 1; i.e., that Avadex is carcinogenic.

D. Estimation of a Common Odds Ratio

When the odds ratios are homogeneous across strata, an important problem is to estimate the common value ψ. Five estimation methods are discussed and their behavior described under both the large and sparse strata models.

The Empiric Logit (Woolf) Estimator

The empiric logit estimator of ψ is a weighted least squares estimate based on the approximate distribution of the empiric logits $\ln(\hat{\psi}_s)$ given in equation (5.5.5). The estimator, also called Woolf's estimator, is denoted $\hat{\psi}_w$ and defined by

$$\ln(\hat{\psi}_w) = \frac{\sum_{s=1}^{S} w_s \ln \hat{\psi}_s}{\sum_{s=1}^{S} w_s} \tag{5.5.12}$$

where $w_s = \hat{v}_s^{-1}$ and $\hat{v}_s$ is defined in (5.5.5). Woolf's estimator is simple to compute, and is consistent, asymptotically normal, and efficient for ψ under the large strata model. However, $\hat{\psi}_w$ need not even be consistent for ψ in the sparse strata case (Breslow (1981)).

The (Unconditional) Maximum Likelihood Estimator

Setting $\psi = \psi_1 = \ldots = \psi_S$ in the likelihood (5.5.2) and simplifying yields:

$$L = \prod_{s=1}^{S} \binom{m_{s1}}{y_{s1}} \binom{m_{s2}}{y_{s2}} \frac{\psi^{m_{s2}-y_{s2}}(1-p_{s1})^{m_{s+}-y_{s+}} p_{s1}^{y_{s+}}}{[p_{s1} + \psi(1-p_{s1})]^{m_{s2}}}. \tag{5.5.13}$$

The value of ψ maximizing the right hand side of (5.5.13), denoted $\hat{\psi}_{mle}$, is the (unconditional) MLE of ψ. Since maximizing L is equivalent to fitting the $\boldsymbol{\lambda}^{123} = \mathbf{0}$ model to the associated $2 \times 2 \times S$ table of success and failures, the IPF algorithm can be used to calculate $\hat{\psi}_{mle}$ (Problem 5.33).

The unconditional MLE is consistent and asymptotically normal in the large strata case, but it need not be consistent in the sparse strata case. Intuitively, as $S \to \infty$ the number of nuisance parameters approaches infinity and this is responsible for the inconsistency of $\hat{\psi}_w$ and $\hat{\psi}_{mle}$ (Problem 5.35).

The Mantel–Haenszel Estimator

The MH estimator of ψ is given by the closed form expression

$$\hat{\psi}_{mh} = \frac{\sum_{s=1}^{S} Y_{s1}(m_{s2} - Y_{s2})/m_{s+}}{\sum_{s=1}^{S} Y_{s2}(m_{s1} - Y_{s1})/m_{s+}}. \tag{5.5.14}$$

It can also be written as

$$\hat{\psi}_{mh} = \sum_{s=1}^{S} \hat{\psi}_s \left[\frac{Y_{s2}(m_{s1} - Y_{s1})/m_{s+}}{\sum_{s=1}^{S} Y_{s2}(m_{s1} - Y_{s1})/m_{s+}} \right]$$

which shows $\hat{\psi}_{mh}$ is a weighted average (with nonnegative data dependent weights) of the $\hat{\psi}_s$, $1 \leq s \leq S$. Since the weights sum to 1 the representation is as a convex combination of the $\hat{\psi}_s$.

The MH estimator, like Woolf's estimator, is easy to compute. It is both consistent and asymptotically normal in the large strata case; however, it is only efficient for the large strata case when $\psi = 1$. Hauck (1979) derives an estimator for the *variance* of $\hat{\psi}_{mh}$ which is consistent under large strata asymptotics. Breslow (1981) shows that $\hat{\psi}_{mh}$ is consistent and asymptotically normal under sparse strata asymptotics and gives an estimator of the variance of $\hat{\psi}_{mh}$ which is consistent in this case. Flanders (1985) and Robins, Breslow, and Greenland (1986) propose estimators of the variance of $\hat{\psi}_{mh}$ which are consistent under either asymptotic model. The Robins et al. estimator of $\text{var}(\hat{\psi}_{mh})$ is

$$\widehat{\text{var}}(\hat{\psi}_{mh}) = (\hat{\psi}_{mh})^2 \left\{ \frac{\sum_{s=1}^{S} P_s R_s}{2R_+^2} + \frac{\sum_{s=1}^{S}(P_s R_s + Q_s R_s)}{2R_+ S_+} + \frac{\sum_{s=1}^{S} Q_s S_s}{2S_+^2} \right\} \tag{5.5.15}$$

where $P_s = (Y_{s1} + m_{s2} - Y_{s2})/m_{s+}$, $Q_s = (Y_{s2} + m_{s1} - Y_{s1})/m_{s+}$, $R_s = Y_{s1}(m_{s2} - Y_{s2})/m_{s+}$, and $S_s = Y_{s2}(m_{s1} - Y_{s1})/m_{s+}$.

The Conditional Maximum Likelihood Estimator

The conditional maximum likelihood estimator $\hat{\psi}_{cmle}$ maximizes

$$L_c(\psi) = \prod_{s=1}^{S} f(y_{s1} \mid y_{s+}; \psi)$$

where $f(\cdot \mid \cdot)$ is defined in (5.2.1). Iterative calculation is required to compute $\hat{\psi}_{cmle}$ and EGRET and StatXact, for example, provide programs to calculate it. Under the large strata model, $\hat{\psi}_{cmle}$ is consistent, asymptotically normal and efficient, while under the sparse strata model it is consistent and asymptotically normal. In the latter case, the asymptotic variance of $\hat{\psi}_{cmle}$ is the same as the asymptotic variance of $\hat{\psi}_{mh}$ and hence can be consistently estimated by (5.5.15).

Bayesian Estimators

Wypij (1986) studies both Bayesian and empirical Bayesian estimators of ψ and more generally of log odds ratio regression coefficients. In the Bayesian case, he considers logistic priors for the parameter $\ln(\psi)$; the resulting estimator has an intuitive interpretation as the conditional MLE ($\hat{\psi}_{cmle}$) based on adding certain pseudo tables to the original data. This is similar to the interpretation of Bayes estimators for the vector of multinomial success probabilities as the MLE based on the original data together with fractional successes and failures. The Bayes estimator is consistent in both asymptotic models and Wypij's simulation studies show that there can be important MSE improvements in both large and small sample settings.

The most frequently used of these five estimators is $\hat{\psi}_{mh}$. However, it is recommended that $\hat{\psi}_{cmle}$ be used when possible, even though it is harder to calculate, because it is fully efficient in the large strata case.

Example 1.2.6 (continued). The analysis in part B of this section shows that it is reasonable to believe the odds ratios in the Avadex data are constant across strata. Applying the four frequentest estimation methods to these data gives $\hat{\psi}_w = 3.14$, $\hat{\psi}_{mle} = 3.11$, $\hat{\psi}_{mh} = 3.08$, and $\hat{\psi}_{cmle} = 3.05$. A 95% confidence interval for ψ based on the asymptotic normality of $\hat{\psi}_{cmle}$ and (5.5.15) is (1.36, 6.79).

Problems

5.1. Show that the following threshold argument leads to the model of equation (5.1.2). Consider an unobserved continuous random variable Z with distribution function $F(\cdot)$ such that the observed random variable Y is 1 if and only if $Z \leq t_0$; here t_0 is an unknown threshold level. Suppose that t_0 is modeled by the linear combination $\mathbf{x}'\boldsymbol{\beta}$ where $\mathbf{x}$ is a known covariate vector associated with response Y and $\boldsymbol{\beta}$ is an unknown parameter vector. Prove that Y satisfies (5.1.2).

5.2. Several parametric families of cumulative distribution functions have been proposed as link function models for the success probabilities of binary regression data $(Y_i, m_i, \mathbf{x}_i)$, $1 \leq i \leq T$. Let $\mathcal{F} = \{F(\cdot \mid \boldsymbol{\theta})\colon \boldsymbol{\theta} \in \Omega\}$ denote a family of distributions; the assumption is

$$p(\mathbf{x}_i) = F(\mathbf{x}_i'\boldsymbol{\beta} \mid \boldsymbol{\theta}) \quad \text{for some} \quad \boldsymbol{\beta} \in \mathbb{R}^k \text{ and } \boldsymbol{\theta} \in \Omega.$$

This problem introduces three such families.

(a) Prentice (1976) essentially proposes using the family of cdf's corresponding to the natural logarithms of F-distributed random variables for $\mathcal{F}$. His proposed family is

$$\mathcal{F}_p = \{F(\cdot \mid \sigma, m_1, m_2)\colon \sigma > 0, m_1 > 0, m_2 > 0\}$$

where $F(z \mid \sigma, m_1, m_2) = \int_{-\infty}^{z} f(w \mid \sigma, m_1, m_2) dw$ and

$$f(w \mid \sigma, m_1, m_2) = \frac{\Gamma(m_1 + m_2)}{\Gamma(m_1)\Gamma(m_2)} \exp\left(\frac{m_1 w}{\sigma}\right)$$
$$\left\{1 + \exp\left(\frac{w}{\sigma}\right)\right\}^{-(m_1+m_2)}.$$

Show that (i) $f(w \mid \sigma, m_1, m_2)$ reduces to the logistic density when $m_1 = m_2 = 1$; (ii) $f(w \mid \sigma, m_1, m_2)$ approaches the normal density when $m_1 = 1$ as $m_2 \longrightarrow \infty$; and (iii) $f(w \mid \sigma, m_1, m_2)$ approaches the double exponential density ($h(t) = e^{-|t|}/2$) as $m_1 \longrightarrow 0$ and $m_2 \longrightarrow 0$.

(b) Aranda-Ordaz (1981) proposes a one-parameter family $\mathcal{F}_{AO}$ of distribution functions:

$$\mathcal{F}_{AO} = \{F(\cdot \mid \lambda) \colon \lambda \in \mathbf{R}\}$$

where

$$F(z \mid \lambda) = \begin{cases} 0, & \lambda z \leq -2 \\ \dfrac{(1+\frac{1}{2}\lambda z)^{1/\lambda}}{(1+\frac{1}{2}\lambda z)^{1/\lambda} + (1-\frac{1}{2}\lambda z)^{1/\lambda}}, & |\lambda z| < 2 \\ 1, & \lambda z \geq 2. \end{cases}$$

Show that $F(\cdot \mid \lambda)$ reduces to the linear distribution when $\lambda = 1$ and approaches the logistic distribution as $\lambda \longrightarrow 0$.

(c) Stukel (1988) considers a two parameter family indexed by $\boldsymbol{\alpha} = (\alpha_1, \alpha_2)'$ which extends the logistic model as follows

$$\text{logit}(p(\mathbf{x})) = h_{\boldsymbol{\alpha}}(\mathbf{x}'\boldsymbol{\beta})$$

where

$$h_{\boldsymbol{\alpha}}(z) = \begin{cases} \alpha_1^{-1}\{\exp(\alpha_1 z) - 1\}, & z \geq 0, \alpha_1 > 0 \\ -\alpha_1^{-1}\ln(1 - \alpha_1 z), & z \geq 0, \alpha_1 < 0 \\ z, & z \geq 0 \text{ and } \alpha_1 = 0, \\ & \text{or } z < 0 \text{ and } \alpha_2 = 0 \\ -\alpha_2^{-1}\{\exp(-\alpha_2 z) - 1\}, & z < 0, \alpha_2 > 0 \\ \alpha_2^{-1}\ln(1 + \alpha_2 z), & z < 0, \alpha_2 < 0. \end{cases}$$

Show that Stukel's model reduces to the logistic model for $\alpha = (0,0)$. Show that $p(\mathbf{x})$ is symmetric about 1/2 when $\alpha_1 = \alpha_2$. Contrast this approach with those taken by Prentice and Aranda-Ordaz.

5.3. Derive the likelihood equations based on the model of equation (5.1.7).

5.4. The data of Example 1.2.7 concern the severity of nausea for cancer patients receiving chemotherapy either with (Cis) or without (~Cis) cisplatinum. The (multinomial) responses Y take values 0 := none, 1, 2, 3, 4, 5 := very severe. Consider a model based on the cumulative logits

$$\ln\left(\frac{P(Y \geq j+1)}{1 - P(Y \geq j+1)}\right) = \beta_{0j} + \beta_{1j} I[\text{Cis}]$$

for $0 \leq j \leq 4$ where $I[\text{Cis}]$ is 1 (0) for patients (not) receiving cisplatinum.

(a) Interpret β_{1j}.

(b) Interpret this model when $\beta_{10} = \beta_{11} = \ldots = \beta_{14}$.

(c) Fit the model as given and the model under (b) to the data of Table 1.2.8. Use the results to test H_0: $\beta_{10} = \beta_{11} = \ldots = \beta_{14}$ versus H_A: (not H_0).

5.5. Suppose $Y_1 \sim B(m_1, p_1)$ is independent of $Y_2 \sim B(m_2, p_2)$ and it is desired to test H_0: $p_1 = p_2$ versus $H_{\neq}$: $p_1 \neq p_2$.

(a) Show the LRT rejects H_0 if and only if

$$2\sum O_{ij} \ln(E_{ij}^0 / E_{ij}^A) \geq \chi^2_{\alpha,1}$$

where the $\{O_{ij}: 1 \leq i, j \leq 2\}$ are the 4 *observed* counts from the 2×2 table:

Y_1	Y_2
$m_1 - Y_1$	$m_2 - Y_2$

the $\{E_{ij}^0: 1 \leq i, j \leq 2\}$ are the *expected* counts under H_0:

$m_1\hat{p}$	$m_2\hat{p}$
$m_1(1-\hat{p})$	$m_2(1-\hat{p})$

($\hat{p} := Y_+/m_+$), and the $\{E_{ij}^A: 1 \leq i, j \leq 2\}$ are the *expected* counts under H_A:

$m_1\hat{p}_1$	$m_2\hat{p}_2$
$m_1(1-\hat{p}_1)$	$m_2(1-\hat{p}_2)$

where $\hat{p}_i := Y_i/m_i$, $i = 1, 2$.

(b) Show that the Wald test can be written as

$$\text{Reject } H_0 \iff (\hat{p}_1 - \hat{p}_2)^2 \geq \chi^2_{\alpha,1} \left\{ \frac{\hat{p}_1(1-\hat{p}_1)}{m_1} + \frac{\hat{p}_2(1-\hat{p}_2)}{m_2} \right\}.$$

Note that, as in the normal means problem (Problem A.2), the right-hand side is a valid estimate of $\text{Var}(\hat{p}_1 - \hat{p}_2)$ under H_0 or $H_{\neq}$.

(c) Prove that the score test can be written as

$$\text{Reject } H_0 \iff (\hat{p}_1 - \hat{p}_2)^2 \geq \chi^2_{\alpha,1} \left\{ \hat{p}(1-\hat{p}) \left(\frac{1}{m_1} + \frac{1}{m_2} \right) \right\}.$$

Again paralleling Problem A.2, the right-hand side is a valid estimate of $\text{Var}(\hat{p}_1 - \hat{p}_2)$ only under H_0.

5.6. Show that the score statistic from Problem 5.5 can also be written as:

$$\frac{(\hat{p}_1 - \hat{p}_2)^2}{(\hat{p}(1-\hat{p})(m_1^{-1} + m_2^{-1}))} = \sum_{i=1}^{2}\sum_{j=1}^{2} \frac{[O_{ij} - E_{ij}^0]^2}{E_{ij}^0} \tag{1}$$

$$= \sum_{i=1}^{2} \frac{m_i(\hat{p}_i - \hat{p})^2}{\hat{p}(1-\hat{p})} \tag{2}$$

$$= \sum_{i=1}^{2} \frac{(Y_i^2/m_i) - (Y_+^2/m_+)}{\hat{p}(1-\hat{p})} \tag{3}$$

where the O_{ij} and E_{ij}^0 are given in part (a) of Problem 5.5. Equation (1) shows that the score statistic is the usual Pearson chi-squared test of equality of cell proportions. Equation (2) shows that it can be viewed as an analog of the one-way ANOVA F-statistic in that it rejects homogeneity of p_1 and p_2 when $\hat{p}_1$ and $\hat{p}_2$ vary too much. Equation (3) can be used for hand computation.

5.7. Suppose $Y_1 \sim B(m_1, p_1)$ is independent of $Y_2 \sim B(m_2, p_2)$ and let $\rho = p_1/p_2$. Consider testing H_0: $\rho = \rho_0$ (given) versus $H_{\neq}$: $\rho \neq \rho_0$ based on the Cressie–Read statistic

$$I^\lambda(\rho_0) = \frac{2}{\lambda(\lambda+1)}\left[\sum\left\{O_{ij}\left(\frac{O_{ij}}{E_{ij}}\right)^\lambda\right\} - 2\right]$$

where the O_{ij} are given in Problem 5.5 and the E_{ij} are the expected counts under H_0 $(\rho = \rho_0)$.

(i) Show that I^1 is the Pearson statistic and I^0 is the likelihood ratio statistic.

(ii) Prove that $\{\rho_0 : I^\lambda(\rho_0) \leq c\}$ is an interval containing $\hat{\rho} = \hat{p}_1/\hat{p}_2$, the MLE of ρ.

[Bedrick (1987) examines a number of large sample intervals of the form (ii) and recommends the choice $\lambda = 2/3$.]

5.8. Suppose $Y_1 \sim B(m_1, p_1)$ is independent of $Y_2 \sim B(m_2, p_2)$ where $0 < p_1, p_2 < 1$ and $m_2/m_1 \to \tau^2$ as $m_1 \to \infty$. Show that

$$\sqrt{m_1}[(\hat{p}_1, \tau\hat{p}_2) - (p_1, \tau p_2)] \to N_2[0_2, \Sigma]$$

where $\Sigma = \text{Diag}[p_1(1-p_1),\ p_2(1-p_2)]$ and $\hat{p}_i = Y_i/m_i$ for $i = 1, 2$. Apply the delta method to show that the large sample distributions of the estimators (i) $\hat{\psi} = \hat{p}_1(1-\hat{p}_2)/(1-\hat{p}_1)\hat{p}_2$ of the odds ratio, (ii) $\hat{\Delta} = \hat{p}_1 - \hat{p}_2$ of the attributable risk, and (iii) $\hat{\rho} = \hat{p}_1/\hat{p}_2$ of the relative risk satisfy

$$\hat{\rho} \overset{app}{\sim} N\left[\rho, \rho^2\left\{\frac{1-p_1}{m_1 p_1} + \frac{1-p_2}{m_1\tau^2 p_2}\right\}\right],$$

$$\ln\{\hat{\rho}\} \overset{app}{\sim} N\left[\ln\{\rho\}, \left\{\frac{1-p_1}{m_1 p_1} + \frac{1-p_2}{m_1 \tau^2 p_2}\right\}\right],$$

$$\hat{\Delta} \overset{app}{\sim} N\left[\Delta, \left\{\frac{p_1(1-p_1)}{m_1} + \frac{p_2(1-p_2)}{m_1 \tau^2}\right\}\right], \text{ and}$$

$$\hat{\psi} \overset{app}{\sim} N\left[\psi, \psi^2\left\{\frac{1}{m_1 p_2(1-p_1)} + \frac{1}{m_1 \tau^2 p_2(1-p_2)}\right\}\right]$$

for large m_1 and $m_2 \simeq m_1 \tau^2$.

5.9. Consider the nephropathic cystosis data of Example 1.2.4. Calculate the score and Wald statistics for the homogeneity hypothesis. Compute point and interval estimates of Δ, ρ, and ψ. Is there evidence that Vitamin C leads to improvement in nephropathic cystosis patients?

5.10. Apply the conditional Cornfield tail limits (5.2.23) to determine a 95% confidence interval of the odds of esophageal cancer among low alcohol consumers to that of esophageal cancer among high alcohol consumers based on the data of Example 1.2.5.

5.11. Consider testing $H_0: p_1 = p_2$ versus $H_>: p_1 > p_2$ based on independent binomial observations $Y_1 \sim B(m_1, p_1)$ and $Y_2 \sim B(m_2, p_2)$. Let $\mathbf{y} = (y_1, y_2)'$ and $\mathbf{m} = (m_1, m_2)'$. The following three tests of H_0 reject the null hypothesis when

$$T(\mathbf{y}, \mathbf{m}) \geq c$$

where the test statistics $T(\cdot,\cdot)$ are given below and the constant $c = c(y_+)$ may depend on y_+.

(1) Fisher's exact test: $T_F = y_1$.

(2) Difference in proportions: $T_D = y_1/m_1 - y_2/m_2$.

(3) Log odds ratio: $T_R = \log(y_1(m_2 - y_2)/(m_1 - y_1)y_2)$.

Although the (unconditional) null distributions of each of these test statistics depend on the (unknown) common value of $p_1 = p_2$, the conditional null distributions given $Y_+ = y_+$ do not. Thus it is common practice to report the P-values

$$p(T) = p(T; \mathbf{y}, \mathbf{m}) = \sum Pr(\mathbf{x})$$

where

$$Pr(\mathbf{x}) = \binom{m_1}{x_1}\binom{m_2}{x_2} / \binom{m_+}{x_+}$$

and the summation is over $\{\mathbf{x} \in \mathcal{T}(\mathbf{y}): T(\mathbf{x}, \mathbf{m}) > T(\mathbf{y}, \mathbf{m})\}$ with $\mathcal{T}(\mathbf{y}) = \{\mathbf{x}: 0 \leq x_i \leq m_i \text{ and integral for } i = 1, 2 \text{ and } x_+ = y_+\}$

denoting the set of all outcomes *consistent* with the observed total y_+. Prove that $p(T_F; \mathbf{y}, \mathbf{m}) = p(T_D; \mathbf{y}, \mathbf{m}) = p(T_R; \mathbf{y}, \mathbf{m})$ for all $\mathbf{y}$. [Davis (1986b)]

5.12. Consider the two-sided version of Problem 5.11 in which $H_0: p_1 = p_2$ is tested against $H_{\neq}: p_1 \neq p_2$. The following five tests reject H_0 when

$$T(\mathbf{y}, \mathbf{m}) \geq c$$

where the test statistics $T(\cdot, \cdot)$ are given below and the constants $c = c(y_+)$ may depend on y_+.

(1) Fisher's exact test: $T_F = -Pr(\mathbf{y})$.

(2) Difference in proportions: $T_D = |y_1/m_1 - y_2/m_2|$.

(3) Log odds ratio: $T_R = |\log(y_1(m_2 - y_2)/(m_1 - y_1)y_2)|$.

(4) Pearson chi-square: $T_P = \sum_{i=1}^{2} \left[\frac{(y_i - S_i)^2}{S_i} + \frac{((m_i - y_i) - F_i)^2}{F_i} \right]$.

(5) Likelihood ratio chi-square:

$$T_L = 2 \sum_{i=1}^{2} \left[y_i \log \left(\frac{y_i}{S_i} \right) + (m_i - y_i) \log \left(\frac{m_i - y_i}{F_i} \right) \right].$$

Here $Pr(\mathbf{y})$ is given in Problem 5.11 and $S_i = m_i(y_+/m_+)$ and $F_i = m_i(1 - y_+/m_+)$ are the expected numbers of successes and failures in the ith binomial population, $i = 1, 2$, under H_0. Thus Fisher's test places all possible outcomes with *smaller* null probability than the observed outcome in the rejection region.

(a) Show that T_D and T_P yield identical conditional P-values for any outcome $\mathbf{Y} = \mathbf{y}$. (The conditional P-value is defined in Problem 5.11.)

(b) Suppose that $m_1 = 5$, $m_2 = 25$ and $y_+ = 9$. There are six outcomes consistent with $Y_+ = 9$; these correspond to values of 0, 1, 2, 3, 4, or 5 for y_1. Show that T_F, T_D, T_R, and T_L order these six outcomes differently. This implies that the conditional P-values associated with these four test statistics need not be consistent.

[Davis (1986b)]

5.13. Consider T sample binomial data with covariates $x_1 \leq \ldots \leq x_T$ satisfying the model $p_i = G(\alpha + \beta x_i)$, $1 \leq i \leq T$. Show that the score test of $H_=: p_1 = \ldots = p_T$ versus the trend alternative $H_\geq$: $p_1 \geq \ldots \geq p_T$ (with at least one inequality holding strictly) is the statistic X^2_{CA} for any monotone twice differentiable function $G(\cdot)$. [Tarone and Gart (1980)]

5.14. Suppose that $Y_1, \ldots, Y_T$ are independent with $Y_i \sim B(m_i, p_i)$ with m_i known and $0 < p_i < 1$ unknown for $1 \le i \le T$. Consider forming simultaneous confidence intervals for the R contrasts $\mathbf{c}_1'\mathbf{p}, \ldots, \mathbf{c}_R'\mathbf{p}$ where $R \le T$, $\mathbf{c}_r = (c_{r1}, \ldots, c_{rT})$ satisfies $\sum_{j=1}^{T} c_{rj} = 0$ for $1 \le r \le R$, and $\mathbf{c}_r'\mathbf{c}_s = 0$ for all $1 \le r \ne s \le R$. Apply the Scheffe projection method to show that

$$\{\mathbf{c}_r'\hat{\mathbf{p}} \pm (\chi^2_{\alpha,R})^{1/2}(\mathbf{c}_r'\hat{\Sigma}\mathbf{c}_r)^{1/2} : 1 \le r \le R\}$$

are asymptotic simultaneous $100(1-\alpha)\%$ confidence intervals where $\hat{\Sigma} = \mathrm{Diag}[\ldots, \hat{p}_i(1-\hat{p}_i)/m_i, \ldots]$.

5.15. In June 1979 a Civil Service entrance examination was given to 36,797 applicants wishing to become police officers. Table 5.P.15 lists the numbers of Blacks, Hispanics, and Whites who took and passed the exam. Let p_B, p_H, and p_W be the probabilities of a Black, Hispanic, and White passing the exam, respectively. One statistical analysis of these data considers Whites as a control group and forms simultaneous confidence intervals for $p_W - p_B$ and $p_W - p_H$. Determine asymptotic 95% simultaneous confidence intervals for these two differences.

[Raab (1980)]

Table 5.P.15. Numbers of Blacks, Hispanics, and Whites Taking and Passing a Civil Service Exam to Become a Police Officer (Reprinted with permission from "Police Strength and Racial Mix Await Decisions in Court Battle" by S. Raals, January 30, 1980, The New York Times. Copyright 1980 by the New York Times Company.)

	Black	Hispanic	White
Took Exam	6,142	5,239	19,798
Passed Exam	1,048	1,047	9,049

5.16. Two approaches have been taken when forming tail limits in situations with nuisance parameters. This problem and Problem 5.17 illustrate the two approaches. Suppose Y follows the beta-binomial distribution based on n trials with parameters α and β. Then Y has probability mass function:

$$\int_0^1 p^y(1-p)^{n-y}\frac{\Gamma(\alpha+\beta)}{\Gamma(\alpha)\Gamma(\beta)}p^{\alpha-1}(1-p)^{\beta-1}\,dp$$

for $y = 0, 1, \ldots, n$. Reparametrize to $\pi := \alpha/(\alpha+\beta)$ which is the marginal probability of success for each trial and $\rho := (1/(\alpha+\beta+1))$ which is the correlation between trials. Consider the problem of forming a confidence interval for π. Let $P[Y = y \mid \pi, \rho]$ denote the probability of y successes in n trials.

(a) Show that

$$A(\pi_0) = \{0, \ldots, n\} \setminus (\underline{R}(\pi_0) \cup \overline{R}(\pi_0))$$

where

$$\underline{R}(\pi_0) = \left\{ y\colon \sup_{0<\rho<1} P[Y \le y \mid \pi_0, \rho] \le \alpha/2 \right\} \quad \text{and}$$

$$\overline{R}(\pi_0) = \left\{ y\colon \sup_{0<\rho<1} P[Y \ge y \mid \pi_0, \rho] \le \alpha/2 \right\}$$

is a $100(1-\alpha)\%$ acceptance region for testing $H_0\colon \pi = \pi_0$ versus $H_{\neq}\colon \pi \neq \pi_0$.

(b) Prove that inversion of the family of acceptance regions $\{A(\pi_0)\colon 0 < \pi_0 < 1\}$ yields the confidence interval $(\underline{\pi}(y), \overline{\pi}(y))$ where $\underline{\pi}(0) = 0$, $\overline{\pi}(n) = 1$, and otherwise $\underline{\pi}(y)$ and $\overline{\pi}(y)$ are the solutions of

$$\inf_{0<\rho<1} P[Y \le y \mid \overline{\pi}(y), \rho] \stackrel{set}{=} \alpha/2 \stackrel{set}{=} \inf_{0<\rho<1} P[Y \ge y \mid \underline{\pi}(y), \rho].$$

Show that $\overline{\pi}(0) = 1 - \alpha/2 = 1 - \underline{\pi}(n)$.

(c) Prove that the intervals possess the invariance property

$$\underline{\pi}(y) = 1 - \overline{\pi}(n-y) \quad \text{for} \quad 0 \le y \le n.$$

[Wypij and Santner (1988)]

5.17. Suppose that $Y_1 \sim B(m_1, p_1)$ is independent of $Y_2 \sim B(m_2, p_2)$ and that it is desired to form a confidence interval for $\psi = \frac{p_1(1-p_2)}{(1-p_1)p_2}$. This problem is perhaps the simplest example of "exact analysis" of a single parameter.

(a) Using (5.2.1), show that conditionally given $Y_+ = t$

$$A(\psi_0) = \{L, \ldots, U\} \setminus (\underline{R}(\psi_0) \cup \overline{R}(\psi_0))$$

is a $100(1-\alpha)\%$ acceptance region for testing $H_0\colon \psi = \psi_0$ versus $H_{\neq}\colon \psi \neq \psi_0$ where

$$L = \max\{0, t - m_2\}, \quad U = \min\{t, m_1\},$$

$$\underline{R}(\psi_0) = \{y_1\colon P_{\psi_0}[Y_1 \le y_1 \mid Y_+ = t] \le \alpha/2\} \text{ and,}$$
$$\overline{R}(\psi_0) = \{y_1\colon P_{\psi_0}[Y_1 \ge y_1 \mid Y_+ = t] \le \alpha/2\}.$$

(Hence the same is true unconditionally.)

(b) Prove that inversion of the family of acceptance regions $\{A(\psi_0): 0 < \psi_0 < \infty\}$ yields the intervals $[\underline{\psi}(\mathbf{y}), \overline{\psi}(\mathbf{y})]$ with $\mathbf{y} = (y_1, y_2)'$ given by:

$$\underline{\psi}(\mathbf{y}) = 0 \quad \text{whenever} \quad \max\{0, y_+ - m_2\} = \min\{m_1, y_+\}$$
$$\text{or} \quad y_1 = \max\{0, y_+ - m_2\}$$
$$\overline{\psi}(\mathbf{y}) = \infty \quad \text{whenever} \quad \max\{0, y_+ - m_2\} = \min\{m_1, y_+\}$$
$$\text{or} \quad y_1 = \min\{m_1, y_+\}$$

and otherwise as solutions of

$$P[Y_1 \geq y_1 \,|\, Y_+ = y_+, \underline{\psi}(\mathbf{y})] = \alpha/2 = P[Y_1 \leq y_1 \,|\, Y_+ = y_+, \overline{\psi}(\mathbf{y})].$$

(c) How does the method described in this problem for dealing with a nuisance parameter differ from that described in Problem 5.16?

5.18. Derive the IRLS equations for computing the MLE for logistic regression data by paralleling the Newton–Raphson development of equation (4.4.1) in Section 4.4.

5.19. Let $R(\boldsymbol{\beta}, \hat{\boldsymbol{\beta}})$ be the MSE of the MLE of $\mathbf{p}(\boldsymbol{\beta})$ and assume that $m_i = 1$ for all responses Y_i. Thus

$$R(\boldsymbol{\beta}, \hat{\boldsymbol{\beta}}) = \sum_{\mathcal{Y}} \left\{ ||\mathbf{p}(\boldsymbol{\beta}) - \mathbf{p}(\hat{\boldsymbol{\beta}})||^2 \exp\left(\sum_{i=1}^{T} y_i \mathbf{x}_i' \boldsymbol{\beta} \right) / Q(\boldsymbol{\beta}) \right\}$$

where $Q(\boldsymbol{\beta}) = \prod_{i=1}^{T} \{1 + \exp[\mathbf{x}_i'\boldsymbol{\beta}]\}$ and $\mathcal{Y}$ is the set of all 2^T vectors of length T with each component either zero or unity (i.e., the set of all possible outcomes $\mathbf{Y} = \mathbf{y}$). Let $\boldsymbol{\beta} \in \mathcal{B}_C$ where

$$\mathcal{B}_C := \{\boldsymbol{\beta}: ||\boldsymbol{\beta}|| = 1, \text{ there exists at least one } \mathbf{y} \in \mathcal{Y} \text{ with}$$
$$(2y_i - 1)\mathbf{x}_i'\boldsymbol{\beta} > 0 \text{ for all } 1 \leq i \leq T\}.$$

In words, $\mathcal{B}_C$ is the set of unit beta vectors for which there exists at least one possible outcome $\mathbf{y} \in \mathcal{Y}$ with components y_i satisfying $y_i = 1$ when $p_i(\boldsymbol{\beta}) > 1/2$ and $y_i = 0$ when $p_i(\boldsymbol{\beta}) < 1/2$.

(a) Show that the complement of $\mathcal{B}_C$ in $\mathbb{R}^k$ has finite cardinality.

(b) Prove that $R(c\boldsymbol{\beta}, \hat{\boldsymbol{\beta}}) \to 0$ as $c \to \infty$ for every $\boldsymbol{\beta} \in \mathcal{B}_C$.

[Duffy and Santner (1989)]

5.20. Suppose data $(Y_i, m_i, \mathbf{x}_i)$, $1 \leq i \leq T$, follow the logistic model

$$\text{logit}(p(\mathbf{x})) = \mathbf{x}'\boldsymbol{\beta}.$$

(a) Derive the likelihood ratio, score and Wald statistics for testing H_0: $\boldsymbol{\beta} = \boldsymbol{\beta}^0$ versus $H_{\neq}$: $\boldsymbol{\beta} \neq \boldsymbol{\beta}^0$ where $\boldsymbol{\beta}^0$ is a given vector.

(b) Suppose $\boldsymbol{\beta}' = ((\boldsymbol{\beta}^1)', (\boldsymbol{\beta}^2)')$ with $\boldsymbol{\beta}^1$ and $\boldsymbol{\beta}^2$ having k_1 and k_2 components, respectively. Show that the score test of H_0: $\boldsymbol{\beta}^1 = \mathbf{0}_{k_1}$ versus H_A: $\boldsymbol{\beta}^1 \neq \mathbf{0}_{k_1}$ rejects H_0 when

$$\{\mathbf{S}(\mathbf{0}_{k_1}, \hat{\boldsymbol{\beta}}^2)\}'\{\mathbf{I}(\mathbf{0}_{k_1}, \hat{\boldsymbol{\beta}}^2)\}^{-1}\{\mathbf{S}(\mathbf{0}_{k_1}, \hat{\boldsymbol{\beta}}^2)\} \geq \chi^2_{\alpha, k_1}$$

where $\hat{\boldsymbol{\beta}}^2$ is the MLE of $\boldsymbol{\beta}^2$ when $\boldsymbol{\beta}^1 = \mathbf{0}_{k_1}$, $S(\cdot, \cdot)$ is the score vector, and $\mathbf{I}(\cdot, \cdot)$ is the Fisher information matrix; i.e., the negative of the matrix of second partial derivatives.

5.21. Show that Tsiatis's (1980) score statistic based on model (5.3.14) for testing H_0: $\gamma_1 = \ldots = \gamma_Q = 0$ versus H_A: not H_0 is

$$A^2 := \{\mathbf{S}(\hat{\boldsymbol{\beta}}, \mathbf{0}_Q)\}'\hat{\mathbf{V}}^-\{\mathbf{S}(\hat{\boldsymbol{\beta}}, \mathbf{0}_Q)\}$$

where $\mathbf{S}(\boldsymbol{\beta}, \boldsymbol{\tau}) := (\ldots, \partial \ln\{L(\boldsymbol{\beta}, \boldsymbol{\tau})\}/\partial \tau_j, \ldots)'$ is $Q \times 1$, $\hat{\boldsymbol{\beta}}$ is the MLE of $\boldsymbol{\beta}$ under H_0, and $\hat{\mathbf{V}}^-$ is the generalized inverse of the $Q \times Q$ matrix $\hat{\mathbf{V}} = \mathbf{V}(\hat{\boldsymbol{\beta}}, \mathbf{0}_Q)$ where $\mathbf{V}(\boldsymbol{\beta}, \boldsymbol{\tau}) := \mathbf{B} - \mathbf{C}\mathbf{D}^{-1}\mathbf{C}$ with

$$\mathbf{B} := -\left[\frac{\partial^2 \ln L(\boldsymbol{\beta}, \boldsymbol{\tau})}{\partial \tau_i \partial \tau_j}\right], Q \times Q,$$

$$\mathbf{C} := -\left[\frac{\partial^2 \ln L(\boldsymbol{\beta}, \boldsymbol{\tau})}{\partial \tau_i \partial \beta_0}\right], Q \times k, \quad \text{and}$$

$$\mathbf{D} := -\left[\frac{\partial^2 \ln L(\boldsymbol{\beta}, \boldsymbol{\tau})}{\partial \beta_i \partial \beta_j}\right], k \times k.$$

5.22. Compare the (asymptotic) variances of $\hat{\beta}_1$ in the two models

$$\text{logit}(p(x)) = \beta_1 x$$

and

$$\text{logit}(p(x)) = \beta_0 + \beta_1 x.$$

What are the qualitative similarities to the analogous results for simple linear regression with and without an intercept?

5.23. Consider the data in Table 5.P.23 from Silvapulle (1981) on 120 individuals cross-classified according to gender, their score on a psychiatric screening questionnaire (the General Health Questionnaire or GHQ), and whether or not they are a psychiatric case based on a standardized interview. The question of interest is whether psychiatric cases can be predicted on the basis of GHQ score.

(i) Fit the simple logistic model with an intercept and a coefficient for GHQ score by both ML and WLS. Analyze and compare the fits.

(ii) Add a gender effect to the model. What happens to the fit?

(iii) To further explore the effect of gender, analyze the males and females separately.

Table 5.P.23. One Hundred and Twenty Patients Classified by Gender, GHQ Score, and the Outcome of a Standardized Psychiatric Interview (Reprinted with permission from M.J. Silvapulle: "On the Existence of Maximum Likelihood Estimator," *Journal of the Royal Statistical Society B,* **43**, 1981, pp. 310–313. Royal Statistical Society, London, England.)

GHQ	Female		Male	
Score	Case	Non-Case	Case	Non-Case
0	2	42	0	18
1	2	14	0	8
2	4	5	1	1
3	3	1	0	0
4	2	1	1	0
5	3	0	3	0
6	1	0	0	0
7	1	0	2	0
8	3	0	0	0
9	1	0	0	0
10	0	0	1	0

5.24. Consider the data of Example 1.2.8 on bald eagle food pirating. Build a logistic model for these data according to the following stepwise procedure.

(a) Start with the intercept-only model.

(b) Fit 3 separate models each with intercept and one of the covariates. Add the factor which has the largest change in loglikelihood. (Compare the changes in loglikelihood with the squares of the z-scores for each model.)

(c) Continue adding main effects and then interactions until no more terms are significant at the .05 level.

(d) Interpret your final model.

5.25. Analyze the prostate data of Example 3.1.3. Does acid phosphatase have predictive ability beyond that contained in the other preoperative variables?

5.26. As mentioned in Section 5.1, the data of Table 5.1.1 are a subset of a larger experiment conducted by Hoblyn and Palmer (1934). Table 5.P.26 provides the complete data; note that besides 2 planting times and 2 cutting lengths, there were 3 diameter categories and 4 harvesting times. Table 5.P.26 lists the number of cuttings surviving out of the 20 cuttings made at each of the $48 = 4 \times 3 \times 2^2$ treatment combinations. Fit the model

$$\begin{aligned} \text{logit}(p(\mathbf{x})) = \beta_0 + \beta_1 I[\text{Length} = \text{long}] \\ + \beta_2 I[\text{Time} = \text{at once}] \\ + \beta_3 I[\text{Diameter} \leq 6\,\text{mm}] + \beta_4 I[\text{Diameter} > 6\,\text{mm}] \\ + \beta_5 I[\text{Cutting made in Dec}] + \beta_6 I[\text{Cutting made in Jan}] \\ + \beta_7 I[\text{Cutting made in Feb}]. \end{aligned} \tag{1}$$

How does (1) compare with a model in which diameter enters linearly? Can the fit be improved by adding interactions?

Table 5.P.26. Number of Surviving Plum Root Stock Cuttings (Out of 20) Planted Under 48 Different Experimental Conditions (Reprinted with permission from Hoblyn and Palmer: "A Complex Experiment in the Propagation of Plum Rootstocks from Root Cuttings," *Journal of Pomology and Horticultural Science,* 1934. Headley Brothers Ltd., Invicta Press, Ashford, Kent, England.)

Planting Time	Length	Diameter mm	Cutting Made in Oct	Dec	Jan	Feb
At Once	Long	4.5	6	5	13	7
		7.7	14	16	15	13
		10.5	18	17	18	14
	Short	4.5	4	4	7	8
		7.5	10	10	8	14
		10.5	11	10	8	13
In Spring	Long	4.5	2	1	1	4
		7.5	6	7	4	9
		10.5	14	15	10	11
	Short	4.5	0	0	0	0
		7.5	2	2	3	3
		10.5	3	10	2	6

5.27. Prove that the estimator $\hat{\boldsymbol{\beta}}^B$ solves the restricted likelihood problem:

$$\text{maximize } \ln L(\boldsymbol{\beta})$$

$$\text{subject to } ||\boldsymbol{\beta} - \boldsymbol{\mu}||^2 \leq C$$

where $L(\boldsymbol{\beta})$ is the likelihood (5.3.4) of $\boldsymbol{\beta}$ and $C := ||\hat{\boldsymbol{\beta}}^B - \boldsymbol{\mu}||^2$. (Hint: Consider the Lagrangian $R(\boldsymbol{\beta}, u) = \ln L(\boldsymbol{\beta}) + u(C - ||\boldsymbol{\beta} - \boldsymbol{\mu}||^2)$ of the restricted likelihood maximization problem with Lagrange multiplier u.)

[Duffy and Santner (1989)]

5.28. Show that the adjusted residuals obtained by applying the definition in (4.4.9) to the enlarged logistic regression data consisting of a $2 \times T$ table of successes and failures, are e_i^a and $-e_i^a$ given in (5.4.12).

5.29. The data in Table 5.P.29 are reported in Finney (1947). They concern a study in human physiology in which a reflex vaso-constriction may occur in the skin of the digits after taking a single deep breath. The response Y_i is the occurrence (1) or non-occurrence (0) of vaso-constriction in the skin of the fingers of a subject after they inhale a certain volume (x_{i1}) of air at a certain average rate (x_{i2}); both explanatory variables are quantitative. There are a total of 39 trials.

(a) Use maximum likelihood estimation to fit the model

$$\text{logit}(p_i) = \beta_0 + \beta_1 \ln(\text{RATE}) + \beta_2 \ln(\text{VOLUME}) \qquad (1)$$

for the probability p_i of the occurrance of vaso-constriction as a function of the rate and volume of air inspired.

(b) Calculate the adjusted residuals e_i^a based on Model (1) and determine whether any points are potential outliers.

(c) Calculate the leverages h_i defined in (5.4.13). Make a coded scatterplot (by Y_i) of the data in ln(RATE) versus ln(VOLUME) space. Which points have the largest leverages?

(d) Do the approximate Cook's distances c_i^a identify any other influential points?

(e) Construct a partial residual plot for each of the variables ln(RATE) and ln(VOLUME). Do the plots suggest any inadequacies in Model (1)?

(f) Three different subjects were involved in this experiment; the first 9 rows of Table 5.P.29 correspond to the first subject, the next 8 rows to the second subject, and the last 22 rows to the third subject. Do the data support the assumption that there are no significant differences between the three subjects?

Table 5.P.29. Occurrence of Vaso-constriction in the Skin of the Digits Following Inspiration of a Certain Volume of Air at a Certain Average Rate (Reprinted with permission from D.J. Finney: "The Estimation from Individual Records of the Relationship Between Dose and Quantal Response," *Biometrika,* **34**, 1947, Biometrika Trust, London, England.)

Volume	Rate	Y
3.7	0.825	1
3.5	1.09	1
1.25	2.5	1
0.75	1.5	1
0.8	3.2	1
0.7	3.5	1
0.6	0.75	0
1.1	1.7	0
0.9	0.75	0
0.9	0.45	0
0.8	0.57	0
0.55	2.75	0
0.6	3	0
1.4	2.33	1
0.75	3.75	1
2.3	1.64	1
3.2	1.6	1
0.85	1.415	1
1.7	1.06	0
1.8	1.8	1
0.4	2	0
0.95	1.36	0
1.35	1.35	0
1.5	1.36	0
1.6	1.78	1
0.6	1.5	0
1.8	1.5	1
0.95	1.9	0
1.9	0.95	1
1.6	0.4	0
2.7	0.75	1
2.35	0.03	0
1.1	1.83	0
1.1	2.2	1
1.2	2	1
0.8	3.33	1
0.95	1.9	0
0.75	1.9	0
1.3	1.625	1

5.30. The data in Table 5.P.30, provided by Keeler (1988), are discussed in Keeler (1985). They concern an experiment to determine the hardiness of two strains of winter wheat (Frederick and Norstar) to thermal stress. Plants were assigned to two watering schemes (normal versus flooded up to the surface of the soil), and cooled to a predetermined temperature. The plants were then removed to a growth room to determine survival by regrowth. The experiment was replicated 4 times using flats of about 10 plants per condition. Table 5.P.30 summarizes the results over the approximately 40 plants used for each variety by watering scheme and temperature combination. The effects of variety and watering method can be incorporated by introducing 0/1 indicator variables

$$V = I[\text{Variety} = \text{Frederick}] \text{ and}$$

$$W = I[\text{Watering Scheme} = \text{normal}].$$

One method of modeling the temperature effect is by means of 5 indicators for the 6 levels of temperature; a second is by a functional parametrization. The following models for the probability of plant death p_i illustrate the two approaches

$$\text{logit}(p_i) = \beta_0 + \beta_1 V_i + \beta_2 W_i + \sum_{i=3}^{7} \beta_i I[\text{Temperature} = (-4 - 2 \times i)°C] \qquad (1)$$

and

$$\text{logit}(p_i) = \beta_0 + \beta_1 V_i + \beta_2 W_i + \beta_3 \ln[-\text{Temperature}]. \qquad (2)$$

(a) Compare the fits of models (1) and (2) to the cold hardiness data.

(b) Use both approaches to explore for the presence of (i) variety-by-treatment interactions, (ii) variety-by-temperature interactions, and (iii) watering method-by-temperature interactions (compute appropriate score tests). State your conclusions.

Table 5.P.30. Cold Hardiness of Winter Wheat: Ratio of Number of Dead Plants to Number of Plants on Test (Reprinted with permission from Lisa C. Keeler: *Genotypic and Environmental Effects on the Cold, Icing, and Flooding Tolerances of Winter Wheat.* M.Sc. thesis, 1985, University of Guelph, Ontario.)

	Norstar		Frederick	
Temp	Normal	Flooded	Normal	Flooded
$-10°C$	1/41	1/41	1/40	3/41
$-12°C$	1/41	2/41	15/41	13/38
$-14°C$	2/41	4/42	36/43	32/41
$-16°C$	7/41	18/40	40/40	40/40
$-18°C$	27/41	38/42	40/40	40/40
$-20°C$	39/42	40/40	40/40	40/40

5.31. Consider the data of Example 1.2.9 on the 23 pre-Challenger space shuttle flights for which post-flight information is available on the booster rockets.

(a) Assuming that the O-rings behave independently, fit a logistic model to these data allowing for an intercept and a temperature effect. Assess fit to your model by making a partial residual plot versus temperature.

(b) Construct a confidence interval for the coefficient of temperature. Compute a point estimate for the probability of experiencing distress in at least one O-ring if the pre-launch temperature is 31°F (the actual launch temperature of the Challenger flight).

(c) Prior to each launch the seals formed by the O-rings were subject to a pressure test and it is possible that conducting this test at high pressure damages the O-rings. What evidence does the data provide concerning this conjecture?

[Dalal, Fowlkes, and Hoadley (1988, 1989)]

5.32 The data in Table 5.P.32 from Crowder (1978) concern a series of experiments in plant biology. In each experiment some seeds are brushed onto a plate covered with a certain extract at a certain dilution level and then the number of germinated seeds is counted. Three seed types, two extracts, and four dilution levels are involved. Crowder notes that there is considerable heterogeneity between plates. Can a logistic regression model account for this? If not, how might you generalize the logit model to better fit data such as these?

Table 5.P.32. Seed Germination Data (Reprinted with permission from M.J. Crowder: "Beta-Binomial ANOVA for Proportions," *Applied Statistics,* **27**, pp. 34–37, 1978. Royal Statistical Society, London, England.)

Number of Germinated Seeds	Total Number of Seeds	Seed Type	Extract Type	Dilution
2	43	A	Bean	1/1
9	51	A	Bean	1/1
9	44	A	Bean	1/1
16	71	A	Bean	1/1
2	24	A	Bean	1/1
0	7	A	Bean	1/1
17	19	A	Bean	1/25
43	56	A	Bean	1/25
79	87	A	Bean	1/25
50	55	A	Bean	1/25
9	10	A	Bean	1/25
11	13	A	Bean	1/625
47	62	A	Bean	1/625
90	104	A	Bean	1/625
46	51	A	Bean	1/625
9	11	A	Bean	1/625
10	39	B	Bean	1/125
23	62	B	Bean	1/125
23	81	B	Bean	1/125
26	51	B	Bean	1/125
17	39	B	Bean	1/125
5	6	B	Cucumber	1/125
53	74	B	Cucumber	1/125
55	72	B	Cucumber	1/125
32	51	B	Cucumber	1/125
46	79	B	Cucumber	1/125
10	13	B	Cucumber	1/125
8	16	C	Bean	1/125
10	30	C	Bean	1/125
8	28	C	Bean	1/125
23	45	C	Bean	1/125
0	4	C	Bean	1/125
3	12	C	Cucumber	1/125
22	41	C	Cucumber	1/125
15	30	C	Cucumber	1/125
52	51	C	Cucumber	1/125
3	7	C	Cucumber	1/125

5.33. Consider the $2 \times 2 \times S$ table of stratified binomial treatment comparisons in Table 5.5.1. Show that

(i) $\psi_1 = \ldots = \psi_S$ holds if and only if $\boldsymbol{\lambda}^{123} = \mathbf{0}$ and

(ii) $\psi_1 = \ldots = \psi_S = 1$ holds if and only if $\boldsymbol{\lambda}^{12} = \boldsymbol{\lambda}^{123} = \mathbf{0}$.

5.34. Suppose stratified data $\{Y_{sj}\}$ follow the binomial model of Section 5.5 with success probabilities $\{p_{sj}\}$ satisfying $\text{logit}(p_{sj}) = \alpha_s + \beta I[j = 2]$, $1 \leq s \leq S$ and $j = 1, 2$. Show that the score statistic for testing H_0: $\beta = 0$ versus $H_{\neq}$: $\beta \neq 0$ has the form

$$\frac{\sum_{s=1}^{S}\{Y_{s1} - Y_{s+}(m_{s1}/m_{s+})^2\}}{\sum_{s=1}^{S}(Y_{s+}(m_{s+} - Y_{s+})m_{s1}m_{s2})/m_{s+}^3}$$

which can be recognized as the CMH statistic (5.5.11) *without* the finite population correction to the variance.

[Day and Byar (1979), Gart and Tarone (1983)]

5.35. Matched pairs experiments lead to $2 \times 2 \times S$ tables for which $m_{sj} = 1$ for all s and both treatments $(j = 1, 2)$. Consider the model

$$\text{logit}(p_{sj}) = \alpha_s + \beta I[j = 1].$$

(a) For a fixed number of pairs S, show that the (unconditional) MLE of β, the common log odds ratio, is

$$\hat{\beta} = 2\ln(n_{10}/n_{01})$$

where n_{ij} is the number of strata s for which $Y_{s1} = i$ and $Y_{s2} = j$.

(b) Prove that $\hat{\beta} \rightarrow 2\beta$ as $S \rightarrow \infty$, which shows that $\hat{\beta}$ is inconsistent. (Hint: Consider the random variables $Z_{ij} = S^{-1}n_{ij} - S^{-1}\sum_{s=1}^{S} P[Y_{s1} = i,\, Y_{s2} = j]$.)

(c) Show that the conditional MLE of the odds ratio $\exp\{\beta\}$ is equal to the Mantel–Haenszel estimator (for the matched pair case only).

5.36. Consider the data of Example 1.2.2 on judicial decisions. Test the homogeneity of the odds ratios for liberal decisions of Johnson versus Nixon appointees across the strata formed by the type of case.

Appendix 1. Some Results from Linear Algebra

This appendix summarizes some basic facts about projection matrices.

Definition. An $n \times n$ matrix of real numbers $\mathbf{P}$ is an orthogonal projection matrix if (i) $\mathbf{P} = \mathbf{P}'$ and (ii) $\mathbf{P} = \mathbf{P}^2$.

The essential intuitive property of a projection operator is (ii) since if $\mathbf{Py}$ is the projection of $\mathbf{y}$ onto a subspace of $\mathbb{R}^n$ then a second projection $\mathbf{P}(\mathbf{Py}) = \mathbf{Py}$ yields $\mathbf{Py}$ again. In addition, property (i) guarantees that the projection is orthogonal in that $\mathbf{y} = \mathbf{Py} + (\mathbf{y} - \mathbf{Py})$ decomposes $\mathbf{y}$ into two orthogonal pieces since

$$\begin{aligned}(\mathbf{Py})'(\mathbf{I}_n - \mathbf{P})\mathbf{y} &= \mathbf{y}'\mathbf{P}'(\mathbf{I}_n - \mathbf{P})\mathbf{y} = \mathbf{y}'\mathbf{P}(\mathbf{I}_n - \mathbf{P})\mathbf{y} \\ &= \mathbf{y}'(\mathbf{P} - \mathbf{P}^2)\mathbf{y} = \mathbf{y}'(\mathbf{P} - \mathbf{P})\mathbf{y} = 0.\end{aligned}$$

Notes.

(a) An $n \times n$ projection matrix $\mathbf{P}$ maps $\mathbb{R}^n$ onto the column space of $\mathbf{P}$, denoted by $\mathcal{C}(\mathbf{P}) = \{\mathbf{Pa}: \mathbf{a} \in \mathbb{R}^n\} = \{\sum_{i=1}^{n} \boldsymbol{\xi}_i a_i: \mathbf{a} \in \mathbb{R}^n\}$ where $\boldsymbol{\xi}_1, \ldots, \boldsymbol{\xi}_n$ are the $(n \times 1)$ column vectors of $\mathbf{P}$.

(b) For each subspace $\mathcal{M} \subset \mathbb{R}^n$ there exists a unique projection matrix $\mathbf{P}: \mathbb{R}^n \xrightarrow{\text{onto}} \mathcal{M}$. If $\mathcal{M} = \mathcal{C}(\mathbf{X})$ for a matrix $\mathbf{X}$ whose columns form a *basis* for $\mathcal{M}$, then $\mathbf{P} = \mathbf{X}(\mathbf{X}'\mathbf{X})^{-1}\mathbf{X}'$ is the formula for this projection matrix (the value of which is independent of $\mathbf{X}$). In the special case when $\mathbf{X}$ has orthonormal columns $\boldsymbol{\eta}_1, \ldots, \boldsymbol{\eta}_k$ then $\mathbf{Pz} = \mathbf{XX}'\mathbf{z} = \sum_{i=1}^{k} \boldsymbol{\eta}_i(\boldsymbol{\eta}_i'\mathbf{z})$, and the terms $\boldsymbol{\eta}_i(\boldsymbol{\eta}_i')\mathbf{z}$ are the projections onto the orthogonal spaces $\mathcal{C}(\boldsymbol{\eta}_i)$, $1 \le i \le k$.

The following lemma gives necessary and sufficient conditions for two vectors to have the same projection. It will be used to state the likelihood equations which arise in maximum likelihood estimation of Poisson, multinomial, and product multinomial means for log linear models.

Lemma A.1.1. *Let* $\mathbf{P}$ *be the projection matrix from* $\mathbb{R}^n$ *onto a linear subspace* $\mathcal{M}$ *of dimension* q *and let* $\boldsymbol{\nu}_1, \ldots, \boldsymbol{\nu}_k$ $(q \le k)$ *span* $\mathcal{M}$. *Then*

$\mathbf{Pw} = \mathbf{Pz}$ *if and only if* $\mathbf{w}'\boldsymbol{\nu}_j = \mathbf{z}'\boldsymbol{\nu}_j$, $1 \leq j \leq k$.

Proof (⇒) A stronger conclusion will be demonstrated; namely that $\mathbf{w}'\boldsymbol{\nu} = \mathbf{z}'\boldsymbol{\nu}$ for any $\boldsymbol{\nu} \in \mathcal{M}$. Fix $\boldsymbol{\nu} \in \mathcal{M}$; then

$$\begin{aligned}
\mathbf{w}'\boldsymbol{\nu} &= \mathbf{w}'\mathbf{P}\boldsymbol{\nu}, \quad \boldsymbol{\nu} \in \mathcal{M} \\
&= \mathbf{w}'\mathbf{P}'\mathbf{P}\boldsymbol{\nu}, \quad \mathbf{P} \text{ is a projection matrix} \\
&= (\mathbf{Pw})'(\mathbf{P}\boldsymbol{\nu}) \\
&= (\mathbf{Pz})'\mathbf{P}\boldsymbol{\nu} \\
&= \mathbf{z}'\boldsymbol{\nu}.
\end{aligned}$$

(⇐) Choose $\boldsymbol{\xi}_1, \ldots, \boldsymbol{\xi}_q$ to be an orthonormal basis for $\mathcal{M}$. Then $\mathbf{Pz} = \sum_{j=1}^{q} \boldsymbol{\xi}_j(\boldsymbol{\xi}_j'\mathbf{z})$ and $\mathbf{Pw} = \sum_{j=1}^{q} \boldsymbol{\xi}_j(\boldsymbol{\xi}_j'\mathbf{w})$. It suffices to show that $\boldsymbol{\xi}_j'\mathbf{z} = \boldsymbol{\xi}_j'\mathbf{w}$ for $j = 1(1)q$. Fix $j \in \{1, \ldots, q\}$ then $\boldsymbol{\xi}_j = \sum_{i=1}^{k} a_i\boldsymbol{\nu}_i$ for some $\{a_i\}_{i=1}^{k}$ and hence $\boldsymbol{\xi}_j'\mathbf{z} = \sum_{i=1}^{k} a_i\boldsymbol{\nu}_i'\mathbf{z} = \sum_{i=1}^{k} a_i\boldsymbol{\nu}_i'\mathbf{w} = \boldsymbol{\xi}_j'\mathbf{w}$. □

Appendix 2. Maximization of Concave Functions

Given $g: \mathbb{R}^p \to \mathbb{R}$, let $\nabla g(\mathbf{w}) := \left(\ldots, \dfrac{\partial g(\mathbf{w})}{\partial w_i}, \ldots\right)'$ be its gradient vector $(p \times 1)$ and $\nabla^2 g(\mathbf{w}) := \left(\dfrac{\partial^2 g(\mathbf{w})}{\partial w_i \partial w_j}\right)$ be its Hessian matrix $(p \times p)$ when the required partials exist.

Example A.2.1. Fix $\mathbf{a} \in \mathbb{R}^p$ and let $g_1(\mathbf{w}) = \mathbf{a}'\mathbf{w} = \sum_{j=1}^p a_j w_j$. Then $\nabla g_1(\mathbf{w}) = \mathbf{a}$ and $\nabla^2 g_1(\mathbf{w})$ is the $p \times p$ matrix of zeros.

Example A.2.2. Fix a matrix $\mathbf{A} \in \mathbb{R}^{p \times p}$ and let $g_2(\mathbf{w}) = \mathbf{w}'\mathbf{A}\mathbf{w}$. Then $\nabla g_2(\mathbf{w}) = 2\mathbf{A}\mathbf{w}$ and $\nabla^2 g_2(\mathbf{w}) = 2\mathbf{A}$.

Two classes of functions important for this text are the concave and strictly concave functions.

Definition. $g: \mathbb{R}^p \to \mathbb{R}$ is (strictly) concave if for all $\mathbf{w}_1, \mathbf{w}_2 \in \mathbb{R}^p$ and $0 < \alpha < 1$,

$$g(\alpha \mathbf{w}_1 + (1-\alpha)\mathbf{w}_2)\ (>) \geq \alpha g(\mathbf{w}_1) + (1-\alpha) g(\mathbf{w}_2).$$

Figure A.2.1 shows an example of a strictly concave function which has a unique maximum and an example of a strictly concave function which fails to attain its maximum. The following proposition states that no other cases can arise.

Proposition A.2.1. *If $g: \mathbb{R}^p \to \mathbb{R}$ is strictly concave and differentiable then the cardinality of $\{\mathbf{w}: \nabla g(\mathbf{w}) = \mathbf{0}_p\}$ is either 0 or 1. If $\nabla g(\mathbf{w}^*) = \mathbf{0}_p$ then $\mathbf{w}^*$ is the unique point such that $g(\mathbf{w}^*) = \sup_{\mathbf{w} \in \mathbb{R}^p} g(\mathbf{w})$.*

The next proposition states a well-known sufficient condition for $g(\cdot)$ to be strictly concave (see e.g., Roberts and Varberg (1973), p. 103).

Proposition A.2.2. *If $g: \mathbb{R}^p \to \mathbb{R}$ has continuous second partials and if for all $\mathbf{w} \in \mathbb{R}^p$, $\nabla^2 g(\mathbf{w})$ is negative definite (i.e., $\mathbf{z}'\nabla^2 g(\mathbf{w})\mathbf{z} < 0$ for all*

$\mathbf{z} \neq \mathbf{0}_p$), *then* $g(\cdot)$ *is strictly concave.*

Example A.2.3. Consider $\mathbf{Y} \sim N_n(\mathbf{X}\boldsymbol{\beta}, \mathbf{I}_n)$ with $\mathbf{X}$ $n \times p$ of rank p. If $L(\boldsymbol{\beta})$ is the likelihood of $\mathbf{Y}$, then apart from constants the log likelihood is

$$\ln L(\boldsymbol{\beta}) = -\frac{1}{2}(\mathbf{Y} - \mathbf{X}\boldsymbol{\beta})'(\mathbf{Y} - \mathbf{X}\boldsymbol{\beta}) \quad \text{and}$$

$$\nabla \ln L(\boldsymbol{\beta}) = \mathbf{X}'\mathbf{Y} - \frac{1}{2}2\mathbf{X}'\mathbf{X}\boldsymbol{\beta}.$$

Then $\nabla \ln L(\hat{\boldsymbol{\beta}}) = \mathbf{0}_p \iff \mathbf{X}'\mathbf{Y} = \mathbf{X}'\mathbf{X}\hat{\boldsymbol{\beta}} \iff \hat{\boldsymbol{\beta}} = (\mathbf{X}'\mathbf{X})^{-1}\mathbf{X}'\mathbf{Y}$ since $\mathbf{X}'\mathbf{X}$ is nonsingular. Furthermore $\hat{\boldsymbol{\beta}}$ is the unique maximum of $\ln L(\boldsymbol{\beta})$ (and hence $L(\boldsymbol{\beta})$) since $\ln L(\boldsymbol{\beta})$ is strictly concave. The latter fact derives from the application of Proposition A.2.2 as $\nabla^2 \ln L(\boldsymbol{\beta}) = -\mathbf{X}'\mathbf{X}$ which is negative definite since for $\mathbf{z} \in \mathbb{R}^p$ with $\mathbf{z} \neq \mathbf{0}_p$, $-(\mathbf{z}'\mathbf{X}'\mathbf{X}\mathbf{z}) = -||\mathbf{X}\mathbf{z}|| < 0$ as $\text{rank}(\mathbf{X}) = p$.

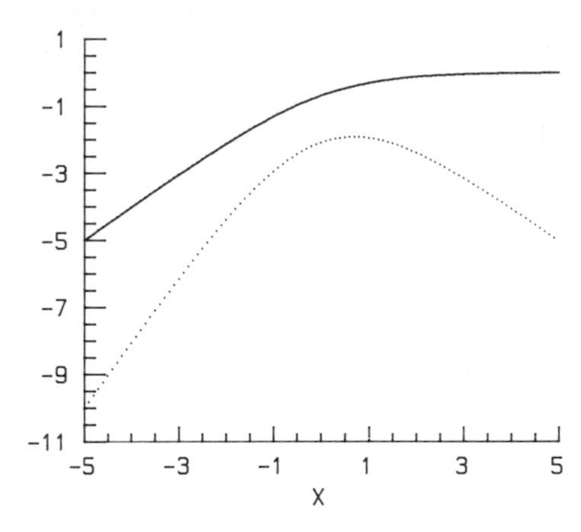

Figure A.2.1. A strictly concave function which attains its maximum ($\cdots$), and a strictly concave function which fails to attain its maximum (—).

Appendix 3. Proof of Proposition 3.3.1 (ii) and (iii)

The proof in this appendix essentially follows Haberman (1973b). To assert uniqueness, it suffices to show that if $\hat{\boldsymbol{\ell}}^m \in \tilde{\mathcal{M}}$ maximizes the product multinomial likelihood among $\boldsymbol{\ell} \in \tilde{\mathcal{M}}$ then $\hat{\boldsymbol{\ell}}^m$ also maximizes the Poisson model likelihood among $\boldsymbol{\ell} \in \mathcal{M}$ and hence must be unique for the Poisson model by Proposition 3.2.1. Furthermore $\hat{\boldsymbol{\ell}}^m$ must also be unique for the product multinomial model by part (i) Proposition 3.3.1.

The proofs of parts (ii) and (iii) of Proposition 3.3.1 require additional notation and two lemmas. Let $\mathcal{N} := \mathcal{C}(\boldsymbol{\nu}_1, \ldots, \boldsymbol{\nu}_S)$ where $\boldsymbol{\nu}_j$ is the 0/1 indicator vector of the set $\mathcal{T}_j$, and let t_j be the cardinality of $\mathcal{T}_j$. Also let $\mathbf{P}_{\mathcal{N}}: \mathbb{R}^n \xrightarrow{\text{onto}} \mathcal{N}$ be the orthogonal projection matrix onto $\mathcal{N}$ and let $\mathcal{N}^{\perp} := \{\mathbf{z} \in \mathbb{R}^n: \mathbf{z} \text{ is orthogonal to } \mathcal{N}\}$ be the orthogonal complement of $\mathcal{N}$. Observe that $\mathcal{N} \subset \mathcal{M}$ since $\{\boldsymbol{\nu}_j: j = 1(1)S\} \subset \mathcal{M}$ by assumption; thus $\mathcal{M}$ has the direct sum decomposition $\mathcal{M} = \mathcal{N} + (\mathcal{M} \cap \mathcal{N}^{\perp})$.

Lemma A.3.1. *The projection* $\mathbf{P}_{\mathcal{N}}$ *is given by*

$$\mathbf{P}_{\mathcal{N}}(\boldsymbol{\ell}) = \sum_{j=1}^{S} \frac{1}{t_j} \left(\sum_{q \in \mathcal{T}_j} \ell_q \right) \boldsymbol{\nu}_j .$$

Proof. The set $\{\boldsymbol{\nu}_1/(t_1)^{1/2}, \ldots, \boldsymbol{\nu}_S/(t_S)^{1/2}\}$ is an orthonormal basis for $\mathcal{N}$ since $\boldsymbol{\nu}_1, \ldots, \boldsymbol{\nu}_S$ are independent and orthogonal. □

Intuitively $\mathbf{P}_{\mathcal{N}}(\boldsymbol{\ell})$ averages the components of $\boldsymbol{\ell}$ in groups corresponding to the $\mathcal{T}_j$. Observe that Lemma A.3.1 and the remarks preceding it yield the formula

$$\mathbf{P}_{\mathcal{M} \cap \mathcal{N}^{\perp}}(\boldsymbol{\ell}) = \boldsymbol{\ell} - \sum_{j=1}^{S} \frac{\boldsymbol{\nu}_j}{t_j} \left(\sum_{q \in \mathcal{T}_j} \ell_q \right) \qquad \text{(A.3.1)}$$

for $\boldsymbol{\ell} \in \mathcal{M}$. The other formula required in the proofs of Proposition 3.3.1

(ii) and (iii) is contained in the next result.

Lemma A.3.2. *If* $\boldsymbol{\ell} \in \mathcal{M} \cap \mathcal{N}^{\perp}$ *then*

$$\boldsymbol{\ell}^* := \boldsymbol{\ell} + \sum_{j=1}^{S} \boldsymbol{\nu}_j \left\{ \ln(m_j) - \ln\left(\sum_{q \in \mathcal{T}_j} \exp\{\ell_q\} \right) \right\} \in \tilde{\mathcal{M}}.$$

Proof. Clearly $\boldsymbol{\ell}^* \in \mathcal{M}$ since $\boldsymbol{\ell} \in \mathcal{M} \cap \mathcal{N}^{\perp}$ and $\boldsymbol{\ell}^* - \boldsymbol{\ell} \in \mathcal{N} \subset \mathcal{M}$. Furthermore for $j = 1(1)S$,

$$\begin{aligned} \sum_{i \in \mathcal{T}_j} \exp\{\ell_i^*\} &= \sum_{i \in \mathcal{T}_j} \exp\left\{ \ell_i + \ln(m_j) - \ln\left(\sum_{q \in \mathcal{T}_j} \exp\{\ell_q\} \right) \right\} \\ &= \frac{m_j}{\sum_{q \in \mathcal{T}_j} \exp\{\ell_q\}} \left(\sum_{i \in \mathcal{T}_j} \exp\{\ell_i\} \right) = m_j \end{aligned}$$

yielding $\boldsymbol{\ell}^* \in \tilde{\mathcal{M}}$. □

Proofs of Proposition 3.3.1 (ii) and (iii). To prove (ii) it suffices to show that $\boldsymbol{\nu}'\mathbf{Y} = \boldsymbol{\nu}' \exp\{\hat{\boldsymbol{\ell}}^m\}$ for all $\boldsymbol{\nu} \in \mathcal{M}$, hence $\mathbf{P}_{\mathcal{M}}\mathbf{Y} = \mathbf{P}_{\mathcal{M}}(\exp\{\hat{\boldsymbol{\ell}}^m\})$ and thus $\hat{\boldsymbol{\ell}}^m$ is the MLE of $\boldsymbol{\ell}$ under $(\mathcal{M}, P)$. Since $\mathcal{M} = \mathcal{N} + (\mathcal{M} \cap \mathcal{N}^{\perp})$, one need only verify that $\boldsymbol{\nu}'\mathbf{Y} = \boldsymbol{\nu}' \exp\{\hat{\boldsymbol{\ell}}^m\}$ for all $\boldsymbol{\nu} \in \mathcal{M} \cap \mathcal{N}^{\perp}$ since $\boldsymbol{\nu}_j'\mathbf{Y} = \sum_{i \in \mathcal{T}_j} Y_i = m_j = \sum_{i \in \mathcal{T}_j} \exp\{\hat{\ell}_i^m\} = \boldsymbol{\nu}_j' \exp\{\hat{\boldsymbol{\ell}}^m\}$ for $j = 1(1)S$ which implies $\boldsymbol{\nu}'\mathbf{Y} = \boldsymbol{\nu}' \exp\{\hat{\boldsymbol{\ell}}\}$ for all $\boldsymbol{\nu} \in \mathcal{N}$.

Let p be the dimension of $\mathcal{M}$. Choose an $n \times (p - S)$ matrix $\mathbf{W}$ such that $\mathcal{C}(\mathbf{W}) = \mathcal{M} \cap \mathcal{N}^{\perp}$. It must be shown that $\mathbf{W}'\mathbf{Y} = \mathbf{W}' \exp\{\hat{\boldsymbol{\ell}}^m\}$. Define $Q(\cdot)$ on $\boldsymbol{\beta} \in \mathbb{R}^{p-S}$ by

$$Q(\boldsymbol{\beta}) := \mathbf{Y}'\mathbf{W}\boldsymbol{\beta} - \sum_{j=1}^{S} m_j \ln\left(\sum_{q \in \mathcal{T}_j} \exp\{\mathbf{w}_q'\boldsymbol{\beta}\} \right)$$

where $\mathbf{w}_1', \ldots, \mathbf{w}_n'$ are the rows of $\mathbf{W}$. In an abuse of notation $Q(\cdot)$ can also be regarded as a function of $\boldsymbol{\ell} \in \mathcal{M} \cap \mathcal{N}^{\perp}$ since $\boldsymbol{\ell} \in \mathcal{M} \cap \mathcal{N}^{\perp}$ if and only if $\boldsymbol{\ell} = \mathbf{W}\boldsymbol{\beta}$ for some $\boldsymbol{\beta} \in \mathbb{R}^{p-S}$ and $Q(\cdot)$ depends on $\boldsymbol{\beta}$ only through $\mathbf{W}\boldsymbol{\beta}$. Hence

$$Q(\boldsymbol{\ell}) = \mathbf{Y}'\boldsymbol{\ell} - \sum_{j=1}^{S} m_j \ln\left(\sum_{q \in \mathcal{T}_j} \exp\{\ell_q\} \right)$$

where $\boldsymbol{\ell} = \mathbf{W}\boldsymbol{\beta}$. It will be shown that

(a) $Q(\cdot)$ is strictly concave in $\boldsymbol{\beta}$, and hence has either 0 or 1 maximizers.

(b) $Q(\mathbf{P}_{\mathcal{M}\cap\mathcal{N}^\perp}[\hat{\boldsymbol{\ell}}^m]) \geq Q(\boldsymbol{\ell})$ for all $\boldsymbol{\ell} \in \mathcal{M} \cap \mathcal{N}^\perp$, hence $\mathbf{P}_{\mathcal{M}\cap\mathcal{N}^\perp}(\hat{\boldsymbol{\ell}}^m)$ is the unique maximizer of Q.

(c) $\nabla Q(\mathbf{P}_{\mathcal{M}\cap\mathcal{N}^\perp}[\hat{\boldsymbol{\ell}}^m]) = \mathbf{W}'(\mathbf{Y} - \exp\{\hat{\boldsymbol{\ell}}^m\})$.

These three statements complete the proof since they imply that

$$\mathbf{O}_{p-S} = \nabla Q(\mathbf{P}_{\mathcal{M}\cap\mathcal{N}^\perp}[\hat{\boldsymbol{\ell}}^m]) = \mathbf{W}'(\mathbf{Y} - \exp\{\hat{\boldsymbol{\ell}}^m\}).$$

Calculation gives

$$\nabla Q(\boldsymbol{\beta}) = \mathbf{W}'\mathbf{Y} - \sum_{j=1}^{S} \sum_{i \in \mathcal{T}_j} \frac{m_j \exp\{\mathbf{w}_i'\boldsymbol{\beta}\}}{(\sum_{q \in \mathcal{T}_j} \exp\{\mathbf{w}_q'\boldsymbol{\beta}\})} \mathbf{w}_i, \quad \text{and}$$

$$\nabla^2 Q(\boldsymbol{\beta}) = -\mathbf{W}'\mathbf{D}\mathbf{W} - \mathbf{W}'\mathbf{D}\left(\sum_{j=1}^{S} \boldsymbol{\nu}_j \boldsymbol{\nu}_j'\right)\mathbf{D}\mathbf{W}$$

where $\mathbf{D}(\boldsymbol{\beta})$ is an $(n \times n)$ diagonal matrix with (positive) ith diagonal element

$$\exp\left\{\mathbf{w}_i'\boldsymbol{\beta} + \ln(m_j) - \ln\left(\sum_{q \in \mathcal{T}_j} \exp\{\mathbf{w}_q'\boldsymbol{\beta}\}\right)\right\}$$

for $i \in \mathcal{T}_j$. To show (a) fixed any nonzero $\mathbf{z} \in \mathbb{R}^{p-S}$ then,

$$\mathbf{z}'\nabla^2 Q(\boldsymbol{\beta})\mathbf{z} = -||\mathbf{D}^{1/2}\mathbf{W}\mathbf{z}||^2 - \sum_{j=1}^{S} ||\boldsymbol{\nu}_j'\mathbf{D}\mathbf{W}\mathbf{z}||^2,$$

where $\mathbf{D} = \mathbf{D}^{1/2}\mathbf{D}^{1/2}$ and $\mathbf{D}^{1/2}$ has positive diagonal elements. The summands $-||\boldsymbol{\nu}_j'\mathbf{D}\mathbf{W}\mathbf{z}||^2 \leq 0$ for $j = 1(1)S$, and $-||\mathbf{D}^{1/2}\mathbf{W}\mathbf{z}||^2 < 0$ since $\mathbf{W}$ has full column rank, $\mathbf{z} \neq \mathbf{0}_{p-S}$, and $\mathbf{D}^{1/2}$ is a positive diagonal matrix; these imply that $\mathbf{z}'\nabla^2 Q(\boldsymbol{\beta})\mathbf{z} < 0$. Hence $\nabla^2 Q(\boldsymbol{\beta})$ is nonnegative definite and $Q(\boldsymbol{\beta})$ is strictly concave in $\boldsymbol{\beta}$. Thus $Q(\boldsymbol{\beta})$ has either no maximizers or a unique maximizer.

By formula (A.3.1)

$$\mathbf{P}_{\mathcal{M}\cap\mathcal{N}^\perp}(\hat{\boldsymbol{\ell}}^m) = \hat{\boldsymbol{\ell}}^m - \sum_{j=1}^{S} \frac{\boldsymbol{\nu}_j}{t_j}\left(\sum_{q \in \mathcal{T}_j} \hat{\ell}_q^m\right).$$

Thus

$$Q(\mathbf{P}_{\mathcal{M}\cap\mathcal{N}^\perp}[\hat{\boldsymbol{\ell}}^m]) = \sum_{j=1}^{S} \sum_{i \in \mathcal{T}_j} Y_i \left(\hat{\ell}_i^m - \frac{1}{t_j} \sum_{q \in \mathcal{T}_j} \hat{\ell}_q^m\right)$$

$$
\begin{aligned}
&- \sum_{j=1}^{S} m_j \ln \left\{ \sum_{q\in\mathcal{T}_j} \exp\left(\hat{\ell}_q^m - \frac{1}{t_j} \sum_{r\in\mathcal{T}_j} \hat{\ell}_r^m \right) \right\} \\
&= \sum_{i=1}^{n} Y_i \hat{\ell}_i^m - \sum_{j=1}^{S} \sum_{i\in\mathcal{T}_j} \frac{Y_i}{t_j} \left(\sum_{q\in\mathcal{T}_j} \hat{\ell}_q^m \right) \\
&\quad + \sum_{j=1}^{S} \frac{m_j}{t_j} \left(\sum_{r\in\mathcal{T}} \hat{\ell}_r^m \right) - \sum_{j=1}^{S} m_j \ln \left(\sum_{q\in\mathcal{T}_j} \exp\{\hat{\ell}_q^m\} \right) \\
&= \sum_{i=1}^{n} Y_i \hat{\ell}_i^m - \sum_{j=1}^{S} m_j \ln(m_j)
\end{aligned}
$$

since $\sum_{q\in\mathcal{T}_j} \exp\{\hat{\ell}_q^m\} = m_j$. To show (b) is must be proved that

$$
\sum_{i=1}^{n} Y_i \hat{\ell}_i^m - \sum_{j=1}^{S} m_j \ln(m_j) \geq \sum_{i=1}^{n} Y_i \ell_i - \sum_{j=1}^{S} m_j \ln \left(\sum_{q\in\mathcal{T}_j} \exp\{\ell_q\} \right)
$$

for any $\boldsymbol{\ell} \in \mathcal{M} \cap \mathcal{N}^{\perp}$. Given $\boldsymbol{\ell} \in \mathcal{M} \cap \mathcal{N}^{\perp}$, form $\boldsymbol{\ell}^* := \boldsymbol{\ell} + \sum_{j=1}^{S} \boldsymbol{\nu}_j \{\ln m_j - \ln(\sum_{q\in\mathcal{T}_j} \exp\{\ell_q\})\}$ as in Lemma A.3.2. Then $\sum_{i=1}^{n} Y_i \hat{\ell}_i^m \geq \sum_{i=1}^{n} Y_i \ell_i^*$ for every $\boldsymbol{\ell}^* = \tilde{\mathcal{M}}$ since $\hat{\boldsymbol{\ell}}^m$ is an MLE under $(\tilde{\mathcal{M}}, PM)$. But,

$$
\begin{aligned}
\sum_{i=1}^{n} Y_i \ell_i^* &= \sum_{j=1}^{S} \sum_{i\in\mathcal{T}_j} Y_i \left\{ \ell_i + \ln m_j - \ln \left(\sum_{q\in\mathcal{T}_j} \exp\{\ell_q \right) \right\} \\
&= \sum_{i=1}^{n} Y_i \ell_i + \sum_{j=1}^{S} m_j \ln(m_j) - \sum_{j=1}^{S} m_j \ln \left(\sum_{q\in\mathcal{T}_j} \exp\{\ell_q\} \right).
\end{aligned}
$$

Substitution gives

$$
\sum_{i=1}^{n} Y_i \hat{\ell}_i^m \geq \sum_{i=1}^{n} Y_i \ell_i + \sum_{j=1}^{S} m_j \ln(m_j) - \sum_{j=1}^{S} m_j \ln \left(\sum_{q\in\mathcal{T}_j} \exp\{\ell_q\} \right)
$$

which is the desired result. Hence $\mathbf{P}_{\mathcal{M}\cap\mathcal{N}^{\perp}}(\hat{\boldsymbol{\ell}}^m)$ is the (unique) maximizer of $Q(\boldsymbol{\beta})$.

Lastly, to prove (c) note that

$$
\nabla Q(\boldsymbol{\ell}) := \nabla Q(\boldsymbol{\beta})|_{\boldsymbol{\ell}=\mathbf{W}\boldsymbol{\beta}} = \mathbf{W}'\mathbf{Y} - \sum_{j=1}^{S} \sum_{i\in\mathcal{T}_j} \frac{m_j \exp\{\ell_i\}}{\sum_{q\in\mathcal{T}_j} \exp\{\ell_q\}} \mathbf{w}_i .
$$

Thus

$$\begin{aligned}
\nabla Q(\mathbf{P}_{\mathcal{M}\cap\mathcal{N}^\perp}(\hat{\boldsymbol{\ell}}^m)) &= \mathbf{W}'\mathbf{Y} - \sum_{j=1}^{S}\sum_{i\in\mathcal{T}_j} \frac{m_j \exp(\hat{\ell}_i^m - t_j^{-1}\sum_{q\in\mathcal{T}_j}\hat{\ell}_q^m)}{\sum_{q\in\mathcal{T}_j}\exp(\hat{\ell}_q^m - t_j^{-1}\sum_{r\in\mathcal{T}_j}\hat{\ell}_r^m)}\mathbf{w}_i \\
&= \mathbf{W}'\mathbf{Y} - \sum_{j=1}^{S}\sum_{i\in\mathcal{T}_j} \frac{\exp(\hat{\ell}_i^m)m_j}{\sum_{q\in\mathcal{T}_j}\exp(\hat{\ell}_q^m)}\mathbf{w}_i \\
&= \mathbf{W}'\mathbf{Y} - \sum_{j=1}^{S}\sum_{i\in\mathcal{T}_j} \exp(\hat{\ell}_i^m)\mathbf{w}_i \\
&= \mathbf{W}'\mathbf{Y} - \mathbf{W}'\exp\{\hat{\boldsymbol{\ell}}^m\}. \qquad \square
\end{aligned}$$

Appendix 4. Elements of Large Sample Theory

This appendix first reviews the likelihood ratio, score, and Wald methods for constructing large sample tests of hypotheses in parametric problems. Then some of the probabilistic tools required to analyze the asymptotic behavior of test statistics and estimators will be sketched.

A. Large Sample Parametric Test Statistics

Suppose $\mathbf{Y} = (Y_1, \ldots, Y_n)'$ has joint cumulative distribution function $F(\mathbf{y}|\boldsymbol{\theta})$ where $\boldsymbol{\theta} \in \Omega \subseteq \mathbb{R}^d$. Given $\omega \subset \Omega$ (strictly), it is desired to test $H_0: \boldsymbol{\theta} \in \omega$ versus H_A: $\boldsymbol{\theta} \in \Omega \backslash \omega$. Two examples illustrating the general testing problem follow and then three large-sample tests of H_0 will be described.

Example A.4.1. Let $\mathbf{Y} \sim \mathcal{M}_t(n, \mathbf{p})$ with unknown $\mathbf{p}$ belonging to the $(t-1)$ dimensional simplex $\mathcal{S}$. Two standard testing problems are those of testing the simple null hypothesis $H_0: \mathbf{p} = \mathbf{p}^0$ versus H_A: not H_0, corresponding to $\omega = \{\mathbf{p}^0\}$, and testing the composite null hypothesis $H_0: \mathbf{p} \in \{(p_1(\boldsymbol{\eta}), \ldots, p_t(\boldsymbol{\eta}))': \boldsymbol{\eta} \in \mathcal{N}\}$ versus H_A: not H_0, corresponding to $\omega = \{(p_1(\boldsymbol{\eta}), \ldots, p_t(\boldsymbol{\eta}))': \boldsymbol{\eta} \in \mathcal{N}\}$.

Example A.4.2. An important class of testing problems are those in which ω can be expressed in the form $\omega := \{\boldsymbol{\theta} \in \Omega: h_1(\boldsymbol{\theta}) = 0, \ldots, h_r(\boldsymbol{\theta}) = 0\}$. The Wald test was specifically devised to test the null hypothesis

$$H_0: \mathbf{h}(\boldsymbol{\theta}) = \mathbf{0}_r \quad \text{versus} \quad H_A: \mathbf{h}(\boldsymbol{\theta}) \neq \mathbf{0}_r,$$

where $\mathbf{h}(\boldsymbol{\theta}) := (h_1(\boldsymbol{\theta}), \ldots, h_r(\boldsymbol{\theta}))'$.

Neyman and Pearson's *likelihood ratio test* (LRT) is the oldest of the three tests considered (Neyman and Pearson (1928)). Let $\hat{\boldsymbol{\theta}}_\Omega$ $(\hat{\boldsymbol{\theta}}_\omega)$ denote the MLE of $\boldsymbol{\theta}$ under Ω (ω), and let $L(\boldsymbol{\theta})$ be the likelihood function of $\mathbf{Y}$. The LRT rejects the null hypothesis if and only if

$$L(\hat{\boldsymbol{\theta}}_\omega)/L(\hat{\boldsymbol{\theta}}_\Omega) \leq C$$

where $\sup_{\boldsymbol{\theta} \in \omega} P_{\boldsymbol{\theta}}[L(\hat{\boldsymbol{\theta}}_\omega)/L(\hat{\boldsymbol{\theta}}_\Omega) \leq C] \stackrel{\text{set}}{=} \alpha$ determines C. If Ω is d-dimensional and ω is s-dimensional, then in the "regular" case (when smoothness conditions assure the asymptotic normality of the MLE and the applicability of the delta method described below), the following null asymptotic result yields the critical region of the LRT: if

$$G^2 := -2\ln(L(\hat{\boldsymbol{\theta}}_\omega)/L(\hat{\boldsymbol{\theta}}_\Omega)), \tag{A.4.1}$$

then $G^2 \xrightarrow{\mathcal{L}} \chi^2_{d-s}$ as $n \to \infty$ for all $\boldsymbol{\theta} \in \omega$. Thus the likelihood ratio test rejects H_0 if and only if $G^2 \geq \chi^2_{\alpha,d-s}$.

Rao (1947) proposed the *score test* (or *Lagrange test*). The test statistic is defined in terms of a vector of "scores."

Definition. The *i*th (*efficient*) *score* of $\mathbf{Y}$ under the model $F(\mathbf{y} \mid \boldsymbol{\theta})$ is $S_i(\boldsymbol{\theta}) := \frac{\partial \ln L(\boldsymbol{\theta})}{\partial \theta_i}$ for $i = 1(1)d$.

The score vector $\mathbf{S}(\boldsymbol{\theta}) := (S_1(\boldsymbol{\theta}), \ldots, S_d(\boldsymbol{\theta}))'$ satisfies $\mathbf{S}(\hat{\boldsymbol{\theta}}_\Omega) = \mathbf{0}_d$. This observation suggests rejecting H_0 when $\mathbf{S}(\cdot)$ evaluated at the *restricted* MLE $\hat{\boldsymbol{\theta}}_\omega$ is "far" from the origin (in $\mathbb{R}^d$). Rao showed that the normalized version

$$X^2 := (\mathbf{S}(\hat{\boldsymbol{\theta}}_\omega))'\mathbf{I}^{-1}(\hat{\boldsymbol{\theta}}_\omega)(\mathbf{S}(\hat{\boldsymbol{\theta}}_\omega)) \tag{A.4.2}$$

of $\mathbf{S}(\hat{\boldsymbol{\theta}}_\omega)$ converges to a χ^2_{d-s} distribution as $n \to \infty$ for any $\boldsymbol{\theta} \in \omega$ under regularity conditions similar to those guaranteeing that G^2 converges to a chi-squared null distribution. Here $\mathbf{I}(\boldsymbol{\theta})$ is the $d \times d$ information matrix with (i,j)th element

$$(\mathbf{I}(\boldsymbol{\theta}))_{ij} = -E_{\boldsymbol{\theta}}\left[\frac{\partial^2 \ln L(\boldsymbol{\theta})}{\partial\theta_i \partial\theta_j}\right].$$

Thus the score test rejects H_0 if and only if $X^2 \geq \chi^2_{\alpha,d-s}$.

Wald (1943) assumes a null hypotheses of the form $\omega = \{\boldsymbol{\theta} \in \Omega \colon \mathbf{h}(\boldsymbol{\theta}) = \mathbf{0}_r\}$. The idea of the test is to reject $H_0 \colon \mathbf{h}(\boldsymbol{\theta}) = \mathbf{0}_r$ when the vector $\mathbf{h}(\hat{\boldsymbol{\theta}}_\Omega)$ is "far" from the null value of zero (in $\mathbb{R}^r$). Wald proposed calibrating $\mathbf{h}(\hat{\boldsymbol{\theta}}_\Omega)$ by

$$W := (\mathbf{h}(\hat{\boldsymbol{\theta}}_\Omega))'(\{\nabla\mathbf{h}(\hat{\boldsymbol{\theta}}_\Omega)\}'\{\mathbf{I}(\hat{\boldsymbol{\theta}}_\Omega)\}^{-1}\{\nabla\mathbf{h}(\hat{\boldsymbol{\theta}}_\Omega)\})^{-1}(\mathbf{h}(\hat{\boldsymbol{\theta}}_\Omega)), \tag{A.4.3}$$

where $\nabla\mathbf{h}(\boldsymbol{\theta})$ is the $d \times r$ matrix with (i,j)th element $(\nabla\mathbf{h}(\boldsymbol{\theta}))_{ij} = (\frac{\partial h_j(\boldsymbol{\theta})}{\partial \theta_i})$. The asymptotic normality of the MLE and an application of the delta method show that $\mathbf{h}(\hat{\boldsymbol{\theta}}_\Omega) \to N_r[\mathbf{0}_r, \{\nabla\mathbf{h}(\boldsymbol{\theta})\}'\{\mathbf{I}(\boldsymbol{\theta})\}^{-1}\{\nabla\mathbf{h}(\boldsymbol{\theta})\}]$ as $n \to \infty$ for any $\boldsymbol{\theta} \in \omega$. Wald's test rejects H_0 if and only if $W \geq \chi^2_{\alpha,r}$. Observe that $r = d - (d - r)$, where $d - r$ is the dimension of ω, so that all three test statistics have the same limiting distribution.

As one comparison, the score test is generally easier to compute than the LRT and Wald tests since the latter two tests both require the calculation of $\hat{\boldsymbol{\theta}}_\Omega$ which is often significantly more difficult to compute than $\hat{\boldsymbol{\theta}}_\omega$.

Additional comparisons of the three tests in the multinomial setting are stated in Section 2.2. To illustrate, the three asymptotic test statistics will be calculated for the simple null hypothesis testing problem of Example A.4.1.

Example A.4.1 (continued). Let $\mathbf{Y} \sim M_t(n, \mathbf{p})$ with unknown $\mathbf{p} \in \mathcal{S}$, and consider testing the simple null hypothesis H_0: $\mathbf{p} = \mathbf{p}^0$ versus the global alternative H_A: $\mathbf{p} \neq \mathbf{p}^0$. Let $\boldsymbol{\theta} := (p_1, \ldots, p_{t-1})'$, $\Omega := \{\mathbf{x} \in \mathbb{R}^{t-1}$: $x_i \geq 0$, $\sum_1^{t-1} x_i \leq 1\}$, and $\omega := \{(p_1^0, \ldots, p_{t-1}^0)'\}$. Note that $d = t-1$, $s = 0$, $\hat{\boldsymbol{\theta}}_\Omega = (Y_1/n, \ldots, Y_{t-1}/n)'$ and $\hat{\boldsymbol{\theta}}_\omega = (p_1^0, \ldots, p_{t-1}^0)'$.

To construct the LRT, observe that $L(\boldsymbol{\theta})$ is proportional to $\prod_{i=1}^t p_i^{y_i}$ where $p_t = 1 - p_1 - \ldots - p_{t-1}$. Thus the LRT rejects H_0 if and only if

$$G^2 = -2 \ln \left(\frac{\prod_{i=1}^t (p_i^0)^{Y_i}}{\prod_{i=1}^t (Y_i/n)^{Y_i}} \right) = 2 \sum_{i=1}^t Y_i \ln \left(\frac{Y_i}{np_i^0} \right) \geq \chi^2_{\alpha,t-1}. \tag{A.4.4}$$

To determine the score test, first note that $\ln L(\boldsymbol{\theta})$ is, aside from constants,

$$\sum_{i=1}^{t-1} Y_i \ln p_i + Y_t \ln(1 - p_1 - \ldots - p_{t-1}).$$

Hence,

$$S_j(\boldsymbol{\theta}) = \frac{\partial \ln L(\boldsymbol{\theta})}{\partial p_j} = \frac{Y_j}{p_j} - \frac{Y_t}{p_t}, \quad 1 \leq j \leq t-1, \quad \text{and}$$

$$\frac{\partial^2 \ln L(\boldsymbol{\theta})}{\partial p_k \partial p_j} = \begin{cases} -\dfrac{Y_j}{p_j^2} - \dfrac{Y_t}{p_t^2}, & k = j \\[2ex] -\dfrac{Y_t}{p_t^2}, & k \neq j. \end{cases}$$

Calculation gives

$$-E_{\boldsymbol{\theta}} \left[\frac{\partial^2 \ln L}{\partial p_k \partial p_j} \right] = \begin{cases} \dfrac{n}{p_j} + \dfrac{n}{p_t}, & k = j \\[2ex] \dfrac{n}{p_t}, & k \neq j, \end{cases}$$

and thus the information matrix is

$$\mathbf{I}(\boldsymbol{\theta}) = n \left[\text{Diag}\left(\frac{1}{p_1}, \ldots, \frac{1}{p_{t-1}} \right) + \frac{1}{p_t} \mathbf{1}_{t-1} \mathbf{1}'_{t-1} \right]. \tag{A.4.5}$$

At the restricted MLE $\hat{\boldsymbol{\theta}}_\omega = (p_1^0, \ldots, p_{t-1}^0)'$, the inverse is

$$\mathbf{I}^{-1}(\hat{\boldsymbol{\theta}}_\omega) = \frac{1}{n} [\text{Diag}(\hat{\boldsymbol{\theta}}_\omega) - \hat{\boldsymbol{\theta}}_\omega (\hat{\boldsymbol{\theta}}_\omega)'].$$

The score test rejects H_0 if and only if

$$X^2 = \left(\ldots, \frac{Y_i}{p_i^0} - \frac{Y_t}{p_t^0}, \ldots\right) \times \left(\frac{1}{n}(\text{Diag}(p_1^0, \ldots, p_{t-1}^0)\right.$$

$$\left. - (p_1^0, \ldots, p_{t-1}^0)'(p_1^0, \ldots, p_{t-1}^0)\right) \times \left(\ldots, \frac{Y_i}{p_i^0} - \frac{Y_t}{p_t^0}, \ldots\right)'$$

$$= \sum_{i=1}^{t} \frac{[Y_i - np_i^0]^2}{np_i^0} \geq \chi^2_{\alpha, t-1} \tag{A.4.6}$$

which is the standard chi-squared test of H_0.

To construct Wald's test, first express ω in constrained form as

$$\omega = \{\theta\colon h_i(\theta) := p_i - p_i^0 = 0, i = 1, \ldots, t-1\}$$

with $r = t - 1 = d$. Thus

$$(\nabla \mathbf{h}(\boldsymbol{\theta}))_{ij} = \frac{\partial h_j(\boldsymbol{\theta})}{\partial p_i} = \begin{cases} 0, & j \neq i \\ 1, & j = i \end{cases}$$

or $\nabla \mathbf{h}(\boldsymbol{\theta}) = \mathbf{I}_{t-1}$. From (A.4.5),

$$\mathbf{I}(\hat{\theta}_\Omega) = n\left[\text{Diag}\left(\frac{1}{\hat{p}_1}, \ldots, \frac{1}{\hat{p}_{t-1}}\right) + \frac{1}{\hat{p}_t}\mathbf{1}_{t-1}\mathbf{1}'_{t-1}\right]$$

where $\hat{p}_i = Y_i/n$. Putting these elements together via A.4.3 shows the Wald test rejects H_0 if and only if

$$W = \sum_{i=1}^{t} \frac{[Y_i - np_i^0]^2}{Y_i} \geq \chi^2_{\alpha, t-1}. \tag{A.4.7}$$

This test is also called the "minimum modified chi squared" test of H_0.

B. Probabilistic Tools

The remainder of this appendix reviews the probabilistic tools necessary to analyze the large-sample properties of tests and estimators. The book *Approximation Theorems of Mathematical Statistics* by Serfling (1980) contains a more detailed study of these results. As an application, the large-sample null and (contiguous) alternative distributions of G^2, W, and X^2 will be derived for the simple null hypothesis multinomial problem of Example A.4.1.

Deterministic O and o

The symbol $\|\mathbf{x}\| := (x_1^2 + \ldots + x_n^2)^{1/2}$ will denote the Euclidean norm of a vector $\mathbf{x} \in \mathbb{R}^n$.

Definition. For vectors $\{\mathbf{a}_n\}_{n\geq 1}$ and positive real numbers $\{b_n\}$, $\mathbf{a}_n = O(b_n)$ means there exists $K > 0$ such that $\|\mathbf{a}_n\|/b_n \leq K$ for all $n \geq 1$; $\mathbf{a}_n = o(b_n)$ means $\|\mathbf{a}_n\|/b_n \to 0$ as $n \to \infty$.

For example, $\mathbf{a}_n = o(1)$ if and only if $\|\mathbf{a}_n\| \to 0$ and $\mathbf{a}_n = o(1/\sqrt{n})$ if and only if $\sqrt{n}\|\mathbf{a}_n\| \to 0$. Many authors extend the above definitions by allowing $\|\mathbf{a}_n\|/b_n$ to be undefined for a finite number of terms.

Deterministic $O(\cdot)$ and $o(\cdot)$ obey many arithmetic operations. As examples, if $\{a_n^1\}$ and $\{a_n^2\}$ are real-valued sequences and $\{b_n^1\}$ and $\{b_n^2\}$ positive real sequences, then

(i) $a_n^1 = O(b_n^1)$, $a_n^2 = O(b_n^2)$ implies $a_n^1 a_n^2 = O(b_n^1 b_n^2)$,

(ii) $a_n^1 = O(b_n^1)$, $a_n^2 = o(b_n^2)$ implies $a_n^1 a_n^2 = o(b_n^1 b_n^2)$, and

(iii) $a_n^1 = o(b_n^1)$, $a_n^2 = o(b_n^2)$ implies $a_n^1 a_n^2 = o(b_n^1 b_n^2)$.

The notions of O, o can be extended to continuous arguments.

Definition. If $\mathbf{f}(\cdot): \mathcal{D} \to \mathbb{R}^k$ and $g(\cdot): \mathcal{D} \to (0, \infty)$ where the domain $\mathcal{D}$ contains an interval about the point $L \in \mathbb{R}^m$ then $\mathbf{f}(\mathbf{x}) = O(g(\mathbf{x}))$ $(o(g(\mathbf{x})))$ as $\mathbf{x} \to L$ means $\mathbf{f}(\mathbf{x}_n) = O(g(\mathbf{x}_n))$ $(o(g(\mathbf{x}_n)))$ for any sequence $\mathbf{x}_n \to L$.

Example A.4.3.

(i) If $f: \mathbb{R} \to \mathbb{R}$ is differentiable at $x = L$ then

$$f(x) = f(L) + f'(L)(x - L) + o(|x - L|),$$

since $f'(L)$ is defined by $\frac{f(x)-f(L)}{x-L} - f'(L) = o(1)$ as $x \to L$.

(ii) If $f: \mathbb{R} \to \mathbb{R}$ is m-times differentiable at $x = L$ then

$$f(x) = \sum_{j=0}^{m} \frac{f^{(j)}(L)(x - L)^j}{j!} + o(|x - L|^m)$$

where $f^{(j)}(\cdot)$ denotes the jth derivative of $f(\cdot)$ at L. Part (ii) follows from (i) by induction.

A review of two modes of convergence of sequences of random variables is now given which will permit an extension of O, o notation to stochastic components.

Modes of Convergence

Let $\{\mathbf{T}_n := (T_{1n}, \ldots, T_{kn})'\}_{n\geq 1}$ be a sequence of random k-vectors, $\mathbf{T} := (T_1, \ldots, T_k)'$ a random k-vector, and $\mathbf{c} \in \mathbb{R}^k$. The notions of convergence in probability and convergence in law of $\{\mathbf{T}_n\}_{n\geq 1}$ are defined below.

Mode 1: Convergence in Probability

Definition. $\mathbf{T}_n$ converges in probability to $\mathbf{c}$, denoted $\mathbf{T}_n \xrightarrow{p} \mathbf{c}$, means for all $\epsilon > 0$, $P[||\mathbf{T}_n - \mathbf{c}|| < \epsilon] \to 1$ as $n \to \infty$.

Convergence in probability concerns only the *marginal* distributions of $\{\mathbf{T}_n\}$. It is straightforward to prove that $\mathbf{T}_n' = (T_{1n}, \ldots, T_{kn})$ converges in probability to $\mathbf{c}' = (c_1, \ldots c_k)$ if and only if there is componentwise convergence $T_{jn} \xrightarrow{p} c_j$ for $1 \leq j \leq k$. The notion of convergence in probability can be extended to allow a random vector as a limit by defining $\mathbf{T}_n \xrightarrow{p} \mathbf{T}$ as $\mathbf{T}_n - \mathbf{T} \xrightarrow{p} 0_k$.

Example A.4.4.

(i) The weak law of large numbers is the classic example of convergence in probability.

(ii) If $\mathbf{T}_n \xrightarrow{p} \mathbf{c}$ as $n \to \infty$ and $\varphi: \mathbb{R}^k \to \mathbb{R}$ is continuous at $\mathbf{c}$, then it can be easily proved from the definition that $\varphi(\mathbf{T}_n) \xrightarrow{p} \varphi(\mathbf{c})$.

Mode 2: Convergence in Law

Definition. The sequence of random vectors $\{\mathbf{T}_n\}$ converges in law (or in distribution or weakly) to $\mathbf{T}$, denoted $\mathbf{T}_n \xrightarrow{\mathcal{L}} \mathbf{T}$, means that for all continuity points $\mathbf{t}$ of the joint distribution function $F_{\mathbf{T}}(\mathbf{t}) := P[T_1 \leq t_1, \ldots, T_k \leq t_k]$ of $\mathbf{T}$,

$$\lim_{n\to\infty} P[T_{1n} \leq t_1, \ldots, T_{kn} \leq t_k] = F_{\mathbf{T}}(\mathbf{t}).$$

Convergence in law provides the justification for approximations such as the normal approximation to the binomial distribution. In general, if for real random variables $\{T_n\}$ and T, $T_n \xrightarrow{\mathcal{L}} T$, and $a < b$ are continuity points of T, then

$$P[a < T_n \leq b] \simeq P[a < T \leq b]$$

for sufficiently large n.

Unlike convergence in probability, convergence in law of $\mathbf{T}_n' = (T_{1n}, \ldots, T_{kn})$ to $\mathbf{T}' = (T_1, \ldots, T_k)$ is not equivalent to the convergence in law of each component T_{jn} to T_j. The former implies the convergence in law of the component random variables, but the converse fails. The relationship between convergence in probability and convergence in law is that $\mathbf{T}_n \xrightarrow{p} \mathbf{c}$

if and only if $\mathbf{T}_n \xrightarrow{\mathcal{L}} \mathbf{T}$ where $\mathbf{T}$ has the degenerate distribution at $\mathbf{c}$. For example, if $k = 1$ then

$$P[T_n \leq t] \to \begin{cases} 0 & \text{if } t < c \\ 1 & \text{if } t > c \end{cases}$$

as $n \to \infty$. In the general case, $\mathbf{T}_n \xrightarrow{p} \mathbf{T}$ implies $\mathbf{T}_n \xrightarrow{\mathcal{L}} \mathbf{T}$ but the converse fails.

Stochastic O and o

Definition. Let $\{\mathbf{T}_n\}_{n\geq 1}$ be a sequence of random k-vectors and $\{b_n\}_{n\geq 1}$ a sequence of positive numbers. Then $\mathbf{T}_n = O_p(b_n)$ means for all $\epsilon > 0$ there exists $K = K(\epsilon) > 0$ such that $P[||\mathbf{T}_n||/b_n < K] \geq 1 - \epsilon$ for all n. The sequence $\mathbf{T}_n = o_p(b_n)$ means for all $\epsilon > 0$ and for all $\eta > 0$ there exists $N = N(\epsilon, \eta)$ such that $P[||\mathbf{T}_n||/b_n < \eta] \geq 1 - \epsilon$ for all $n \geq N$; equivalently, $||\mathbf{T}_n||/b_n \xrightarrow{p} 0$.

As in the deterministic case, the definitions of O_p and o_p are relaxed by some authors to allow finitely many n to violate the probability requirements. Special cases of the definitions are $o_p(1)$ which is convergence in probability and $O_p(1)$ which is called "bounded in probability."

Example A.4.5.

(i) If $\mathbf{T}_n \xrightarrow{\mathcal{L}} \mathbf{T}$ then $\mathbf{T}_n = O_p(1)$.

(ii) Suppose $\{T_n\}$ is real-valued and $(T_n - \mu_n)/\sigma_n \xrightarrow{\mathcal{L}} T$ for some $\{\mu_n\}$ and $\{\sigma_n\}$. Then $(T_n - \mu_n)/\sigma_n = O_p(1)$ by (i) which implies $T_n = \mu_n + O_p(\sigma_n)$.

As in the case of deterministic O, o there is an arithmetic that O_p, o_p obey. For example, if $\{X_n^1\}$ and $\{X_n^2\}$ are real-valued random sequences and $\{b_n^1\}$ and $\{b_n^2\}$ are positive real sequences, then

(i) $X_n^1 = O_p(b_n^1)$, $X_n^2 = O_p(b_n^2)$ implies $X_n^1 X_n^2 = O_p(b_n^1 b_n^2)$,

(ii) $X_n^1 = O_p(b_n^1)$, $X_n^2 = o_p(b_n^2)$ implies $X_n^1 X_n^2 = o_p(b_n^1 b_n^2)$, and

(iii) $X_n^1 = o_p(b_n^1)$, $X_n^2 = o_p(b_n^2)$ implies $X_n^1 X_n^2 = o_p(b_n^1 b_n^2)$.

Establishing Convergence in Law

As convergence in law is the most important mode of convergence for this text, the appendix concludes with a review of some methods for establishing it:

(1) apply the definition of $\xrightarrow{\mathcal{L}}$,

(2) apply a central limit theorem,

(3) identify the generating function, and

(4) argue from continuity.

Method 1: Apply the definition of $\xrightarrow{\mathcal{L}}$. In some problems the brute force application of the definition can be used to establish $\xrightarrow{\mathcal{L}}$.

Example A.4.6. Let $X_1, \ldots, X_n$ be iid random variables distributed $U[0, \theta]$ and $X_{(n)} := \max\{X_1, \ldots, X_n\}$. Then

$$P[X_{(n)} \leq t] = \begin{cases} 0 & \text{for } t \leq 0 \\ (t/\theta)^n & \text{for } 0 < t < \theta \\ 1 & \text{for } \theta \leq t. \end{cases}$$

It is clear from this expression that $X_{(n)} \xrightarrow{p} \theta$ as $n \to \infty$, i.e.,

$$P[X_{(n)} \leq t] \to \begin{cases} 0 & \text{if } t < \theta \\ 1 & \text{if } t > \theta. \end{cases}$$

Hence $X_{(n)} \xrightarrow{\mathcal{L}} X$ which is degenerate at θ. To obtain a nondegenerate distribution the difference $(X_{(n)} - \theta)$ must be "blown up"; in this case

$$T_n := n(\theta - X_{(n)}) \xrightarrow{\mathcal{L}} T \sim \text{Exponential with mean } \theta$$

as $n \to \infty$. To prove this, again apply the definition to obtain

$$\begin{aligned} P[T_n \leq t] &= 1 - P[n(\theta - X_{(n)}) \geq t] \\ &= 1 - P[X_{(n)} \leq \theta - t/n] \\ &= 1 - \begin{cases} 0, & \theta - t/n \leq 0 \\ (1 - t/n\theta)^n, & 0 < \theta - t/n < \theta \\ 1, & \theta \leq \theta - t/n \end{cases} \\ &= \begin{cases} 1 & \text{if } t \geq n\theta \\ 1 - (1 - t/n\theta)^n & \text{if } 0 < t < n\theta \\ 0 & \text{if } t \leq 0. \end{cases} \end{aligned}$$

So, $t \leq 0$ implies $F_n(t) := P[T_n \leq t] = 0$, while if $t > 0$ and n is sufficiently large, $F_n(t) = 1 - (1 - t/n\theta)^n \to 1 - \exp\{-t/\theta\}$ which gives the desired conclusion.

Method 2: Apply a Central Limit Theorem. The normal approximation to the binomial distribution discussed in Section 1.3 provides one application of the central limit theorem. A multivariate version of the central limit theorem will now be stated (Serfling (1980), p. 28) followed by an application.

Theorem A.4.1 (Multivariate Central Limit Theorem). *Let $\{\mathbf{T}_n\}$ be a sequence of iid k-vectors with mean $\boldsymbol{\mu}$ and variance–covariance matrix $\boldsymbol{\Sigma}$, and set*

$$\overline{\mathbf{T}}_n := \frac{1}{n}\left(\sum_{j=1}^{n} T_{1j}, \sum_{j=1}^{n} T_{2j}, \ldots, \sum_{j=1}^{n} T_{kj}\right)'.$$

Then $\sqrt{n}(\overline{\mathbf{T}}_n - \boldsymbol{\mu}) \xrightarrow{\mathcal{L}} N_k(\mathbf{0}_k, \boldsymbol{\Sigma})$.

Example A.4.1 (continued). Since $\mathbf{Y}_n \sim M_t(n, \mathbf{p})$, $\mathbf{Y}_n := \sum_{j=1}^{n} \mathbf{T}_j$ where $\{\mathbf{T}_j\}_{j=1}^{n}$ are the iid $M_t(1, \mathbf{p})$ component trial outcomes. Thus $E[\mathbf{T}_j] = \mathbf{p}$ and the variance–covariance matrix of $\mathbf{T}_j$ is $\boldsymbol{\Sigma}(\mathbf{p}) = \text{Diag}(\mathbf{p}) - \mathbf{p}\mathbf{p}'$. The multivariate central limit theorem asserts

$$\sqrt{n}\left(\frac{\mathbf{Y}_n}{n} - \mathbf{p}\right) \xrightarrow{\mathcal{L}} N_t[\mathbf{0}_t, \boldsymbol{\Sigma}(\mathbf{p})].$$

Method 3: Identify the Generating Function

Definition. The moment generating function (mgf) of the random vector $\mathbf{Y}$, is defined by

$$\psi(\mathbf{s}) := E[\exp\{\mathbf{s}'\mathbf{Y}\}]$$

provided $\psi(\mathbf{s})$ is finite for all $\mathbf{s}$ in some hyperrectangle containing the origin.

For any $\mathbf{Y}$, $\psi(\mathbf{0}) = 1$, but $\psi(\mathbf{s})$ can be infinite for all $\mathbf{s} \neq \mathbf{0}$ (e.g., if the components of $\mathbf{Y}$ are independent Cauchy distributions). The existence of a moment generating function for $\mathbf{Y}$ is a strong property—it implies that $\mathbf{Y}$ has moments of all orders. Mgf's are important in studying convergence in law because of the following result describing the correspondence between convergence of mgf's and convergence in law.

Theorem A.4.2. *Suppose $\{\mathbf{T}_n\}_{n=1}^{\infty}$ have moment generating functions $\psi_n(\mathbf{s})$ existing on $||\mathbf{s}|| < K$ for some $K > 0$, and $\mathbf{T}$ has a moment generating function $\psi(\mathbf{s})$ also existing on $||\mathbf{s}|| < K$. If $\psi_n(\mathbf{s}) \to \psi(\mathbf{s})$ for $||\mathbf{s}|| < K$, then $\mathbf{T}_n \xrightarrow{\mathcal{L}} \mathbf{T}$.*

Example A.4.7. Suppose $\mathbf{Y}_n \sim M_t(n, \mathbf{p}_n = \mathbf{p}^0 + \mathbf{u}/\sqrt{n})$ where $\mathbf{u} = (u_1, \ldots, u_t)'$ is a fixed vector satisfying $\sum_{i=1}^{t} u_i = 0$ and $\mathbf{p}^0 \in \mathcal{S}$, then $\mathbf{T}_n := \sqrt{n}(\mathbf{Y}_n/n - \mathbf{p}^0) \xrightarrow{\mathcal{L}} N_t[\mathbf{u}, \boldsymbol{\Sigma}(\mathbf{p}^0)]$, where $\boldsymbol{\Sigma}(\mathbf{p}^0) = \text{Diag}(\mathbf{p}^0) - \mathbf{p}^0\mathbf{p}^{0\prime}$. Note that the mean of $\mathbf{T}_n$ is

$$E[\mathbf{T}_n] = E\left[\sqrt{n}\left(\frac{\mathbf{Y}_n}{n} - \mathbf{p}^0\right)\right] = \sqrt{n}\left(\frac{n(\mathbf{p}^0 + \mathbf{u}/\sqrt{n})}{n} - \mathbf{p}^0\right) = \mathbf{u}$$

for all n, which is exactly the mean of the limiting normal distribution. The sequence of cell probabilities $\mathbf{p}_n$ is said to be contiguous to $H_0\colon \mathbf{p} = p^0$ since $\mathbf{p}_n \to \mathbf{p}^0$ as $n \to \infty$. Observe that the multivariate central limit thoerem does not apply unless $\mathbf{u} = \mathbf{0}_t$ since the $\mathbf{Y}_n$ have different distributions as n changes. Instead, convergence of the sequence of mgf's of $\mathbf{T}_n$ will be shown. Let $\mathbf{s}' = (s_1, \ldots, s_t)$ and consider

$$\begin{aligned}\psi_n(\mathbf{s}) &= E\left[\exp\{\sqrt{n}\mathbf{s}'((\mathbf{Y}_n/n) - \mathbf{p}^0)\}\right] \\ &= \left(\sum_{j=1}^{t}\left(p_j^0 + \frac{u_j}{\sqrt{n}}\right)\exp\{(1/\sqrt{n})(s_j - \mathbf{s}'\mathbf{p}^0)\}\right)^n \\ &= \left(1 + \frac{1}{n}[\mathbf{s}'\mathbf{u} + \frac{1}{2}\mathbf{s}'\boldsymbol{\Sigma}(\mathbf{p}^0)\mathbf{s}] + o\left(\frac{1}{n}\right)\right)^n \\ &\to \exp\left\{\mathbf{s}'\mathbf{u} + \frac{1}{2}\mathbf{s}'\boldsymbol{\Sigma}(\mathbf{p}^0)\mathbf{s}\right\},\end{aligned}$$

which is the mgf of the $N_t[\mathbf{u}, \boldsymbol{\Sigma}(p^0)]$ distribution. The second equality follows by considering $\mathbf{Y}_n$ as the sum of n iid $M_t(1, \mathbf{p}_n)$ random t-vectors and the third equality from the facts $e^x = 1 + x + x^2/2 + o(x^2)$, $p^0 \in \mathcal{S}$, and $\sum_{i=1}^t u_i = 0$.

Method 4: Argue from Continuity. A result which has broad application is the Mann–Wald theorem.

Theorem A.4.3 (Mann–Wald). *Suppose that* $\{\mathbf{T}_n\}_{n=1}^{\infty}$ *is a sequence of* k*-vectors satisfying* $\mathbf{T}_n \xrightarrow{\mathcal{L}} \mathbf{T}$ *for some* $\mathbf{T}$, *and that* $\mathbf{g}\colon \mathbb{R}^k \to \mathbb{R}^m$ *satisfies*

$$P[\mathbf{T} \in \{\mathbf{x} \in \mathbb{R}^k\colon \mathbf{g} \text{ is discontinuous at } \mathbf{x}\}] = 0, \tag{A.4.8}$$

then $\mathbf{g}(\mathbf{T}_n) \xrightarrow{\mathcal{L}} \mathbf{g}(\mathbf{T})$.

When $\mathbf{g}(\cdot)$ is continuous the condition (A.4.8) is immediate. Thus, for example, suppose $k = 2$ and

$$(T_{1n}, T_{2n}) \xrightarrow{\mathcal{L}} (T_1, T_2)$$

as $n \to \infty$ where all random variables are real. Then by Theorem A.4.3, $T_{1n} + T_{2n} \xrightarrow{\mathcal{L}} T_1 + T_2$ (take $g(x_1, x_2) = x_1 + x_2$) and $T_{1n}T_{2n} \xrightarrow{\mathcal{L}} T_1T_2$ (take $g(x_1, x_2) = x_1x_2$).

Example A.4.7 (continued). Recall that $\mathbf{Y}_n \sim M_t(n, \mathbf{p}_n)$ where $\mathbf{p}_n = \mathbf{p}^0 + \mathbf{u}/\sqrt{n}$ with $\sum_{i=1}^t u_i = 0$ and $\mathbf{p}^0 \in \mathcal{S}$. Then

$$\sum_{i=1}^{t} \frac{(Y_{in} - np_i^0)^2}{np_i^0} \xrightarrow{\mathcal{L}} \chi_{t-1}^2(\mathbf{u}'\mathbf{A}\mathbf{u}), \tag{A.4.9}$$

where $\mathbf{A} = [\text{Diag}(\mathbf{p}^0)]^{-1}$. Note that for $\mathbf{u} = \mathbf{0}_t$ ($\mathbf{p}_n = \mathbf{p}^0$ for all n), (A.4.9) shows that Pearson's chi-squared statistic converges in law to a central chi-square distribution.

To prove (A.4.9), observe that

$$\sum_{i=1}^{t} \frac{(Y_{in} - np_i^0)^2}{np_i^0} = \sqrt{n}\left[\frac{\mathbf{Y}_n}{n} - \mathbf{p}^0\right]' \mathbf{A} \left[\frac{\mathbf{Y}_n}{n} - \mathbf{p}^0\right] = g\left(\frac{\mathbf{Y}_n}{n} - \mathbf{p}^0\right)$$

where $g: \mathbb{R}^t \to \mathbb{R}$ is defined by $g(\mathbf{x}) := \mathbf{x}'\mathbf{A}\mathbf{x}$. Since $g(\cdot)$ is continuous the Mann–Wald theorem implies

$$\sum_{i=1}^{t} \frac{(Y_{in} - np_i^0)^2}{np_i^0} \xrightarrow{\mathcal{L}} g(\mathbf{T}) = \mathbf{T}'\mathbf{A}\mathbf{T} \tag{A.4.10}$$

where $\mathbf{T} \sim N_t[\mathbf{u}, \boldsymbol{\Sigma}(\mathbf{p}^0)]$. To complete the proof recall that if $\mathbf{Z} \sim N_t[\boldsymbol{\mu}, \boldsymbol{\Sigma}]$, $\mathbf{A}$ is a symmetric matrix, and $\mathbf{A}\boldsymbol{\Sigma}$ is idempotent, then $\mathbf{Z}'\mathbf{A}\mathbf{Z} \sim \chi_d^2(\boldsymbol{\mu}'\mathbf{A}\boldsymbol{\mu})$ where $d = \text{rank}(\mathbf{A}\boldsymbol{\Sigma}) = \text{Tr}(\mathbf{A}\boldsymbol{\Sigma})$. In (A.4.10), $\mathbf{A} = [\text{Diag}(\mathbf{p}^0)]^{-1}$ and thus $\mathbf{A}\boldsymbol{\Sigma} = \mathbf{A}[\mathbf{A}^{-1} - \mathbf{p}^0(\mathbf{p}^0)'] = \mathbf{I}_t - \mathbf{1}_t(\mathbf{p}^0)'$. Also $(\mathbf{A}\boldsymbol{\Sigma})(\mathbf{A}\boldsymbol{\Sigma}) = (\mathbf{I}_t - \mathbf{1}_t(\mathbf{p}^0)')(\mathbf{I}_t - \mathbf{1}_t(\mathbf{p}^0)') = \mathbf{I}_t - \mathbf{1}_t(\mathbf{p}^0)' - \mathbf{1}_t(\mathbf{p}^0)' + \mathbf{1}_t(\mathbf{p}^0)'\mathbf{1}_t(\mathbf{p}^0)' = \mathbf{I}_t - \mathbf{1}_t(\mathbf{p}^0)' = \mathbf{A}\boldsymbol{\Sigma}$, and $d = \text{Tr}(\mathbf{I}_t - \mathbf{1}_t(\mathbf{p}^0)') = \text{Tr}(\mathbf{I}_t) - \text{Tr}(\mathbf{1}_t(\mathbf{p}^0)') = t - \sum_{i=1}^t p_i^0 = t - 1$ which gives the desired result.

The Mann–Wald theorem requires convergence in law of the joint distribution of $\mathbf{T}_n$ and, in general, the conclusion need not follow if only the componentwise distributions of $\mathbf{T}_n$ converge in law. For example if T_{1n} are iid $N(0,1)$, and $T_{2n} = -T_{1n}$, then $T_{1n} \xrightarrow{\mathcal{L}} T_1 \sim N(0,1)$ and $T_{2n} \xrightarrow{\mathcal{L}} T_2 \sim N(0,1)$ but $T_{1n} + T_{2n} = 0$ does not converge to a $N(0,2)$ distribution. In general, the distribution of a function $g(\mathbf{T})$ depends on the joint distribution of the components of $\mathbf{T}$ rather than their marginal distributions. However, there is one important case in which convergence of the components is sufficient for joint convergence in law and hence for the Mann–Wald theorem to hold. The following result is sometimes called Slutsky's theorem.

Theorem A.4.4. *If $\{\mathbf{T}_n\}$ and $\{\mathbf{W}_n\}$ are sequences of random vectors such that $\mathbf{T}_n \xrightarrow{\mathcal{L}} \mathbf{T}$ and $\mathbf{W}_n \xrightarrow{p} \mathbf{c}$, where $\mathbf{c}$ is a constant, then*

$$(\mathbf{T}_n, \mathbf{W}_n) \xrightarrow{\mathcal{L}} (\mathbf{T}, \mathbf{W})$$

where $(\mathbf{T}, \mathbf{W})$ has a degenerate joint distribution along the line $\mathbf{w} = \mathbf{c}$.

Example A.4.8. If $\mathbf{T}_n \xrightarrow{\mathcal{L}} \mathbf{T}$, $\mathbf{W}_n \xrightarrow{p} \mathbf{0}$ then $\mathbf{T}_n + \mathbf{W}_n \xrightarrow{\mathcal{L}} \mathbf{T}$.

A final type of continuity argument, known as the "delta method," is used to determine the limiting distribution of a transformation of a known convergent sequence. The method is first stated in the univariate case and then in general.

Theorem A.4.5. *If* $\sqrt{n}(T_n - \mu) \xrightarrow{\mathcal{L}} N(0, \sigma^2)$ *with* $\sigma^2 > 0$ *and* $g: \mathbb{R} \to \mathbb{R}$ *is differentiable at* μ *then*

$$\sqrt{n}(g(T_n) - g(\mu)) \xrightarrow{\mathcal{L}} N(0, [g'(\mu)]^2\sigma^2).$$

Proof. The proof is an application of Theorem A.4.4 and O_p, o_p arithmetic. By Theorem A.4.4, it suffices to show that

$$W_n := \sqrt{n}(g(T_n) - g(\mu)) - \sqrt{n}(T_n - \mu)g'(\mu)$$

satisfies $W_n = o_p(1)$ since

$$X_n := \sqrt{n}(T_n - \mu)g'(\mu) \xrightarrow{\mathcal{L}} N(0, [g'(\mu)]^2\sigma^2)$$

and

$$X_n + W_n = \sqrt{n}(g(T_n) - g(\mu)).$$

Define the following function,

$$h(x) = \begin{cases} \dfrac{g(x) - g(\mu)}{x - \mu} - g'(\mu), & x \neq \mu \\ \\ 0, & x = \mu. \end{cases}$$

Observe that $h(x) \to 0 = h(\mu)$ as $x \to \mu$; i.e., $h(\cdot)$ is continuous at $x = \mu$. Recall that $\sqrt{n}(T_n - \mu) \xrightarrow{\mathcal{L}} N(0, \sigma^2)$ implies $T_n \xrightarrow{P} \mu$ and $\sqrt{n}(T_n - \mu) = O_p(1)$. Thus $h(T_n) \xrightarrow{P} 0 = h(\mu)$; that is, $h(T_n) = o_p(1)$. Then, $W_n = h(T_n)\sqrt{n}(T_n - \mu) = o_p(1)O_p(1) = o_p(1)$ which completes the proof. □

Example A.4.9. Let $\{Y_n\}_{n \geq 1}$ be iid Bernoulli random variables with success probability p, $0 < p < 1$. By the central limit theorem,

$$\sqrt{n}(\hat{p}_n - p) \xrightarrow{\mathcal{L}} N(0, p(1-p))$$

where $\hat{p}_n := (\sum_1^n Y_j)/n$. Let $g(x) = x^m$, $m \geq 1$. Then $g'(x) = mx^{m-1}$ and by Theorem A.4.5,

$$\sqrt{n}(\hat{p}_n^m - p^m) \xrightarrow{\mathcal{L}} N(0, m^2p^{2m-1}(1-p)).$$

The proof of the delta method consists of showing that only the linear terms in the Taylor series expansion of $g(T_n)$ about $g(\mu)$ are relevant in the analysis of $\sqrt{n}(g(T_n) - g(\mu))$. In a similar way the multivariate version of Theorem A.4.5 requires that the function defining the transformation have an analogous two term expansion about the mean with remainder going to zero; i.e., the transformation has total differential at the mean. A well-known sufficient condition for this is continuity of the first partials of the transformation.

Theorem A.4.6. *Let* $\sqrt{n}(\mathbf{T}_n - \boldsymbol{\mu}) \xrightarrow{\mathcal{L}} N_k[\mathbf{0}_k, \boldsymbol{\Sigma}]$, *where* $\mathbf{T}_n$ *and* $\boldsymbol{\mu}$ *are k-dimensional. Let* $\mathbf{g}: \mathbb{R}^k \to \mathbb{R}^s$ *have components* $\mathbf{g}(\mathbf{x}) = (g_1(\mathbf{x}), \ldots, g_s(\mathbf{x}))'$ *and suppose each* $g_i(\cdot)$ *has continuous first partial derivatives at* $\mathbf{x} = \boldsymbol{\mu}$. *Then*

$$\sqrt{n}(\mathbf{g}(\mathbf{T}_n) - \mathbf{g}(\boldsymbol{\mu})) \xrightarrow{\mathcal{L}} N_s[\mathbf{0}_s, \nabla\mathbf{g}(\boldsymbol{\mu})\boldsymbol{\Sigma}(\nabla\mathbf{g}(\boldsymbol{\mu}))']$$

where $\nabla\mathbf{g}(\boldsymbol{\mu})$ *is the* $s \times k$ *matrix defined by*

$$(\nabla\mathbf{g}(\boldsymbol{\mu}))_{ij} := \left(\frac{\partial g_i(\boldsymbol{\mu})}{\partial \mu_j}\right)$$

for $1 \le i \le s$ *and* $1 \le j \le k$.

Problems

A.1. For a given linear subspace $\mathcal{M} \subset \mathbb{R}^n$, prove that there exists a unique orthogonal projection matrix $\mathbf{P}$ which maps $\mathbb{R}^n$ onto $\mathcal{M}$.

A.2. Suppose $Y_1, \ldots, Y_n$ are iid $N(\mu, \sigma^2)$ where $\mu \in \mathbb{R}^1$ and $\sigma^2 > 0$. Consider testing H_0: $\mu = \mu_0$ (given) versus $H_{\neq}$: $\mu \neq \mu_0$. Show that

(i) the score test of H_0 versus $H_{\neq}$ is:

$$\text{Reject } H_0 \iff \frac{n(\overline{Y} - \mu_0)^2}{\hat{\sigma}_0^2} \ge \chi^2_{\alpha,1}$$

where $\hat{\sigma}_0^2 = \sum_{i=1}^n (Y_i - \mu_0)^2/n$ is the MLE of σ^2 under H_0 and

(ii) the Wald test of H_0 is:

$$\text{Reject } H_0 \iff \frac{n(\overline{Y} - \mu_0)^2}{\hat{\sigma}^2} \ge \chi^2_{\alpha,1}$$

where $\hat{\sigma}^2 = \sum_{i=1}^n (Y_i - \overline{Y})^2/n$ is the MLE of σ^2 under $H_{\neq}$.

(The difference between the two tests is that the score test uses an estimate of σ^2 valid only under H_0 while the Wald test uses an estimate of σ^2 valid under either H_0 or $H_{\neq}$.)

A.3. Suppose $\mathbf{Y} = (Y_1, \ldots, Y_n)'$ has joint cdf $F(\mathbf{y} \mid \boldsymbol{\theta})$ where $\boldsymbol{\theta} \in \Omega \subseteq \mathbb{R}^d$; it is desired to test H_0: $\boldsymbol{\theta} \in \omega \subseteq \Omega$ versus H_A: $\boldsymbol{\theta} \in \Omega \setminus \omega$. Suppose further that $\mathbf{T}: \mathbb{R}^d \to \mathbb{R}^d$ is 1–1 so that $\boldsymbol{\psi} = \mathbf{T}(\boldsymbol{\theta})$ gives an alternative parametrization of the distribution of $\mathbf{Y}$. Which of the likelihood ratio, score, and Wald statistics are invariant with respect to the parametrization?

A.4. If $X_n = O_p(1/\sqrt{n})$, show $X_n = o_p(1)$.

A.5. Show that if X_n is the sum of n iid random variables with common mean μ and variance σ^2, then $X_n = n\mu + O_p(\sqrt{n})$.

A.6. Show that O_p, o_p obey the stochastic arithmetic laws:

$$\begin{aligned}
&\text{(i) } X_n^1 = O_p(b_n^1),\ X_n^2 = O_p(b_n^2) \text{ implies } X_n^1 X_n^2 = O_p(b_n^1 b_n^2),\\
&\text{(ii) } X_n^1 = O_p(b_n^1),\ X_n^2 = o_p(b_n^2) \text{ implies } X_n^1 X_n^2 = o_p(b_n^1 b_n^2),\\
&\text{(iii) } X_n^1 = o_p(b_n^1),\ X_n^2 = o_p(b_n^2) \text{ implies } X_n^1 X_n^2 = o_p(b_n^1 b_n^2).
\end{aligned}$$

A.7. If $X_n \xrightarrow{\mathcal{L}} X$ and $Y_n \xrightarrow{P} c$, then show $(X_n, Y_n) \xrightarrow{\mathcal{L}} (X, Y)$ where (X, Y) is degenerate on the line $Y = c$ with cdf

$$F_{X,Y}(x,y) = \begin{cases} 0, & y < c \\ F_X(x), & y \geq c, \end{cases}$$

where $F_X(x)$ is the cdf of X.

A.8. Suppose $Y \sim B(n,p)$ where $0 < p < 1$. Show that the Arcsine transformation of $\hat{p} = Y/n$, (Arcsine$\{[\hat{p}]^{1/2}\}$), has variance independent of p.

References

Agresti, A. (1984). *Analysis of Ordinal Categorical Data.* New York: John Wiley & Sons.

Aitchison, J. and Aitken, C.G.G. (1976). Multivariate Binary Discrimination by the Kernel Method. *Biometrika* **63**, 413–420.

Aitkin, M. (1978). The Analysis of Unbalanced Cross-Classifications. *J. Roy. Statist. Soc. (A)* **141**, 195–223.

Aitkin, M. (1979). A Simultaneous Test Procedure for Contingency Table Models. *J. Roy. Statist. Soc. (C)* **28**, 233–242.

Aitkin, M. (1980). A Note on the Selection of Log-Linear Models. *Biometrics* **36**, 173–178.

Alam, K. (1971). Selection from Poisson Processes. *Ann. Inst. Statist. Math.* **23**, 411–418.

Alam, K. (1979). Estimation of Multinomial Probabilities. *Ann. Statist.* **7**, 282–283.

Alam, K. and Thompson, J.R. (1972). On Selecting the Least Probable Multinomial Event. *Ann. Math. Statist.* **43**, 1981–1990.

Alam, K. and Thompson, J.R. (1973). A Problem of Ranking and Estimation with Poisson Processes. *Technometrics* **15**, 801–808.

Albert, A. and Anderson, J.A. (1984). On the Existence of Maximum Likelihood Estimates in Logistic Regression Models. *Biometrika* **71**, 1–10.

Albert, J.H. (1981a). Pseudo-Bayes Estimation of Multinomial Proportions. *Commun. Statist. - Theory Meth.* **10**, 1587–1611.

Albert, J.H. (1981b). Simultaneous Estimation of Poisson Means. *J. Mult. Anal.* **11**, 400–417.

Albert J.H. (1987). Empirical Bayes Estimation in Contingency Tables. *Commun. Statist. - Theory Meth.* **16**, 2459–2485.

Albert, J.H. and Gupta, A.K. (1982). Mixtures of Dirichlet Distributions and Estimation in Contingency Tables. *Ann. Statist.* **10**, 1261–1268.

Albert, J.H. and Gupta, A.K. (1983a). Bayesian Estimation Methods for 2×2 Contingency Tables Using Mixtures of Dirichlet Distributions. *J. Amer. Statist. Assoc.* **78**, 708–717.

Albert, J.H. and Gupta, A.K. (1983b). Estimation in Contingency Tables using Prior Information. *J. Roy. Statist. Soc. (B)* **45**, 60–69.

Altham, P.M.E. (1971). The Analysis of Matched Proportions. *Biometrika* **58**, 561–576.

Amemiya, T. (1980). The n^{-2}-order Mean Squared Errors of the Maximum Likelihood and the Minimum Logit Chi-square Estimators. *Ann. Statist.* **8**, 488–505. (Correction: *Ann. Statist.* **12**, 783.)

Amemiya, T. (1985). *Advanced Econometrics.* Cambridge, Massachusetts: Harvard University Press.

Anbar, D. (1983). On Estimating the Difference Between Two Probabilities with Special Reference to Clinical Trials. *Biometrics* **39**, 257–262.

Anbar, D. (1984). Confidence Bounds for the Difference Between Two Probabilities. *Biometrics* **40**, 1176.

Andersen, E.B. (1970). Asymptotic Properties of Conditional Maximum Likelihood Estimators. *J. Roy. Statist. Soc. (B)* **32**, 283–301.

Anderson, J.A. and Richardson, S.C. (1979). Logistic Discrimination and Bias Correction in Maximum Likelihood Estimation. *Technometrics* **21**, 71–78.

Andrews, D.W.K. (1988a). Chi-Square Diagnostic Tests for Econometric Models: Introduction and Applications. *J. of Econometrics,* **37**, 135–156.

Andrews, D.W.K. (1988b). Chi-Square Diagnostic Tests for Econometric Models: Theory. *Econometrika* **56**, 1419–1453.

Angers, C. (1974). A Graphical Method to Evaluate Sample Sizes for the Multinomial Distribution. *Technometrics* **16**, 469–471.

Angus, J.E. and Schaefer, R.E. (1984). Improved Confidence Statements for the Binomial Parameter. *The American Statistician* **38**, 189–191.

Anscombe, F.J. (1956). On Estimating Binomial Response Relations. *Biometrika* **43**, 461–464.

Aranda-Ordaz, F.J. (1981). On Two Families of Transformations to Additivity for Binary Response Data. *Biometrika* **68**, 357–363.

Armesan, P. (1955). Tables for Significance Tests of 2 × 2 Contingency Tables. *Biometrika* **42**, 494–511.

Armitage, P. (1955). Tests for Linear Trends in Proportions and Frequencies. *Biometrics* **11**, 375–386.

Armitage, P. (1966). The Chi-Square Test for Heterogeneity of Proportions After Adjustment for Stratification. *J. Roy. Statist. Soc. (B)* **28**, 150–163.

Armitage, P. (1975). *Sequential Medical Trials,* 2nd Ed. New York: Halsted Press.

Asmussen, S. and Edwards, D. (1983). Collapsibility and Response Variables in Contingency Tables. *Biometrika* **70**, 657–678.

Atkinson, E.N. and Brown, B.W. (1985). Confidence Limits for Probability of Response in Multistage Phase II Clinical Trials. *Biometrics* **41**, 714–744.

Baglivo, J., Olivier, D., and Pagano, M. (1988). Methods for the Analysis of Contingency Tables with Large and Small Cell Counts. *J. Amer. Statist. Assoc.* **83**, 1006–1013.

Baptista, J. and Pike, M.C. (1977). Exact Two-Sided Confidence Limits for the Odds Ratio in a 2×2 Table. *J. Roy. Statist. Soc. (C)* **26**, 214–220.

Bartlett, M.A. (1935). Contingency Table Interactions. *J. Roy. Statist. Soc. Suppl.* **2**, 248–252.

Beal, S.L. (1987). Asymptotic Confidence Intervals for the Difference Between Two Binomial Parameters for Use with Small Samples. *Biometrics* **43**, 941–950.

Bechhofer, R.E. (1954). A Single-Sample Multiple Decision Procedure for Ranking Means of Normal Populations with Known Variances. *Ann. Math. Statist.* **25**, 16–39.

Bechhofer, R.E. (1985). An Optimal Sequential Procedure for Selecting the Best Bernoulli Process—A Review. *Nav. Res. Log. Quart.* **32**, 665–674.

Bechhofer, R.E., Elmagraby, S., and Morse, N. (1959). A Single-Sample Multiple-Decision Procedure for Selecting the Multinomial Event Which Has the Highest Probability. *Ann. Math. Statist.* **30**, 102–119.

Bechhofer, R.E. and Goldsman, D.M. (1985). Truncation of the Bechhofer–Kiefer–Sobel Sequential Procedure for Selecting the Multinomial Event Which Has the Largest Probability. *Comm. Statist. - Simul. Comput.* **14**, 283–315.

Bechhofer, R.E. and Goldsman, D.M. (1986). Truncation of the Bechhofer–Kiefer–Sobel Sequential Procedure for Selecting the Multinomial Event Which Has the Largest Probability (II): Extended Tables and an Improved Procedure. *Comm. Statist. - Simul. Comput.* **15**, 829–851.

Bechhofer, R.E., Kiefer, J. and Sobel, M. (1967). Sequential Identification and Ranking Procedures. Statistical Research Monographs, 3, IMS and University of Chicago Press, Chicago.

Bechhofer, R.E. and Kulkarni, R. (1984). Closed Sequential Procedures for Selecting the Multinomial Events Which Have the Largest Probabilities. *Comm. Statist. - Theor. Meth.* **13**, 829–851.

Bedrick, E.J. (1987). A Family of Confidence Intervals for the Ratio of Two Binomial Proportions. *Biometrics* **43**, 993–998.

Belsley, D.A., Kuh, E., and Welsch, R.E. (1980). *Regression Diagnostics.* New York: J. Wiley and Sons, Inc.

Benedetti, J.K. and Brown, M.B. (1978). Strategies for the Selection of Log Linear Models. *Biometrics* **34**, 680–686.

Bennett, B.M. and Horst, C. (1966). *Supplement to Tables for Testing Significance in a* 2×2 *Contingency Table.* London: Cambridge University Press.

Bennett, B.M. and Hsu, P. (1960). On the Power Function of the Exact Test for the 2×2 Contingency Table. *Biometrika* **47**, 393–398. (Correction: *Biometrika* **48**, 475.)

Bennett, B.M. and Kaneshire, C. (1974). On the Small-Sample Properties of the Mantel–Haenszel Test for Relative Risk. *Biometrika* **61**, 233–236.

Berengut, D. and Petkau, A.J. (1979). On Testing the Equality of Two Proportions. Technical Report 267, Department of Statistics, Stanford University, Stanford, California.

Berger, A., Wittes, J., and Gold, R. (1979). On the Power of the Cochran–Mantel–Haenszel Test and Other Approximately Optimal Tests for Partial Association. Technical Report B-03, Division of Biostatistics, Columbia University, New York, New York.

Berger, J.O. (1985). *Statistical Decision Theory and Bayesian Analysis* (2nd edition). New York: Springer-Verlag.

Berger, R.O. (1980). Minimax Subset Selection for the Multinomial Distribution. *J. Statist. Pl. Inf.* **4**, 391–402.

Berkson, J. (1944). Application of the Logistic Function to Bioassay. *J. Amer. Statist. Assoc.* **39**, 357–365.

Berkson, J. (1955). Maximum Likelihood and Minimum Chi-Square Estimates of the Logistic Function. *J. Amer. Statist. Assoc.* **50**, 130–162.

Berry, D.A. and Christensen, R. (1979). Empirical Bayes Estimation of a Binomial Parameter via Mixtures of Dirichlet Processes. *Ann. Statist.* **7**, 558–568.

Bhapkar, V.P. (1966). A Note on the Equivalence of Two Test Criteria for Hypotheses in Categorical Data. *J. Amer. Statist. Assoc.* **61**, 228–235.

Birch, M.W. (1963). Maximum Likelihood in Three-Way Contingency Tables. *J. Roy. Statist. Soc. (B)* **25**, 220–233.

Birch, M.W. (1964). The Detection of Partial Association, I: The 2×2 Case. *J. Roy. Statist. Soc. (B)* **26**, 313–324.

Birch, M.W. (1965). The Detection of Partial Association, II: The General Case. *J. Roy. Statist. Soc. (B)* **27**, 111–124.

Bishop, Y.M.M., Fienberg, S.E., and Holland, P.W. (1975). *Discrete Multivariate Analysis.* Cambridge, Massachusetts: The MIT Press.

Blyth, C. (1986). Approximate Binomial Confidence Limits. *J. Amer. Statist. Assoc.* **81**, 843–855.

Blyth, C. and Hutchinson, D. (1960). Tables of Neyman–Shortest Unbiased Confidence Intervals for the Binomial Parameter. *Biometrika* **47**, 381–391.

Blyth, C. and Still, H. (1983). Binomial Confidence Intervals. *J. Amer. Statist. Assoc.* **78**, 108–116.

Boschloo, R.D. (1970). Raised Conditional Level of Significance for the 2×2 Table When Testing for the Equality of Two Probabilities. *Statistica Neerlandica* **21**, 1–35.

Bowman, A.W., Hall, P., and Titterington, D.M. (1984). Cross-Validation in Nonparametric Estimation of Probabilities and Probability Densities. *Biometrika* **71**, 341–351.

Brand, R.J., Pinnock, D.E., and Jackson, K.L. (1973). Large Sample Confidence Bands for the Logistic Response Curve and Its Inverse. *The American Statistician* **27**, 157–160.

Breiman, L. and Friedman, J.H. (1985). Estimating Optimal Transformations for Multiple Regression and Correlation (and discussion). *J. Amer. Statist. Assoc.* **80**, 580–619.

Breslow, N.E. (1976). Regression Analysis of the Log Odds Ratio: A Method for Retrospective Studies. *Biometrics* **32**, 409–416.

Breslow, N.E. (1981). Odds Ratio Estimators When the Data are Sparse. *Biometrika* **68**, 73–84.

Breslow, N.E. (1984). Extra-Poisson Variation in Log-Linear Models. *J. Roy. Statist. Soc. (C)* **33**, 38–44.

Breslow, N.E. and Day, N.E. (1980). *Statistical Methods in Cancer Research, Vol. 1: The Analysis of Case-Control Studies.* IARC Scientific Publications No. 32, International Agency of Research on Cancer, Lyon, France.

Brier, S.S., Zacks, S., and Marlow, W.H. (1986). An Application of Empirical Bayes Techniques to the Simultaneous Estimation of Many Probabilities. *Nav. Res. Log. Quart.* **33**, 77–90.

Brown, B.W. (1980). Prediction Analyses for Binary Data. In *Biostatistics Casebook* (R. Miller, Jr., B. Efron, B.W. Brown, Jr., L. Moses, eds.). New York: J. Wiley and Sons, Inc., 3–18.

Brown, C.C. and Muenz, L.R. (1976). Reduced Mean Square Error Estimation in Contingency Tables. *J. Amer. Statist. Assoc.* **76**, 176–182.

Brown, L.D. and Farrell, R.H. (1985a). All Admissible Linear Estimators of a Multivariate Poisson Mean. *Ann. Statist.* **13**, 282–294.

Brown, L.D. and Farrell, R.H. (1985b). Complete Class Theorems for Estimation of Multivariate Poisson Means and Related Problems. *Ann. Statist.* **13**, 706–726.

Brown, M.B. (1974). Identification of the Sources of Significance in Two-Way Contingency Tables. *J. Roy. Statist. Soc. (C)* **23**, 405–413.

Brown, M.B. (1976). Screening Effects in Multidimensional Contingency Tables. *J. Roy. Statist. Soc. (C)* **25**, 37–46.

Brown, P.J. and Rundell, P.W.K. (1985). Kernel Estimates for Categorical Data. *Technometrics* **27**, 293–299.

Buhrman, J.M. (1977). Tests and Confidence Intervals for the Difference and Ratio of Two Probabilities. *Biometrika* **64**, 160–162.

Bulick, S., Montgomery, K.L., Fetterman, J., and Kent, A. (1976). Use of Library Materials in Terms of Age. *J. Amer. Soc. Info. Sci.* **27**, 175–178.

Bunke, O. (1985). An Adaptive Smoothing Estimator for Probabilities in Contingency Tables. *Statistics* **16**, 55–62.

Burman, P. (1987). Smoothing Sparse Contingency Tables. *Sankyā* (*A*) **49**, 24–36.

Burrell, Q.L. and Cane, V.R. (1982). The Analysis of Library Data. *J. Roy. Statist. Soc. (A)* **145**, 439–471.

Cacoullos, T. and Sobel, M. (1966). An Inverse Sampling Procedure for Selecting the Most Probable Event in a Multinomial Distribution. *Proceedings of the International Symposium on Multivariate Analysis.* New York: Academic Press, Inc., 423–455.

Cameron, A.C. and Trivedi, P.K. (1986). Econometric Models Based on Count Data: Comparisons and Applications of Some Estimators and Tests. *J. Appl. Econ.* **1**, 29–53.

Carp, R.A. and Rowland, C.K. (1983). *Policymaking and Politics in the Federal District Courts.* Knoxville, TN: The University of Tennessee Press.

Carroll, R.J. and Lombard, F. (1985). A Note on N Estimators for the Binomial Distribution. *J. Amer. Statist. Assoc.* **80**, 423–426.

Carroll, R.J., Spiegelman, C.H., Lan, K.K., Bailey, K.T., and Abbott, R.D. (1984). On Errors-in-Variables for Binary Regression Models. *Biometrika* **71**, 19–25.

Carter, W.H., Chinchili, V.M., Wilson, J.D., Campbell, E.D., Kessler, F.K., and Carchman, R.A. (1986). An Asymptotic Confidence Region for the ED_{100p} from the Logistic Response Surface for a Combination of Agents. *The American Statistician* **40**, 124–128.

Casagrande, J.T., Pike, M.C., and Smith, P.G. (1978). The Power Function of the "Exact" Test for Comparing Two Binomial Distributions. *J. Roy. Statist. Soc. (C)* **27**, 176–180.

Casella, G. (1986). Refining Binomial Confidence Intervals. *Canad. J. Statist.* **14**, 113–129.

Casella, G. (1987). Refining Poisson Confidence Intervals. Technical Report BU-903-M, Biometrics Unit, Cornell University.

Casella, G. and Strawderman, W. (1980). Estimating a Bounded Normal Mean. *Ann. Statist.* **9**, 870–878.

Ceder, L., Thorngren, K.G., and Wallden, B. (1980). Prognostic Indicators and Early Home Rehabilitation in Elderly Patients with Hip Fractures. *Clin. Orth. Rel. Res.* **152**, 173–184.

Chacko, V.J. (1966). Modified Chi-Square Tests for Ordered Alternatives. *Sankhyā (B)*, **28**, 185–190.

Chapman, D. and Nam, J. (1968). Asymptotic Power of Chi-Square Tests for Linear Trends in Proportions. *Biometrics* **24**, 317–327.

Chen, P. (1987). Comparison of Multinomial Cells with a Control. *Statistics and Decisions* **5**, 33–45.

Chernoff, H. and Lehmann, E. (1954). The Use of Maximum Likelihood Estimation in χ^2 Tests for Goodness-of-Fit. *Ann. Math. Statist.* **25**, 579–586.

Chvatal, V. (1980). *Linear Progamming.* New York: W.H. Freeman Company.

Clevenson, M.L. and Zidek, J.V. (1975). Simultaneous Estimation of the Means of Independent Poisson Laws. *J. Amer. Statist. Assoc.* **70**, 698–705.

Clopper, C.J. and Pearson, E. (1934). The Use of Confidence or Fiducial Limits Illustrated in the Case of Binomial. *Biometrika* **26**, 404–413.

Cochran, W.G. (1952). The Chi-Squared Test of Goodness of Fit. *Ann. Math. Statist.* **23**, 315–345.

Cochran, W.G. (1954). Some Methods for Strengthening the Common χ^2 Tests. *Biometrics* **10**, 417–451.

Collins, B.J. and Margolin, B.H. (1985). Testing Goodness of Fit for the Poisson Assumption When Observations Are Not Identically Distributed. *J. Amer. Statist. Assoc.* **80**, 411–418.

Cook, R.D. (1977). Detection of Influential Observations in Linear Regression. *Technometrics* **19**, 15–18.

Cook, R.D. (1979). Influential Observations in Linear Regression. *J. Amer. Statist. Assoc.* **74**, 169–174.

Cook, R.D. and Weisberg, S. (1982). *Residuals and Influence in Regression.* New York: Chapman Hall.

Copas, J.B. (1972). Empirical Bayes Methods and the Repeated Use of a Standard. *Biometrika* **59**, 349–360.

Copas, J.B. (1988). Binary Regression Models for Contaminated Data (with discussion). *J. Roy. Statist. Soc. (B)* **50**, 225–265.

Cornfield, J. (1956). A Statistical Problem Arising from Retrospective Studies. *Proceedings of the 3rd Berkeley Symposium, Berkeley, California* **4**, 135–148. Berkeley: University of California Press.

Cox, D.R. (1970). *The Analysis of Binary Data.* London: Metheun.

Cox, D.R. and Lauh, E. (1967). A Note on the Graphical Analysis of Multidimensional Contingency Tables. *Technometrics* **9**, 481–488.

Cox, D.R. and Snell, E.J. (1968). A General Definition of Residuals. *J. Roy. Statist. Soc. (B)* **30**, 248–265.

Cox, D.R. and Snell, E.J. (1989). *The Analysis of Binary Data* (2nd edition). London: Metheun.

Cressie, N. and Read, T.R.C. (1984). Multinomial Goodness-of-Fit Tests. *J. Roy. Statist. Soc. (B)* **46**, 440–464.

Crook, J.F. and Good, I.J. (1980). On the Application of Symmetric Dirichlet Distributions and Their Mixtures to Contingency Tables, Part II. *Ann. Statist.* **8**, 1198–1218.

Crow, E.L. (1956). Confidence Limits for a Proportion. *Biometrika* **43**, 423–435.

Crow, E.L. and Gardner, R.S. (1959). Confidence Intervals for the Expectation of a Poisson Variable. *Biometrika* **46**, 441–453.

Crowder, M.J. (1978). Beta-binomial ANOVA for Proportions. *J. Roy. Statist. Soc. (C)* **27**, 34–37.

Csiszár, I. (1975). *I*-Divergence Geometry of Probability Distributions and Minimization Problems. *Ann. Prob.* **3**, 146–158.

Czado, C. (1988). *Link Misspecification and Data Selected Transformations in Binary Regression Models.* Ph.D. Thesis, School of Op. Res. and Ind. Eng., Cornell University.

Dalal, S.R., Fowlkes, E.B., and Hoadley, B.A. (1988). The Pre-Challenger Prediction of Space Shuttle Failure. Bellcore Technical Memorandum, Bell Communications Research, Morristown, N.J.

Dalal, S.R., Fowlkes, E.B., and Hoadley, B. (1989). Risk Analysis of Space Shuttle: Pre-Challenger Prediction of Failure. To appear in *J. Am. Statist. Assoc.*

Dalal, S.R. and Klein, R.W. (1988). A Flexible Class of Discrete Choice Models. *Marketing Science* **7**, 232–251.

Darroch, J.N., Lauritzen, S.L., and Speed, T.P. (1980). Markov Fields and Log-Linear Interaction Models for Contingency Tables. *Ann. Statist.* **8**, 522–539.

Darroch, J.N. and Ratcliff, D. (1972). Generalized Iterative Scaling for Loglinear Models. *Ann. Math. Statist.* **43**, 1470–1480.

Davey, P. (1979). Log-Linear Models. In *Interactive Statistics* (D. McNeil, ed.). North Holland, 147–152.

Davies, O.L. (1961). *Statistical Methods in Research and Production.* London: Oliver and Boyd.

Davis, L.J. (1984). Comments on a Paper by T. Amemiya on Estimation in a Dichotomous Logit Regression Model. *Ann. Statist.* **12**, 778–782. (Correction: *Ann. Statist.* **13**, 1629.)

Davis, L.J. (1985). Consistency and Asymptotic Normality of the Minimum Logit Chi-Squared Estimator When the Number of Design Points is Large. *Ann. Statist.* **13**, 947–957.

Davis, L.J. (1986a). Whittemore's Notion of Collapsibility in Multidimensional Contingency Tables. *Comm. Statist. - Theor. Meth.* **15**, 2241–2554.

Davis, L.J. (1986b). Exact Tests for 2 × 2 Contingency Tables. *The American Statistician* **40**, 139–141.

Davis, L.J. (1988a). Intersection Union Tests for Testing Strict Collapsibility in Three Dimensional Contingency Tables. Presented at Spring, 1988 Biometrics Meeting (Boston).

Davis, L.J. (1988b). Collapsibility of Likelihood Ratio Tests in Multidimensional Contingency Tables. Preprint.

Day, N.E. and Byar, D.P. (1979). Testing Hypotheses in Case-Control Studies—Equivalence of Mantel–Haenszel Statistics and Logit Score Tests. *Biometrics* **35**, 623–630.

Dean, C. and Lawless, J.F. (1989). Tests for Detecting Everdispersion in Poisson Regression Models. *J. Am. Statist. Assoc.* **84**, 467–472.

Deming, W.E. and Stephan, F. (1940). On a Least Squares Adjustment of a Sampled Frequency Table When the Expected Marginal Totals are Known. *Ann. Math. Statist.* **11**, 427–444.

Dempster, A.P., Laird, N.M., and Rubin, D.B. (1977). Maximum Likelihood from Incomplete Data via the EM Algorithm (with discussion). *J. Roy. Statist. Soc. (B),* 1–38.

Dorn, H.F. (1954). The Relationship of Cancer of the Lung and the Use of Tobacco. *The American Statistician* **8**, 7–13.

Drinkwater, R.W. and Hastings, N.A.J. (1967). An Economic Replacement Model. *Oper. Res. Quart.* **18**, 121–138.

Drost, F.C., Kallenberg, W.C.M., Moore, D.S., and Oosterholf, J. (1987). Asymptotic Error Bounds for Power Approximations to Multinomial Tests of Fit. Technical Report 640, Faculty of Applied Mathematics, University of Twente, The Netherlands.

Drost, F.C., Kallenberg, W.C.M., Moore, D.S., and Oosterholf, J. (1989). Power Approximations to Multinomial Tests of Fit. *J. Am. Statist. Assoc.* **84**, 130–141.

Ducharme, G.R. and Lepage, Y. (1986). Testing Collapsibility in Contingency Tables. *J. Roy. Statist. Soc. (B)* **48**, 197–205.

Duffy, D.E. (1986). *Alternative Methods of Estimation in Logistic Regression.* Ph.D. Thesis, School of Operations Research and Industrial Engineering, Cornell University, Ithaca, New York.

Duffy, D.E. (1987). Consistency of an Adaptively Restricted Maximum Likelihood Estimator in Logistic Regression Models. Bellcore Technical Memorandum. Bell Communications Research, Morristown, N.J.

Duffy, D.E. (1988). Adaptive Estimation for Logistic Regression Models. Contributed talk presented at the XIVth International Biometric Conference, July 1988, Namur, Belgium.

Duffy, D.E. (1989). On Continuity-Corrected Residuals in Logistic Regression. Bellcore Technical Memorandum. Bell Communications Research, Morristown, N.J.

Duffy, D.E. and Santner, T.J. (1987a). Estimating Logistic Regression Probabilities. In *Proceedings of the Fourth Purdue Symposium on Statistics and Decision Theory* 1. New York: Springer-Verlag, 177–194.

Duffy, D.E. and Santner, T.J. (1987b). Confidence Intervals for a Binomial Parameter Based on Multistage Tests. *Biometrics* **43**, 81–93.

Duffy, D.E. and Santner, T.J. (1989). On the Small Sample Properties of Restricted Maximum Likelihood Estimators for Logistic Regression Models. *Comm. Statist. - Theor. Meth.* **18**, 959–989.

Eberhardt, K.R. and Fligner, M.A. (1977). A Comparison of Two Tests for Equality of Two Proportions. *The American Statistician* **31**, 151–155.

Edwards, A.W.F. and Fraccaro, M. (1960). Distribution and Sequences of Sexes in a Selected Sample of Swedish Families. *Ann. Human Genet.* **24**, 245–252.

Edwards, D. and Havranek, T. (1985). A Fast Procedure for Model Search in Multidimensional Contingency Tables. *Biometrika* **72**, 339–351.

Edwards, D. and Kreiner, S. (1983). The Analysis of Contingency Tables by Graphical Models. *Biometrika* **70**, 553–565.

EGRET Statistical Software, Statistics and Epidemiology Research Corporation, 909 Northeast 43rd Street, Suite 310, Seattle, WA 98105.

El-Sayyad, G.M. (1973). Bayesian and Classical Analysis of Poisson Regression. *J. Roy. Statist. Soc. (B)* **35**, 445–451.

Fahrmeir, L. and Kaufmann, H. (1985). Consistency and Asymptotic Normality of the Maximum Likelihood Estimator in Generalized Linear Models. *Ann. Statist.* **13**, 342–368. (Correction: *Ann. Statist.* **14**, 1643.)

Fahrmeir, L. and Kaufmann, H. (1986). Asymptotic Inference in Discrete Response Models. To appear in *Statistische Hefte* **27**.

Farewell, V. (1982). A Note on Regression Analysis of Ordinal Data with Variability of Classification. *Biometrika* **69**, 533–538.

Feder, P.I. (1968). On the Distribution of the Log Likelihood Ratio Test Statistic When the True Parameter is "Near" the Boundaries of the Hypothesis Regions. *Ann. Statist.* **39**, 2044–2055.

Feigl, P. (1978). A Graphical Aid for Determining Sample Size When Comparing Two Independent Proportions. *Biometrics* **34**, 111–122.

Fenech, A. and Westfall, P. (1988). The Power Function of Conditional Log-Linear Model Tests. *J. Amer. Statist. Assoc.* **83**, 198–203.

Fienberg, S.E. (1969). Preliminary Graphical Analysis and Quasi-Independence for Two-Way Contingency Tables. *J. Roy. Statist. Soc. (C)* **18**, 153–168.

Fienberg, S.E. (1980). *The Analysis of Cross-Classified Categorical Data* (2nd edition). Cambridge, Massachusetts: The MIT Press.

Fienberg, S.E. and Holland, P.W. (1970). Methods for Eliminating Zero Counts in Contingency Tables. In *Random Counts and Structures* (G.P. Patil, ed.). University Park: Pennsylvania State University Press, 233–260.

Fienberg, S.E. and Holland, P.W. (1973). Simultaneous Estimation of Multinomial Cell Probabilities. *J. Amer. Statist. Assoc.* **68**, 683–691.

Finney, D.J. (1947). The Estimation from Individual Records of the Relationship Between Dose and Quantal Response. *Biometrika* **34**, 320–334.

Finney, D.J. (1971). *Probit Analysis* (3rd edition). Cambridge: Cambridge University Press.

Finney, D.J., Latscha, R., Bennett, B.M., Hsu, P., and Pearson, E.S. (1963). *Tables for Testing Significance in a* 2×2 *Contingency Table.* Cambridge: Cambridge University Press.

Fisher, R.A. (1935). The Logic of Inductive Inference. *J. Roy. Statist. Soc.* **48**, 39–82.

Fisher, R.A. (1940). The Precision of Discriminant Functions. *Ann. Eugenics* **10**, 422–429.

Fisher, R.A. (1962). Confidence Limits for a Cross-Product Ratio. *Austral. J. Statist.* **4**, 41.

Fix, E.J., Hodges, J.L., and Lehmann, E.L. (1959). The Restricted Chi-Square Test. In *Probability and Statistics: The Harald Cramer Volume.* (U. Grenander, ed.). Stockholm: Almqvist and Wiksell, 92–107.

Flanders, W.D. (1985). A New Variance Estimator for the Mantel–Haenszel Odds Ratio. *Biometrics* **41**, 637–642.

Fleiss, J. (1979). Confidence Intervals for the Odds Ratio in Case-Control Studies: The State of the Art. *J. Chronic Dis.* **32**, 69–77.

Fleming, T.R. (1982). One Sample Multiple Testing Procedures for Phase II Clinical Trials. *Biometrics* **38**, 143–151.

Fowlkes, E.B. (1987). Some Diagnostics for Binary Logistic Regression Via Smoothing. *Biometrika* **74**, 503–515.

Fowlkes, E.B., Freeny, A.E., and Landwehr, J.M. (1988). Evaluating Logtistic Models for Large Contingency Tables. *J. Amer. Statist. Assoc.* **83**, 611–622.

Freeman, M.R. and Tukey, J.W. (1950). Transformation Related to the Angular and Square Root Transformations. *Ann. Math. Statist.* **21**, 607–611.

Frome, E.L. (1983). The Analysis of Rates Using Poisson Regression Models. *Biometrics* **39**, 665–674.

Frome, E.L., Kutner, M.H., and Beauchamp, J.J. (1973). Regression Analysis of Poisson-Distributed Data. *J. Amer. Statist. Assoc.* **68**, 935–940.

Fujino, Y. (1980). Approximate Binomial Confidence Limits. *Biometrika* **67**, 677–681.

Fujino, Y. and Okuno, T. (1984). The Minimax Average Confidence Limits for a Binomial Probability – One-Sided Case. *Rep. Statist. Appl. Res.* JUSE **31**, 1–7.

Gabriel, K.R. (1969). Simultaneous Test Procedures—Some Theory of Multiple Comparisons. *Ann. Math. Statist.* **40**, 224–250.

Gafarian, A.V. (1964). Confidence Bands in Straight Line Regression. *J. Amer. Statist. Assoc.* **59**, 182–213.

Gail, M. (1973). The Determination of Sample Size for Trials Involving Several Independent 2×2 Tables. *J. Chronic Dis.* **26**, 669–673.

Gail, M. (1974). Power Computations for Designing Comparative Poisson Trials. *Biometrics* **30**, 231–237.

Gail, M. and Gart, J.J. (1973). The Determination of Sample Sizes for Use With the Exact Conditional Test in 2×2 Comparative Trials. *Biometrics* **29**, 441–448.

Garside, G.R. and Mack, C. (1976). Actual Type I Error Probabilities for Various Tests in the Homogeneity Case of the 2×2 Contingency Table. *The American Statistician* **30**, 18–21.

Gart, J.J. (1964). The Analysis of Poisson Regression with an Application in Virology. *Biometrika* **51**, 517–521.

Gart, J.J. (1971). The Comparisons of Proportions: A Review of Significance Tests, Confidence Intervals, and Adjustments for Stratification. *Rev. Interntl. Statist. Inst.* **29**, 148–169.

Gart, J.J. and Nam, J.-N. (1988). Approximate Interval Estimation of the Ratio of Binomial Parameters: A Review and Corrections for Skewness. *Biometrics* **44**, 323–338.

Gart, J.J., Pettigrew, H., and Thomas, D. (1985). The Effect of Bias, Variance Estimation, Skewness and Kurtosis of the Empirical Logit on Weighted Least Squares Analyses. *Biometrika* **72**, 179–190.

Gart, J.J., Pettigrew, H., and Thomas, D. (1986). Further Results on the Effect of Bias, Variance Estimation, and Non-normality of the Empirical Logit on Weighted Least Squares Analyses. *Comm. Statist. - Theor. Meth.* **15**, 755–782.

Gart, J.J. and Tarone, R.E. (1983). The Relation Between Score Tests and Approximately UMPU Tests in Exponential Models Common in Biometry. *Biometrics* **39**, 781–786.

Gart, J.J. and Thomas, D.G. (1972). Numerical Results on Approximate Confidence Limits for the Odds Ratio. *J. Roy. Statist. Soc. Ser. (B)* **34**, 441–447.

Gart, J.J. and Thomas, D.G. (1982). The Performance of Three Approximate Confidence Limit Methods for the Odds Ratio. *Amer. J. Epidem.* **115**, 453–470.

Gart, J.J. and Zweifel, J.R. (1967). On the Bias of the Logit and its Variance with Application to Quantal Bioassay. *Biometrika* **54**, 181–187.

Garwood, F. (1936). Fiducial Limits for the Poisson Distribution. *Biometrika* **46**, 441–453.

Gavor, D.P. and O'Muircheartaigh, I.G. (1987). Robust Empirical Bayes Analyses of Event Rates. *Technometrics* **29**, 1–15.

Gbur, E.F.E. (1979). Analysis of Variance with Poisson Responses. *Comm. Statist. - Theor. Meth.* **A8**, 433–445.

Geisser, S. (1984). On Prior Distributions for Binary Trials. *The American Statistician* **38**, 244–251.

Ghosh, B.K. (1979). A Comparison of Some Approximate Confidence Intervals for the Binomial Parameter. *J. Amer. Statist. Assoc.* **74**, 894–900.

Ghosh, M. (1983). Estimates of Multiple Poisson Means: Bayes and Empirical Bayes. *Statist. and Dec.* **1**, 183–195.

Ghosh, M., Hwang, J.T., and Tsui, K.W. (1983). Construction of Improved Estimators in Multiparameter Estimation for Discrete Exponential Families. *Ann. Statist.* **11**, 351–367.

Ghosh, M. and Parsian, A. (1981). Bayes Minimax Estimation of Multiple Poisson Parameters. *J. Mult. Anal.* **11**, 280–288.

Ghosh, M. and Yang, M.C. (1988). Simultaneous Estimation of Poisson Means Under Entropy Loss. *Ann. Statist.* **16**, 278–291.

Gibbons, J.D., Olkin, I., and Sobel, M. (1977). *Selecting and Ordering Populations.* New York: John Wiley and Sons, Inc.

Gilula, Z. (1986). Grouping and Association in Contingency Tables: An Exploratory Canonical Correlation Approach. *J. Amer. Statist. Assoc.* **81**, 773–779.

Gold, R.Z. (1963). Tests Auxiliary to χ^2 Tests in a Markov Chain. *Ann. Statist.* **34**, 56–74.

Good, I.J. (1956). On the Estimation of Small Frequencies in Contingency Tables. *J. Roy. Statist. Soc. (B)* **18**, 113–124.

Good, I.J. (1965). *The Estimation of Probabilities: An Essay on Modern Bayesian Methods.* Cambridge, Massachusetts: The MIT Press.

Good, I.J. (1967). A Bayesian Significance Test for Multinomial Distributions. *J. Roy. Statist. Soc. (B)* **29**, 399–431.

Good, I.J. (1976). On the Application of Symmetric Distributions and Their Mixtures to Contingency Tables. *Ann. Statist.* **4**, 1159–1189.

Good, I.J. (1983). The Robustness of a Hierarchical Model for Multinomials and Contingency Tables. In *Statistical Inference, Data Analysis, and Robustness* (G. Box, T. Leonard, and C.-F. Wu, eds.). New York: Academic Press.

Good, I.J., Gover, T.N., and Mitchell, G.J. (1970). Exact Distributions for Chi-Square and for the Likelihood-Ratio Statistics for the Equiprobable Multinomial Distribution. *J. Amer. Statist. Assoc.* **65**, 267–283.

Goodman, L.A. (1964). Simultaneous Confidence Limits for Cross-Product Ratios in Contingency Tables. *J. Roy. Statist. Soc. (B)* **26**, 86–102.

Goodman, L.A. (1965). On Simultaneous Confidence Intervals for Multinomial Proportions. *Technometrics* **7**, 247–254.

Goodman, L.A. (1969). On Partitioning Chi-Squared and Detecting Partial Association in Three-Way Contingency Tables. *J. Roy. Statist. Soc. (B)* **31**, 486–498.

Goodman, L.A. (1971a). The Analysis of Multidimensional Contingency Tables: Stepwise Procedures and Direct Estimation Methods for Building Models for Multiple Classifications. *Technometrics* **13**, 33–61.

Goodman, L.A. (1971b). Partitioning of Chi-Square, Analysis of Marginal Contingency Tables, and Estimation of Expected Frequencies in Multi-Dimensional Contingency Tables. *J. Amer. Statist. Assoc.* **66**, 339–344.

Goodman, L.A. (1973). Guided and Unguided Methods for the Selection of Models for a Set of T Multidimensional Contingency Tables. *J. Amer. Statist. Assoc.* **68**, 165–175.

Goodman, L.A. (1984). *The Analysis of Cross-Classified Data Having Ordered Categories.* Cambridge, Massachusetts: Harvard University Press.

Greenwood, Major and Yule, G.U. (1920). An Inquiry into the Nature of Frequency Distributions Representative of Multiple Happenings With Particular Reference to the Occurrence of Multiple Attacks of Disease or of Repeated Accidents. *J. Roy. Statist. Soc.* **83**, 255–279.

Griffin, B. and Krutchkoff, R. (1971). Optimal Linear Estimates: An Empirical Bayes Version with Application to the Binomial Distribution. *Biometrika* **58**, 195–201.

Griffiths, D.A. (1973). Maximum Likelihood Estimation for the Beta-Binomial Distribution and an Application to the Household Distribution of the Total Number of Cases of a Disease. *Biometrics* **29**, 637–648.

Grizzle, J.E., Starmer, C.F., and Koch, G.G. (1969). Analysis of Categorical Data by Linear Models. *Biometrika* **28**, 137–156.

Guerrero, V.M. and Johnson, R.A. (1982). Use of the Box–Cox Transformation With Binary Response Models. *Biometrika* **69**, 309–314.

Gupta, S.S. (1956). *On a Decision-Rule for a Problem in Ranking Means.* Ph.D. Thesis, University of North Carolina, Chapel Hill, North Carolina.

Gupta, S.S. (1965). On Some Multiple Decision (Selection and Ranking) Rules. *Technometrics* **7**, 225–245.

Gupta, S.S., Leong, Y.K. and Wong, W.-Y. (1979). On Subset Selection Procedures for Poisson Populations. *Bull. Malaysia Math. Soc.* **2**, 89–110.

Gupta, S.S. and Nagel, K. (1967). On Selection and Ranking Procedures and Order Statistics from the Multinomial Distributions. *Sankhyā (B)* **29**, 1–34.

Gupta, S.S. and Panchapakesan, S. (1979). *Multiple Decision Procedures: Theory and Methodology of Selecting and Ranking Populations.* New York: John Wiley and Sons, Inc.

Gupta, S.S. and Sobel, M. (1960). Selecting a Subset Containing the Best of Several Binomial Populations. In *Contributions to Probability and Statistics.* Stanford University Press, 224–248.

Gupta, S.S. and Wong, W.-Y. (1977). On Subset Selection Procedures for Poisson Processes and Some Applications to the Binomial and Multinomial Problems. *Op. Research Verfahren* (eds. Henn et al.), Verlag Anton Hain, Meisenheim am Glan, Germany, 49–70.

Gutmann, S. (1982). Stein's Paradox is Impossible in Problems with Finite Sample Space. *Ann. Statist.* **10**, 1017–1020.

Haber, M. (1987). A Comparison of Some Conditional and Unconditional Exact Tests for 2 × 2 Contingency Tables. *Comm. Statist. - Simul. Comput.* **16**, 999–1013.

Haberman, S.J. (1973a). The Analysis of Residuals in Cross-Classified Tables. *Biometrics* **29**, 205–220.

Haberman, S.J. (1973b). Log-Linear Models for Frequency Data: Sufficient Statistics and Likelihood Equations. *Ann. Statist.* **1**, 617–632.

Haberman, S.J. (1974). *The Analysis of Frequency Data.* Chicago: The University of Chicago Press.

Haberman, S.J. (1976). Generalized Residuals for Log-Linear Models. In *Proceedings of the Ninth Intl. Biometrics Conf. Boston.* The Biometric Society, **1**, 104–122.

Haberman, S.J. (1977a). Maximum Likelihood Estimates in Exponential Response Models. *Ann. Statist.* **5**, 815–841.

Haberman, S.J. (1977b). Log-Linear Models and Frequency Tables With Small Expected Cell Counts. *Ann. Statist.* **5**, 1148–1169.

Haberman, S.J. (1978). *Analysis of Qualitative Data.* Vols. 1 and 2. Orlando: Academic Press, Inc.

Haldane, J.B.S. (1955). The Estimation and Significance of the Logarithm of a Ratio of Frequencies. *Ann. Hum. Genet.* **20**, 309–311.

Haldane, J.B.S. (1957). Almost Unibased Estimates of Functions of Frequencies. *Sankyā* **17**, 201–208.

Hall, P. (1981). On Nonparametric Multivariate Binary Discrimination. *Biometrika* **68**, 287–294.

Hall, P. and Titterington, D.M. (1987). On Smoothing Sparse Multinomial Data. *Austral. J. Statist.* **29**, 19–37.

Halperin, J. (1982). Maximally Selected Chi Square Statistics for Small Samples, *Biometrics* **38**, 1017–1023.

Hart, J.S., George, S.L., Frei, B., Budey, G.P., Nickerson, R.C., and Freireich, E.J. (1977). Prognostic Significance of Pretreatment Proliferative Activity in Adult Acute Leukemia. *Cancer* **39**, 1603–1617.

Haseman, J. (1978). Exact Sample Sizes for Use With the Fisher–Irwin Test for 2 × 2 Tables. *Biometrics* **34**, 106–109.

Haseman J. and Kupper, L.L. (1979). Analysis of Dichotomous Response Data from Certain Toxicological Experiments. *Biometrics* **35**, 281–293.

Hastie, T. and Tibshirani, R. (1987). Non-parametric Logistic and Proportional Odds Regression. *J. Roy. Statist. Soc. (C)* **36**, 260–276.

Hauck, W.W. (1979). The Large Sample Variance of the Mantel–Haenszel Estimator of a Common Odds Ratio. *Biometrics* **35**, 817–819.

Hauck, W.W. (1983). A Note on Confidence Bands for the Logistic Response Curve. *The American Statistician* **37**, 158–160.

Hauck, W.W. (1989). Odds Ratio Inference from Stratified Samples. *Comm. Statist. - Theor. Meth.* **18**, 767–800.

Hauck, W.W. and Anderson, S. (1985). On Testing for Bioequivalence. *Biometrics* **41**, 561–563.

Hauck, W.W., Anderson, S., and Leahy, F. (1982). Finite-Sample Properties of Some Old and Some New Estimators of a Common Odds Ratio from Multiple 2 × 2 Tables. *J. Amer. Statist. Assoc.* **77**, 145–152.

Hauck, W.W. and Donner, A. (1977). Wald's Test as Applied to Hypotheses in Logit Analysis. *J. Amer. Statist. Assoc.* **72**, 851–853.

Havranek, T. (1984). A Procedure for Model Search in Multidimensional Contingency Tables. *Biometrics* **40**, 95–100.

Hayam, G.E., Govindarajulu, Z., and Leone, F.C. (1973). Tables of the Cumulative Non-Central Chi-Square Distribution. In *Selected Tables in Mathematical Statistics* **1** (H.L. Harter and D.B. Owen, eds.). Providence, Rhode Island: Institute of Mathematical Statistics, 1–78.

Hinde, J. (1982). Compound Poisson Regression Models. In *GLIM 82: Proceedings of the International Conference in Generalized Linear Models* (R. Gilchrist, ed.). Berlin: Springer-Verlag, 109–121.

Hoadley, B. (1981). The Quality Measurement Plan (QMP). *Bell Sys. Tech. J.* **60**, 215–273.

Hoaglin, D.C. (1980). A Poissonness Plot. *The American Statistician* **34**, 146–149.

Hoaglin, D.C. and Welsch, R. (1978). The Hat Matrix in Regression and ANOVA. *The American Statistician* **32**, 17–22.

Hoblyn, T.N. and Palmer, R.C. (1934). A Complex Experiment in the Propagation of Plum Rootstocks from Root Cuttings. *J. of Pomology and Horticultural Science* **7**, 36–55.

Hochberg, Y. and Tamhane, A. (1987). *Multiple Comparison Procedures.* New York: John Wiley and Sons, Inc.

Hoel, D.G. and Yanagawa, T. (1986). Incorporating Historical Controls in Testing for a Trend in Proportions. *J. Amer. Statist. Assoc.* **81**, 1095–1099.

Hoerl, A. and Kennard, R. (1970). Ridge Regression: Biased Estimation for Non-Orthogonal Problems. *Technometrics* **12**, 53–63.

Hosmer, D.W. and Lemeshow, S. (1980). Goodness-of-fit Tests for the Multiple Logistic Regression Model. *Comm. Statist. - Theor. Meth.* **9**, 1043–1069.

Hudson, H.M. (1985). Adaptive Estimators for Simultaneous Estimation of Poisson Means. *Ann. Statist.* **13**, 246–261.

Hudson, H.M. and Tsui, K.W. (1981). Simultaneous Poisson Estimators for a Priori Hypotheses About Means. *J. Amer. Statist. Assoc.* **76**, 182–187.

Hwang, J.T. (1982). Improving Upon Standard Estimators in Discrete Exponential Families with Applications to Poisson and Negative Binomial Cases. *Ann. Statist.* **10**, 857–867.

Ighodaro, A. (1980). *Ridge and James-Stein Methods for Contingency Tables.* Ph.D. Thesis, School of Operations Research and Industrial Engineering, Cornell University, Ithaca, New York.

Ighodaro, A. and Santner, T.J. (1982). Ridge Type Estimators of Multinomial Cell Probabilities. In *Statistical Design Theory and Related Topics* III, Vol. 2. New York: Academic Press.

Ighodaro, A., Santner, T.J., and Brown, L. (1982). Admissibility and Complete Class Results for the Multinomial Estimation Problem with Entropy and Squared Error Loss. *J. Mult. Anal.* **12**, 469–479.

Innes, J.R.M., Ulland, B.M., Valerio, M.G., Petrucelli, L., Fishbein, L., Hart, E.R., Pallotta, A.J., Bates, R.R., Falk, H.L., Gart, J.J., Klein, M.,

Mitchell, I., and Peters, J. (1969). Bioassay of Pesticides and Industrial Chemicals for Tumorigenicity in Mice: A Preliminary Note. *J. Nat. Cancer Inst.* **42**, 1101–1114.

James, A.T. and Stein, C. (1961). Estimation with Quadratic Loss. In *Proceedings of the Fourth Berkeley Symposium* **1**, 361–379.

Jaynes, E.T. (1968). Prior Probabilities. *IEEE Trans. Syst. Sci. Cyber. SSC*-**4**, 227–241.

Jeffreys, H. (1961). *Theory of Probability.* Oxford: Oxford University Press.

Jennings, D.E. (1986a). Judging Inference Adequacy in Logistic Regression. *J. Amer. Statist. Assoc.* **81**, 471–476.

Jennings, D.E. (1986b). Outliers and Residual Distributions in Logistic Regression. *J. Amer. Statist. Assoc.* **81**, 987–990.

Jennison, C. and Turnbull, B.W. (1983). Confidence Intervals for a Binomial Parameter Following a Multistage Test with Application to MIL-STD 105D and Medical Trials. *Technometrics* **25**, 49–58.

Jewell, N. (1984). Small Sample Bias of Point Estimators of the Odds Ratio from Matched Sets. *Biometrics* **40**, 421–435.

Johnson, B.M. (1971). On the Admissible Estimators for Certain Fixed Sample Binomial Problems. *Ann. Math. Statist.* **42**, 1579–1587.

Johnson, W. (1985). Influence Measures for Logistic Regression: Another Point of View. *Biometrika* **72**, 59–65.

Jones, M.P., O'Gorman, T.W., Lemke, J.H., and Woolson, R.F. (1989). A Monte Carlo Investigation of Homogeneity Tests of the Odds Ratio Under Various Sample Size Configurations. *Biometrics* **45**, 171–181.

Jorgenson, D.W. (1961). Multiple Regression Analysis of a Poisson Process. *J. Amer. Statist. Assoc.* **56**, 235–245.

Kaplan, S. (1983). On a "Two Stage" Bayesian Procedure for Determining Failure Rates From Experimental Data. *IEEE Transactions on Power Apparatus and Systems* PAS-102, 195–202.

Kaplan, W. (1952). *Advanced Calculus.* Reading, MA: Addison-Wesley.

Karlin, S. and Taylor, H. (1975). *A First Course in Stochastic Processes.* New York: Academic Press.

Keeler, L.C. (1985). *Genotypic and Environmental Effects on the Cold, Icing, and Flooding Tolerances of Winter Wheat.* M.Sc. Thesis, University of Guelph, Ontario.

Keeler, L.C. (1988). personal communication.

Kesten, H. and Morse, N. (1959). A Property of the Multinomial Distribution. *Ann. Statist.* **30**, 120–127.

Kettenring, J.R. (1982). Canonical Analysis. In *Encyclopedia of Statistical Sciences* **1**, 354–365.

Klein, R.W. and Spady, R.H. (1989). An Efficient Semiparametric Estimator for Discrete Choice Models. Bellcore Technical Memorandum, Bell Communications Research, Morristown, N.J.

Knight, R.L. and Skagen, S.K. (1988). Agonistic Asymmetry and the Foraging of Bald Eagles. *Ecology* **69**, 1188–1194.

Knoke, D. (1976). *Change and Continuity in American Politics: The Social Bases of Political Parties.* Baltimore, MD: The Johns Hopkins University Press.

Koehler, K.J. (1986). Goodness-of-Fit for Log-Linear Models in Sparse Contingency Tables. *J. Amer. Statist. Assoc.* **81**, 483–493.

Koehler, K.J. and Larntz, K. (1980). An Empirical Investigation of Goodness-of-Fit Statistics for Sparse Multinomials. *J. Amer. Statist. Assoc.* **75**, 336–344.

Korff, F.A., Taback, M.A.M., and Beard, J.H. (1952). A Coordinated Investigation of a Food Poisoning Outbreak. *Public Health Reports* **67** (6), 909–913.

Kotze, T.J. van and Hawkins, D.M. (1984). The Identification of Outliers in Two-Way Contingency Tables Using 2 × 2 Subtables. *J. Roy. Statist. Soc. (C)* **33**, 215–233.

Kulkarni, R.V. and Kulkarni, V.G. (1987). Optimal Bayes Procedures for Selecting the Better of Two Bernoulli Populations. *J. Statist. Planning Inference* **15**, 311–330.

Kullback, S. (1971). Marginal Homogeneity of Multidimensional Contingency Tables. *Ann. Math. Statist.* **42**, 594–606.

Kupper, L., Portier, C., Hogan, M., and Yamamoto, E. (1986). The Impact of Litter Effects on Dose-Response Modeling in Teratology. *Biometrics* **42**, 85–98.

Kwei, L. (1983). *Tests of Goodness-of-Fit with Random Cells for Conditional Distributions.* Ph.D. Thesis, Dept. of Statistics, University of California, Berkeley, CA.

Lacampagne, C.B. (1979). *An Evaluation of the Women and Mathematics (WAM) Program and Associated Sex-Related Differences in the Teaching, Learning, and Counseling of Mathematics.* Ed.D. Thesis, Teachers College, Columbia University, New York, New York.

Laird, N.M. (1978a). Empirical Bayes Methods for Two-Way Contingency Tables. *Biometrika* **65**, 581–590.

Laird, N.M. (1978b). Nonparametric Maximum Likelihood Estimation of a Mixing Distribution. *J. Amer. Statist. Assoc.* **73**, 805–811.

Landwehr, J.M., Pregibon, D., and Shoemaker, A.C. (1984). Graphical Methods for Assessing Logistic Regression Models (with discussion). *J. Amer. Statist. Assoc.* **79**, 61–83.

Larntz, K. (1978). Small Sample Comparisons of Exact Levels for Chi-Square Goodness-of-Fit Statistics. *J. Amer. Statist. Assoc.* **73**, 253–263.

Lawal, H.B. (1984). Comparisons of the X^2, Y^2 Freeman–Tukey and William's Improved G^2 Test Statistics in Small Samples of One-Way Multinomials. *Biometrika* **71**, 415–418.

Lawless, J.F. (1987a). Negative Binomial and Mixed Poisson Regression. *Canad. J. Statist.* **15**, 209–225.

Lawless, J.F. (1987b). Regression Methods for Poisson Process Data. *J. Amer. Statist. Assoc.* **82**, 808–815.

LeCessie, S. and van Houwelingen, J.C. (1989). Goodness of Fit Tests for Binary Regression Models Based on Smoothing Methods. Technical Rpt. 7, Dept. of Medical Statistics, University of Leiden, The Netherlands.

Lee, A.H. (1987). Diagnostic Displays for Assessing Leverage and Influence in Generalized Linear Models. *Aust. J. Statist.* **29**, 233–243.

Lee, A.H. (1988). Assessing Partial Influence in Generalized Linear Models. *Biometrics* **44**, 71–77.

Lee, C.C. (1987). Chi-Squared Tests For and Against an Order Restriction on Multinomial Parameters. *J. Amer. Statist. Assoc.* **82**, 611–618.

Lee, J.A.H. (1963). Seasonal Variations on Leukemia Incidence. *Brit. Med. Journal* **2**, 623.

Lee, J.C. and Sabavala, D. (1982). Some Further Results on Bayesian Methods in the Stochastic Modeling of Buying Behavior. *Business and Economics Statistics Section Proceedings of the Amer. Statist. Assoc.* 6–9.

Lee, J.C. and Sabavala, D. (1987). Bayesian Estimation and Prediction for the Beta-Binomial Model. *J. Bus. Statist.* **5**, 357–367.

Lee, E.T. (1974). Computer Programs for Linear Logistic Regression Analysis. *Computer Programs in Biomedicine* **4**, 82–97.

Lee, L.F. (1986). Specification Tests for Poisson Regression Models. *Inter. Econ. Rev.* **27**, 689–706.

Lee, S.K. (1977). On the Asymptotic Variances of $\hat{u}$ Terms in Log-linear Models of Multidimensional Contingency Tables. *J. Amer. Statist. Assoc.* **72**, 412–419.

Lee, Y.J. (1977). Maximum Tests of Randomness Against Ordered Alternatives: The Multinomial Distribution Case. *J. Amer. Statist. Assoc.* **72**, 673–675.

Lehmann (1986). *Testing Statistical Hypotheses* (2nd edition). New York: John Wiley and Sons, Inc.

Lenk, P. (1987). Bayesian, Predictive Distributions Under Multinomial Sampling. Technical Report No. 87-88, Graduate School of Business Administration, New York University.

Leonard, T. (1972). Bayesian Methods for Binomial Data. *Biometrika* **59**, 581–589.

Leonard, T. (1973). A Bayesian Method for Histograms. *Biometrika* **60**, 297–308.

Leonard, T. (1975). Bayesian Estimation Methods for Two-Way Contingency Tables. *J. Roy. Statist. Soc. (B)* **37**, 23–37.

Leonard, T. (1977a). A Bayesian Approach to Some Multinomial Estimation and Pretesting Problems. *J. Amer. Statist. Assoc.* **72**, 869–874.

Leonard, T. (1977b). Bayesian Simultaneous Estimation for Several Multinomial Distributions. *Comm. Statist. - Theor. Meth.* **6**, 619–630.

Li, S.-H., Simon, R., and Gart, J.J. (1979). Small-Sample Properties of the Mantel–Haenszel Test. *Biometrics* **66**, 181–183.

Liang, K.-Y. and Self, S. (1985). Tests for Homogeneity of Odds Ratio when the Data are Sparse. *Biometrika* **72**, 353–358.

Liang, K.-Y. and Zeger, S. (1986). Longitudinal Data Analysis for Discrete and Continuous Outcomes. *Biometrics* **42**, 121–130.

Lindley, D.V. (1964). The Bayesian Analysis of Contingency Tables. *Ann. Math. Statist.,* 1622–1643.

Lindsay, B.G. (1983). Efficiency of the Conditional Score in a Mixture Setting. *Ann. Statist.* **11**, 486–497.

Maddala, G.A. (1983). *Limited-Dependent and Qualitative Variables in Econometrics.* Cambridge: Cambridge University Press.

Madsen, M. (1976). Statistical Analysis of Multiple Contingency Tables—Two Examples. *Scand. J. Statist.* **3**, 97–106.

Mantel, N. and Haenszel, W. (1959). Statistical Aspects of the Analysis of Data from Retrospective Studies of Disease. *J. Nat. Cancer Inst.* **22**, 719–748.

Marhoul, J.C. (1984). A Model for Large Sparse Contingency Tables. Tech. Report No. 13, Stanford Linear Accelerator Center, Stanford University.

Maritz, J.S. (1969). Empirical Bayes Estimation for the Poisson Distribution. *Biometrika* **56**, 349–359.

Maritz, J.S. and Lian, M.G. (1974). Empirical Bayes Estimation of the Binomial Parameter. *Biometrika* **61**, 517–523.

McCullagh, P. (1980). Regression Models for Ordinal Data. *J. Roy. Statist. Soc. (B)* **42**, 109–142.

McCullagh, P. (1986). The Conditional Distribution of Goodness-of-Fit Statistics for Discrete Data. *J. Amer. Statist. Assoc.* **81**, 104–107.

McCullagh, P. and Nelder, J.A. (1983). *Generalized Linear Models.* London: Chapman and Hall.

McDonald, L.L., Davis, B.M., and Milliken, G.A. (1977). A Nonrandomized Unconditional Test for Comparing Two Proportions in 2 × 2 Contingency Tables. *Technometrics* **19**, 145–157.

Mee, R. (1984). Confidence Bounds for the Difference Between Two Probabilities. *Biometrics* **40**, 1175–1176.

Mehta, C.R. and Patel, N.R. (1983). A Network Algorithm for Performing Fisher's Exact Test in $r \times c$ Contingency Tables. *J. Amer. Statist. Assoc.* **78**, 427–434.

Mehta, C.R., Patel, N.R., and Gray, R. (1985). Computing an Exact Confidence Interval for the Common Odds Ratio in Several 2×2 Contingency Tables. *J. Am. Statist. Assoc.* **80**, 969–973.

Mehta, C.R., Patel, N.R., and Senchaudhuri, P. (1988). Importance Sampling for Estimating Exact Probabilities in Permutational Inference. *J. Am. Statist. Assoc.* **83**, 999–1005.

Mehta, C.R. and Walsh, S.J. (1989). Comparison of Exact, Mid-P, and Mantel–Haenszel Confidence Intervals for the Common Odds Ratio Across Several 2×2 Contingency Tables. Dept. of Biostatistics Technical Report, Harvard University.

Meittinen, O. and Nurminen, M. (1985). Comparative Analysis of Two Rates. *Statistics in Medicine* **4**, 213–226.

Mendel, G. (1967). *Experiments in Plant Hybridization.* Cambridge, Massachusetts: Harvard University Press.

Meyer, M.M. (1982). Transforming Contingency Tables. *Ann. Statist.* **10**, 1172–1181.

Miller, R. and Siegmund, D. (1982). Maximally Selected Chi Square Statistics. *Biometrics* **38**, 1011–1016.

MIL-STD 105D (1963). *Military Standard Sampling Procedures and Tables for Inspection by Attributes.* Washington, D.C.

Mōhner, M. (1986). A Comparative Study of Estimators for Probabilities in Contingency Tables. *Statistics* **17**, 557–568.

Moore, D.S. (1971). A Chi-Square Statistic with Random Cell Boundaries. *Ann. Math. Statist.* **42**, 147–156.

Moore, D.S. and Spruill, M.C. (1975). Unified Large-Sample Theory of General Chi-Squared Statistics for Tests of Fit. *Ann. Statist.* **3**, 599–616.

Moore, L.M. and Beckman, R.J. (1988). Approximate One-Sided Tolerance Bounds on the Number of Failures Using Poisson Regression. *Technometrics* **30**, 283–290.

Moran, P.A.P. (1970). On Asymptotically Optimal Tests of Composite Hypotheses. *Biometrika* **57**, 47–55.

Nam, J. (1987). A Simple Approximation for Calculating Sample Sizes for Detecting Linear Trend in Proportions. *Biometrics* **43**, 701–705.

Nazaret, W.A. (1987). Bayesian Log Linear Estimates for Three-Way Contingency Tables. *Biometrika* **74**, 401–410.

Nelder, J.A. (1984). Models for Rates with Poisson Errors. *Biometrics* **40**, 1159–1162.

Nelder, J.A. and Wedderburn, R.W.M. (1972). Generalized Linear Models. *J. Roy. Statist. Soc. (A)* **135**, 370–384.

Neyman, J. and Pearson, E.S. (1928). On the Use and Interpretation of Certain Test Criteria for Purposes of Statistical Inference. *Biometrika* **20A**, 175–240 and 263–294.

Novick, M.R., Lewis, C., and Jackson, P.H. (1973). The Estimation of Proportions in m Groups. *Psychometrika* **38**, 19–46.

Odoroff, C.L. (1970). A Comparison of Minimum Logit Chi-Square Estimation and Maximum Likelihood Estimation in $2 \times 2 \times 2$ and $3 \times 2 \times 2$ Contingency Tables: Tests for Interaction. *J. Amer. Statist. Assoc.* **65**, 1617–1631.

Oler, J. (1985). Noncentrality Parameters in Chi-Squared Goodness-of-Fit Analyses with an Application to Loglinear Procedures. *J. Amer. Statist. Assoc.* **80**, 181–189.

Olkin, I., Petkau, A.J., and Zidek, J.V. (1981). A Comparison of n Estimators for the Binomial Distribution. *J. Amer. Statist. Assoc.* **76**, 637–642.

Olkin, I. and Sobel, M. (1979). Admissible and Minimax Estimation for the Multinomial Distribution and for K Independent Binomial Distributions. *Ann. Statist.* **7**, 284–290.

Ord, J.K. (1967). Graphical Methods for a Class of Discrete Distributions. *J. Roy. Stat. Soc. (A)* **130**, 232–238.

O'Sullivan, F., Yandell, B., and Raynor, W., Jr. (1986). Automatic Smoothing of Regression Functions in Generalized Linear Models. *J. Amer. Statist. Assoc.* **81**, 96–103.

Panchapakesan, S. (1971). On Subset Selection Procedures for the Most Probable Event in a Multinomial Distribution. *Statistical Decision Theory and Related Topics.* New York: Academic Press, Inc., 275–298.

Park, M. (1985). A Graphic Representation of a Three-Way Contingency Table: Simpson's Paradox and Correlation. *The American Statistician* **39**, 53–54.

Pauling, L. (1971). The Significance of the Evidence about Ascorbic Acid and the Common Cold. *Proc. Acad. Sci. USA* **68**, 2678–2681.

Peizer, D.B. and Pratt, F. (1968). A Normal Approximation for Binomial, F, Beta, and Other Common Related Tail Probabilities, I. *J. Amer. Statist. Assoc.* **63**, 1416–1456.

Peng, J.C.M. (1975). Simultaneous Estimation of the Parameters of Independent Poisson Distributions, Technical Report 78, Department of Statistics, Stanford University, Stanford, California.

Piegorsch, W. and Casella, G. (1988). Confidence Bands for Logistic Regression with Restricted Predictor Variables. *Biometrics* **44**, 739–750.

Piegorsch, W., Weinberg, C.R., and Margolin, B.H. (1988). Exploring Simple Independent Action in Multifactor Tables of Proportions. *Biometrics* **44**, 595–603.

Pirie, W.R. and Hamden, M.A. (1972). Some Revised Continuity Corrections for Discrete Distributions. *Biometrics* **28**, 693–701.

Platefield, W.M. (1981). An Efficient Method of Generating Random $R \times C$ Tables with Given Row and Column Totals. *J. Roy. Statist. Soc. (C)* **30**, 91–97.

Potthoff, R. and Whittinghill, M. (1966). Testing for Homogeneity II. The Poisson Distribution. *Biometrika* **53**, 183–190.

Pregibon, D. (1980). Goodness of Link Tests for Generalized Linear Models. *J. Roy. Statist. Soc. (C)* **29**, 15–24.

Pregibon, D. (1981). Logistic Regression Diagnostics. *Ann. Statist.* **9**, 705–724.

Pregibon, D. (1982a). Score Tests in GLIM With Applications. In *Lecture Notes in Statistics, No. 14, GLIM.82: Proceedings of the International Conference on Generalized Linear Models* (R. Gilchrist, ed.). New York: Springer-Verlag.

Pregibon, D. (1982b). Resistant Fits for Some Commonly Used Logistic Models with Medical Applications. *Biometrics* **38**, 485–498.

Prentice, R.L. (1976). A Generalization of the Probit and Logit Methods for Dose Response Curves. *Biometrics* **32**, 761–768.

Prentice, R.L. (1988). Correlated Binary Regression with Covariates Specific to Each Binary Observation. *Biometrics* **44**, 1033–1048.

Prentice, R.L. and Pyke, R. (1979). Logistic Disease Incidence Models and Case-Control Studies. *Biometrika* **66**, 403–411.

Presidential Commission on the Space Shuttle Challenger Accident (1986). Report to the President by the Presidential Commission on the Space Shuttle Challenger Accident. June 6, Washington, D.C.

Quesenberry, C.P. and Hurst, D.C. (1964). Large Sample Simultaneous Confidence Intervals for Multinomial Proportions. *Technometrics* **6**, 191–195.

Quine, S. (1975). *Achievement Orientation of Aboriginal and White Australian Adolescents.* Ph.D. Thesis, Australian National University.

Raab, S. (1980). Police Strength and Racial Mix await Decisions in Court Battle. *New York Times* **129**, Section B, col. 1.

Raftery, A. (1988). Inference for the Binomial N Parameter: A Hierarchical Bayes Approach. *Biometrika* **75**, 223–228.

Ramey, J.T. and Alam, K. (1979). A Sequential Procedure for Selecting the Most Probable Multinomial Event. *Biometrika* **66**, 171–173.

Rao, C.R. (1947). Large Sample Tests of Statistical Hypotheses Concerning Several Parameters with Applications to Problems of Estimation. *Proc. Camb. Phil. Soc.* **44**, 50–57.

Rao, C.R. (1965). *Linear Statistical Inference and its Applications.* New York: John Wiley.

Rao, K.C. and Robson, D.S. (1974). A Chi-Square Statistic for Goodness-of-Fit Tests Within the Exponential Family. *Comm. Statist. - Theor. Meth.* **3**, 1139–1153.

Read, T.R. and Cressie, N.A. (1988). *Goodness-of-Fit Statistics for Discrete Multivariate Data.* New York: Springer-Verlag.

Roberts, A.W. and Varberg, D.E. (1973). *Convex Functions.* New York: Academic Press.

Robertson, T. (1978). Testing For and Against an Order Restriction on Multinomial Parameters. *J. Amer. Statist. Assoc.* **73**, 197–202.

Robertson, T. and Wright, F.T. (1983). On Approximation of the Level Probabilities and Associated Distributions in Order Restricted Inference. *Biometrika* **70**, 597–606.

Robertson, T., Wright, F.T., and Dykstra, R.L. (1988). *Order Restricted Statistical Inference.* Chichester: John Wiley and Sons, Inc.

Robins, J., Breslow, N., and Greenland, S. (1986). Estimators for the Mantel–Haenszel Variance Consistent in Both Sparse Data and Large Strata Limiting Models. *Biometrics* **42**, 311–323.

Rubin, D.B. and Schenker, N. (1986). Logit-Based Interval Estimation for Binomial Data Using the Jeffreys' Prior. Preprint.

Rudas, T. (1986). A Monte Carlo Comparison of the Small Sample Behavior of the Pearson, the Likelihood Ratio, and the Cressie–Read Statistics. *J. Statist. Comput. Simul.* **24**, 107–120.

Rutherford, E. and Geiger, H. (1910). The Probability Variations in the Distribution of α Particles. *Phil. Mag.* Sixth Ser., **20**, 698–704.

Sabavala, D. and Lee, J.C. (1981). Bayesian Methods in the Stochastic Modeling of Buying Behavior. *Business and Economic Statistics Section Proceedings of the Amer. Statist. Assoc.*, 455–456.

Sakamoto, Y. and Akaike, H. (1978). Analysis of Cross Classified Data by AIC. *Ann. Inst. Statist. Math. (B)* **30**, 185–197.

Sanchez, S. (1987). A Modified Least-Failures Sampling Procedure for Bernoulli Subset Selection. *Comm. Statist. - Theor. Meth.* (*A*) **16**, 1051–1065.

Santner, T.J. and Duffy, D.E. (1986). A Note on A. Albert and J.A. Anderson's Conditions for the Existence of Maximum Likelihood Estimates in Logistic Regression Models. *Biometrika* **73**, 755–758.

Santner, T.J. and Snell, M.K. (1980). Small-Sample Confidence Intervals for $p_1 - p_2$ and p_1/p_2 in 2×2 Contingency Tables. *J. Amer. Statist. Assoc.* **75**, 386–394.

Santner, T.J. and Yamagami, S. (1988). Invariant Small-Sample Confidence Intervals for the Difference of Two Success Probabilities. Technical Report 698, School of Operations Research, Cornell University.

Schaefer, R.L. (1983). Bias Correction in Maximum Likelihood Logistic Regression. *Statistics in Medicine* **2**, 71–78.

Schaefer, R.L. (1986). Alternative Estimators in Logistic Regression when the Data are Collinear. *J. Statist. Comput. Simul.* **25**, 75–91.

Schaefer, R.L., Roi, L.D., and Wolfe, R.A. (1984). A Ridge Logistic Estimator. *Commun. Statis. - Theor. Meth.* **13**, 99–113.

Scheffe, H. (1959). *The Analysis of Variance*. New York: John Wiley and Sons, Inc.

Schneider, J.A., Schlesselman, J.J., Mendoza, S.A., Orloff, S., Thoene, J.G., Kroll, W.A., Godfrey, A.D., and Schulman, J.D. (1979). Ineffectiveness of Ascorbic Acid Therapy in Nephropathic Cystinosis. *New Eng. Jour. Med.* **300**, 756–759.

Segaloff, A. (1961). Progress Report: Results of Studies by the Cooperative Breast Cancer Group—1956–60. *Can. Chem. Rep.* **11**, 109–141.

Serfling, R.J. (1980). *Approximation Theorems of Mathematical Statistics*. New York: John Wiley & Sons.

Sewell, W. and Shah, V. (1968). Social Class, Parental Encouragement and Educational Aspirations. *Amer. J. Social.* **73**, 559–572.

Shapiro, S.H. (1982). Collapsing Contingency Tables—A Geometric Approach. *The American Statistician* **36**, 43–46.

Silvapulle, M.J. (1981). On the Existence of Maximum Likelihood Estimators for the Binomial Response Models. *J. Roy. Statist. Soc. (B)* **43**, 310–313.

Simar, L. (1976). Maximum Likelihood Estimation of a Compound Poisson Process. *Ann. Statist.* **4**, 1200–1209.

Simonoff, J. (1983). A Penalty Function Approach to Smoothing Large Sparse Contingency Tables. *Ann. Statist.* **11**, 208–218.

Simonoff, J. (1985). An Improved Goodness-of-Fit Statistic for Sparse Multinomials. *J. Amer. Statist. Assoc.* **80**, 671–677.

Simonoff, J. (1986). Jackknifing and Bootstrapping Goodness-of-Fit Statistics in Sparse Multinomials. *J. Amer. Statist. Assoc.* **81**, 1005–1011.

Simonoff, J. (1987). Probability Estimation Via Smoothing in Sparse Contingency Tables with Ordered Categories. *Statist. and Prob. Letters* **5**, 55–63.

Simonoff, J. (1988). Detecting Outlying Cells in Two-Way Contingency Tables via Backwards-Stepping. *Technometrics* **30**, 339–345.

Simpson, C.H. (1951). The Interpretation of Interaction in Contingency Tables. *J. Roy. Statist. Soc. (B)* **13**, 238–241.

Slakter, M.J. (1966). Comparative Validity of the Chi-Squared and Two Modified Chi-Square Goodness-of-Fit Tests for Small but Equal Expected Frequencies. *Biometrika* **53**, 619–622.

Sobel, M. (1963). Single Sample Ranking Problems with Poisson Populations. Technical Report 19, University of Minnesota, Minneapolis, Minnesota.

Sobel, M. and Huyett, M. (1957). Selecting the Best One of Several Binomial Populations. *Bell System Tech. J.* **36**, 537–576.

StatXact Statistical Software. Cytel Software Corporation, 137 Erie Street, Cambridge, MA 02139.

Stefanski, L.A. and Carroll, R.J. (1985). Covariate Measurement Error in Logistic Regression. *Ann. Statist.* **13**, 1335–1351.

Stefanski, L.A., Carroll, R.J., and Ruppert, D. (1986). Optimally Bounded Score Functions for Generalized Linear Models with Applications to Logistic Regression. *Biometrika* **73**, 413–424.

Stein, C. (1956). Inadmissibility of the Usual Estimator for the Mean of a Multivariate Normal Distribution. In *Proceedings of the Third Berkeley Symposium, Math. Statist. Prob.*, **1**, University of California Press, 197–206.

Stein, C. (1973). Estimation of the Mean of a Multivariate Distribution. In *Proceedings of The Prague Symp. Asymptotic Statist.*, 345–381.

Sterne, T. (1954). Some Remarks on Confidence or Fiducial Limits. *Biometrika* **41**, 275–278.

Stewart, L. (1987). Hierarchical Bayesian Analysis Using Monte Carlo Integration: Computing Posterior Distributions When There Are Many Possible Models. *The Statistician* **36**, 211–219.

Stiratelli, R., Laird, N., and Ware, J. (1984). Random-Effects Models for Serial Observations with Binary Response. *Biometrics* **40**, 961–971.

Stone, M. (1974). Cross-Validation and Multinomial Prediction. *Biometrika* **61**, 509–515.

Storer, B.E. and Kim, C. (1988). Exact Small Sample Properties of Some Test Statistics for Comparing Two Binomial Proportions. Technical Report 41, Department of Statistics, University of Wisconsin-Madison.

Stuart, A. (1953). The Estimation and Comparison of Strengths of Association in Contingency Tables. *Biometrika* **40**, 105–110.

Stukel, T. (1988). Generalized Logistic Models. *J. Amer. Statist. Assoc.* **83**, 426–431.

Suissa, S. and Shuster, J.J. (1985). Exact Unconditional Sample Sizes for the 2×2 Binomial Trial. *J. Roy. Statist. Soc. (A)* **148**, 317–327.

Sutherland, M., Fienberg, S.E., and Holland, P. (1974). Combining Bayes and Frequency Approaches to Estimate a Multinomial Parameter. In *Studies in Bayesian Econometrics and Statistics* (S.E. Fienberg and A. Zellner, eds.). Amsterdam: North Holland, 585–617.

Tamura, R.N. and Young, S.S. (1987). A Stabilized Moment Estimator for the Beta-Binomial Distribution. *Biometrics* **43**, 813–824.

Tarone, R.E. (1982). The Use of Historical Control Information in Testing for a Trend in Proportions. *Biometrics* **38**, 215–220.

Tarone, R.E. and Gart, J.J. (1980). On the Robustness of Combined Tests for Trends in Proportions. *J. Amer. Statist. Assoc.* **75**, 110–116.

Taylor, J.M.G. (1988). The Cost of Generalizing Logistic Regression. *J. Am. Statist. Assoc.* **83**, 1078–1083.

Thisted, R. (1988). *Elements of Statistical Computing: Numerical Computation.* New York: Chapman and Hall.

Thomas, D.G. (1971). Exact Confidence Limits for the Odds Ratio in a 2×2 Table. *Appl. Statist.* **20**, 105–110.

Thomas, D.G. and Gart, J.J. (1977). A Table of Exact Confidence Limits for Differences and Ratios of Two Proportions and Their Odds Ratios. *J. Amer. Statist. Assoc.* **72**, 73–76.

Thompson, S.K. (1987). Sample Size for Estimating Multinomial Proportions. *The American Statistician* **41**, 42–46.

Titterington, D.M. (1980). A Comparative Study of Kernel-Based Density Estimates for Categorical Data. *Technometrics* **22**, 259–268.

Tsiatis, A.A. (1980). A Note on a Goodness-of-Fit Test for the Logistic Regression Model. *Biometrika* **67**, 250–251.

Tsui, K.W. (1979). Multiparameter Estimation of Discrete Exponential Distributions. *Can. J. Statist.* **7**, 193–200.

Tsui, K.W. (1981). Simultaneous Estimation of Several Poisson Parameters Under Squared Error Loss. *Ann. Inst. Statist. Math. (A)* **33**, 215–223.

Tsui, K.W. (1984). Robustness of Clevenson–Zidek-Type Estimators. *J. Amer. Statist. Assoc.* **79**, 152–157.

Tsui, K.W. (1986). Further Developments on the Robustness of Clevenson–Zidek-Type Means Estimators. *J. Amer. Statist. Assoc.* **81**, 176–180.

Tsui, K.W. and Press, S.J. (1982). Simultaneous Estimation of Several Poisson Parameters Under k-Normalized Squared Error Loss. *Ann. Statist.* **10**, 93–100.

Tuyns, A.J., Pequinot, G., and Jensen, O.M. (1977). Le Cancer de L'Oesophage en Ille-et-Vilaine en Fonction des Niveaux de Consommation d'Alcool et de Tabac: Des Risques qui se Multiplient. *Bull. Cancer* **64**, 45–60.

Ury, H.K. (1982). Hauck's Approximate Large-Sample Variance of the Mantel–Haenszel Estimator. *Biometrics* **38**, 1094–1095.

van Dijk, H.K. (1987). Some Advances in Bayesian Estimation Methods Using Monte Carlo Integration. Report 8704/A, Erasmus University, Rotterdam, The Netherlands.

Venzon, D. and Moolgavkar, S.H. (1988). Origin-Invariant Relative Risk Functions for Case-Control and Survival Studies. *Biometrika* **75**, 325–333.

Vos, J.W.E. (1978). Confidence Intervals for a Binomial Parameter. ISO/TC 69/SC 2/N 165.

Vos, J.W.E. (1979). Average Confidence Intervals. ISO/TC 69/SC 2/N 175.

Wagner, C.H. (1982). Simpson's Paradox in Real Life. *The American Statistician* **36**, 46–48.

Wald, A. (1943). Tests of Statistical Hypotheses Concerning Several Parameters When the Number of Parameters is Large. *Trans. of Am. Math. Soc.* **54**, 426–482.

Walter, S.D. (1975). The Distribution of Levin's Measure of Attributable Risk. *Biometrika* **62**, 371–375.

Walter, S.D. (1976). The Estimation and Interpretation of Attributable Risk in Health Research. *Biometrics* **32**, 829–849.

Walter, G.G. and Hamdani, G.G. (1987). Empiric Bayes Estimation of Binomial Probability. *Comm. Statist. - Theor. Meth.* **16**, 559–577.

Wang, P.C. (1985). Adding a Variable in Generalized Linear Models. *Technometrics* **27**, 273–276.

Weber, D.C. (1971). Accident Rate Potential: An Application of Multiple Regression Analysis of a Poisson Process. *J. Amer. Statist. Assoc.* **66**, 285–288.

Weiss, L. (1975). The Asymptotic Distribution of the Likelihood Ratio in Some Nonstandard Cases. *J. Amer. Statist. Assoc.*, 70.

Wermuth, N. (1976). Model Search Among Multiplicative Models. *Biometrics* **32**, 253–263.

Wermuth, N. and Lauritzen, S.L. (1983). Graphical and Recursive Models for Contingency Tables. *Biometrika* **70**, 537–552.

Whitehead, J. (1983). *The Design and Analysis of Clinical Trials.* New York: Halsted Press.

Whittaker, J. and Aitkin, M. (1978). A Flexible Strategy for Fitting Complex Log-Linear Models. *Biometrics* **34**, 487–495.

Whittemore, A.S. (1978). Collapsibility of Multidimensional Contingency Tables. *J. Roy. Statist. Soc. (B)* **40**, 328–340.

Williams, D.A. (1984). Residuals in Generalized Linear Models. In *Proceedings of the 12th International Biometric Conference,* 59–68.

Williams, D.A. (1987). Generalized Linear Model Diagnostics Using the Deviance and Single Case Deletions. *J. Roy. Statist. Soc. (C)* **36**, 181–191.

Wittes, J. and Wallenstein, S. (1987). The Power of the Mantel–Haenszel Test. *J. Amer. Statist. Assoc.* **82**, 1104–1109.

Woolf, B. (1955). On Estimating the Relationship Between Blood Group and Disease. *Ann. Hum. Genet.* **19**, 251–253.

Wypij, D. (1986). *Estimation Methods for Grouped Binary Data.* Ph.D. Thesis, School of Operations Research and Industrial Engineering, Cornell University, Ithaca, New York.

Wypij, D. (1987). Pseudotable Methods for the Analysis of 2 × 2 Tables. Department of Biostatistics Technical Report, Harvard University.

Wypij, D. and Santner, T.J. (1988). Interval Estimation of the Marginal Probability of Success for the Beta-Binomial Distribution. To appear in *J. Statist. Comp. Simul.*

Yarnold, J.K. (1970). The Minimum Expectation in χ^2 Goodness-of-Fit Tests and the Accuracy of Approximations for the Null Distribution. *J. Amer. Statist. Assoc.* **65**, 864–886.

Zacks, S., Brier, S.S., and Marlow, W.H. (1988). A Simulation Study of Empirical Bayes Estimators of Multiple Correlated Probability Vectors. *Nav. Res. Log. Quart.* **35**, 237–246.

Zellner, A. and Rossi, P.E. (1984). Bayesian Analysis of Dichotomous Quantal Response Models. *J. Econometrics* **25**, 365–393.

Zelterman, D. (1987). Goodness-of-Fit Tests for Large Sparse Multinomial Distributions. *J. Amer. Statist. Assoc.* **82**, 624–629.

List of Notation

ACE	alternating conditional expectation (algorithm)
ANOVA	analysis of variance
Arcsine(x)	inverse sine function evaluated at x
$Be(\alpha,\beta)$	beta distribution with parameters $\alpha > 0$, $\beta > 0$
$B(n,p)$	binomial distribution with n trials and success probability p
CMH	Cochran–Mantel–Haenszel (test)
$\mathbf{c}$	center of the $(t-1)$-dimensional simplex; $= (1/t,\ldots,1/t)'$
cdf	cumulative distribution function
$\cos(w)$	cosine of w
$\mathcal{C}(\mathbf{X})$	column space of the $n \times p$ matrix $\mathbf{X}$; $= \{\mathbf{X}\boldsymbol{\beta}: \boldsymbol{\beta} \in \mathbb{R}^p\}$
CS	correct selection
Cov(X,Y)	covariance between random variables X and Y
df	degrees of freedom
$\mathcal{D}$	class of direct loglinear models
$D(\boldsymbol{\beta})$	Diag$(\ldots, m_i p_i(\boldsymbol{\beta})[1 - p_i(\boldsymbol{\beta})], \ldots)$
d-NSEL	**d**-normalized squared error loss
$\mathcal{D}_t(\boldsymbol{\beta})$	Dirichlet distribution with t cells and parameter $\boldsymbol{\beta} = (\beta_1,\ldots,\beta_t)'$ $(> \mathbf{0}_t)$
$D(w)$	digamma function evaluated at w; $= \frac{d}{dw}\ln(\Gamma(w+1))$
Diag($\mathbf{x}$)	diagonal matrix with vector $\mathbf{x}$ on the diagonal
e	base of the natural logarithm
$\mathbf{e}^P$	Pearson residuals
$\mathbf{e}^r$	raw residuals
$\mathbf{e}^a$	adjusted residuals
$\mathbf{e}^d$	deviance residuals
$\mathbf{e}^{ad}$	adjusted deviance residuals
$\mathbf{e}^o$	outlier residuals
EB	empirical Bayes
EL	entropy loss
EM	expectation-maximization (algorithm)
$\exp\{\mathbf{x}\}$	$(e^{x_1},\ldots,e^{x_n})'$
$E[X]$	expectation of the random variable X
$E[X \mid Y]$	conditional expectation of X given Y
$f(y \mid t; \psi)$	conditional probability of $Y_1 = y$ given $Y_1 + Y_2 = t$ for independent binomial random variables Y_1 and Y_2
F_{ν_1,ν_2}	F distribution with ν_1 and ν_2 degrees of freedom

F_{α,ν_1,ν_2}	upper α percentile of the F distribution with ν_1 and ν_2 degrees of freedom
$\mathcal{G}$	class of graphical loglinear models
$G^2(\mathcal{M})$	LRT of the LLM $\mathcal{M}$ versus the (unrestricted) global alternative
$G^2(\mathcal{M}' \mid \mathcal{M})$	LRT of the nested submodel $\mathcal{M}'$ of the LLM $\mathcal{M}$ versus the alternative $\mathcal{M} \setminus \mathcal{M}'$
$\Gamma(\alpha, \beta)$	gamma distribution with parameters $\alpha > 0$ and $\beta > 0$
$\Gamma(w)$	gamma function evaluated at $w > 0$; $= \int_0^\infty e^{-u} u^{x-1} du$
HLLM	hierarchical loglinear model
$\mathcal{H}$	class of hierarchical loglinear models
$i = 1(1)n$	$i \in \{1, 2, \ldots, n\}$
IPF	Iterative Proportional Fitting (algorithm)
IRLS	iteratively reweighted least squares
$I[A]$	indicator function of event A
I^λ	Cressie–Read goodness-of-fit statistic
$\mathbf{I}(\boldsymbol{\theta})$	Fisher information matrix evaluated at $\boldsymbol{\theta}$
$\mathbf{I}_n$	$n \times n$ identity matrix
$\mathbf{I}$	identity matrix of appropriate dimension
$\mathcal{I}$	index set $\{1, \ldots, n\}$
k-NSEL	k-normalized squared error loss
$\boldsymbol{\ell}$	vector of natural logarithms of the mean of $E[Y_i]$, $i \in \mathcal{I}$
$L_S(\cdot, \cdot)$	squared error loss
LLM	loglinear model
$\ln(w)$	logarithm (base e) of $w > 0$
$\text{logit}(w)$	$= \ln\left(\frac{w}{1-w}\right)$, $0 < w < 1$
LRT	likelihood ratio test
LS	large sparse (asymptotics)
$(\tilde{\mathcal{M}}, M)$	loglinear model with $\boldsymbol{\ell} \in \tilde{\mathcal{M}}$ and multinomial sampling
$(\mathcal{M}, P)$	loglinear model with $\boldsymbol{\ell} \in \mathcal{M}$ and Poisson sampling
$(\tilde{\mathcal{M}}, PM)$	loglinear model with $\boldsymbol{\ell} \in \tilde{\mathcal{M}}$ and product multinomial sampling
mgf	moment generating function
MLE	maximum likelihood estimator(s)
MPLE	maximum penalized likelihood estimator(s)
MSE	mean squared error
M_k	HLLM containing *all* k-factor interactions
$M_t(n, \mathbf{p})$	t-cell multinomial distribution with n trials and vector of cell probabilities $\mathbf{p} = (p_1, \ldots, p_t)'$
$m \bmod n$	remainder when m is divided by n for integers m and n
$NB(\alpha, \beta)$	negative binomial distribution with parameters $\alpha > 0$ and $\beta > 0$
$N(\mu, \sigma^2)$	normal distribution with mean μ and variance σ^2
$\phi(w)$	standard normal density; $= \frac{1}{\sqrt{2\pi}} \exp(-w^2/2)$
$\Phi(x)$	standard normal distribution evaluated at x; $= \int_{-\infty}^{x} \phi(w) dw$
$P[A]$	probability of event A

$P[A \mid B]$	conditional probability of event A given event B
$P(\lambda)$	Poisson distribution with rate $\lambda > 0$
PB	pseudo Bayes
pdf	probability density function
$\text{rank}(\mathbf{X})$	rank of matrix $\mathbf{X}$
RSEL	relative squared error loss
$\mathcal{S}$	$(t-1)$-dimensional simplex; $= \{\mathbf{x} \in \mathbb{R}^t : x_i \geq 0 \text{ for } 1 \leq i \leq t \text{ and } \sum_{i=1}^t x_i = 1\}$
SEL	squared error loss
$\sin(w)$	the sine of w
$Tr(A)$	trace of the square matrix $\mathbf{A} = (a_{ij})$; $= \sum_i a_{ii}$
UMAU	uniformly most accurate unbiased (confidence intervals)
UMP	uniformly most powerful
UMPU	uniformly most powerful unbiased
UMVUE	uniform minimum variance unbiased estimator
$\text{Var}(X)$	variance of X
WLS	weighted least squares
$X \perp Y$	X is independent of Y
$X \perp Y \mid Z$	X and Y are conditionally independent given Z
$X \sim Y$	X and Y have the same distribution
$X \overset{app}{\sim} Y$	X has approximately the same distribution as Y
χ^2_n	chi squared distribution with n degrees of freedom
$\chi^2_{\alpha,n}$	upper α percentile of the χ^2_n distribution
$\chi^2_n(\delta)$	non-central chi-square distribution with n degrees of freedom and noncentrality parameter δ
$\chi^2_{\alpha,n}(\delta)$	upper α percentile of the $\chi^2_n(\delta)$ distribution
$X^2(\mathcal{M})$	Pearson's chi-squared test of fit of the LLM $\mathcal{M}$ versus the global alternative
$\hat{Y}$	MLE of Y
z_α	upper α percentile of the standard normal distribution
zpa	zero partial association
$\nabla \mathbf{f}(\mathbf{x})$	matrix of first partials of $\mathbf{f}(\mathbf{x})$
$\nabla^2 \mathbf{f}(\mathbf{x})$	(Hessian) matrix of second partials of $\mathbf{f}(\mathbf{x})$
$\mathbf{0}_n$	vector of zeroes of length n
$\mathbf{0}$	vector of zeroes of appropriate length
$\mathbf{0}_{n,m}$	$n \times m$ matrix of zeroes
$\mathbf{1}_n$	vector of n ones
$\mathbf{1}$	vector of ones of appropriate length
$\mathbf{1}_{n,m}$	$n \times m$ matrix of ones
$\overline{x}$	$\frac{1}{n} \sum_{i=1}^n x_i$
$\lVert\mathbf{x}\rVert$	norm of $\mathbf{x}$; $= (\sum_i x_i^2)^{1/2}$
$x := y$	x defined by y
$x \overset{\text{set}}{=} y$	the quantity of interest is defined to be that value which solves the equation

x^+	$\max\{0, x\}$
$\mathbf{x} > \mathbf{0}$	$x_i > 0$ for all i
$\lvert x \rvert$	$\max\{x, -x\}$
$\times_{i=1}^{n} A_i$	Cartesian product of the sets $A_1, A_2, \ldots, A_n$
$\emptyset$	emptyset
$\sim A$	complement of A
$A \setminus B$	set difference; $= A \cap (\sim B)$
$\xrightarrow{\mathcal{L}}$	convergence in law
$\xrightarrow{p}$	convergence in probability

Index to Data Sets

Accidents in the Work Force. Greenwood and Yule (1920).
Number of accidents over a 3-month period for 414 workers.
Table 1.3.1: 20.
Example 1.3.1: 19.

Acute Lymphatic Leukemia Incidence. Lee (1963).
Numbers of cases of acute lymphatic leukemia recorded by the British Cancer Registry during 1946–1960.
Table 2.3.3: 95.
Example 2.3.3: 94, 97.

Acute Myeloblastic Leukemia Remissions. Lee (1974).
Remission status of 27 acute myeloblastic leukemia cases and two explanatory variables.
Table 5.3.1: 231.
Example 5.3.1: 230, 243–246, 251, 256–260.

Alcohol Consumption and Esophogeal Cancer. Tuyns et al. (1977).
A retrospective 2×2 table.
Table 1.2.6: 8.
Example 1.2.5: 7–8.
Problem 5.10: 273.

Avadex and Tumor Development. Innes et al. (1969).
A 2^4 table with one binary response variable and 3 binary explanatory variables.
Table 1.2.7: 8.
Table 5.5.2: 264.
Example 1.2.6: 8, 208, 263–264, 266, 269.

Breast Cancer Treatment. Segaloff (1961).
Single sample binomial response data.
Example 2.1.1: 26, 27–28, 31–32.

Cell Differentiation. Piegorsch, Weinberg, and Margolin (1988).
Number of cells undergoing differentiation after exposure to one of two agents.
Table 3.P.5: 140.
Problem 3.5: 140–141.

Cisplatinum Use and Severity of Nausea. Farewell (1982).
A 6 × 2 table with a 6 level ordinal response and a binary explanatory variable.
Table 1.2.8: 9.
Example 1.2.7: 8–9.
Problem 2.16: 104.
Problem 5.4: 270–271.

Civil Service Examinations to Become a Police Officer. Raab (1980).
Classification by race of individuals passing a civil service examination to become a police officer.
Table 5.P.15: 275.
Problem 5.15: 275.

Cold Hardiness of Winter Wheat. Keeler (1988).
Binary regression data in which the response is the survival of two varieties of wheat under various environmental conditions.
Table 5.P.30: 284.
Problem 5.30: 283.

Convictions of Same Sex Twins of Criminals. Fisher (1935).
Nunber of convictions of same sex monozygotic and dizygotic twins of criminals.
Table 5.2.6: 223.
Example 5.2.4: 223, 225.

Educational Aspirations of Wisconsin High School Students. Sewell and Shah (1968).
A 4 × 4 × 2 × 2 × 2 table of survey results of education plans and potential explanatory variables of Wisconsin high school seniors.
Table 4.P.8: 199.
Problem 4.8: 197–198.

Equipment Breakdowns. Jorgenson (1961).
Number of breakdowns of a complex piece of electronic equipment over a 9-week period.
Table 3.P.4: 140.
Problem 3.4: 136.

Familial Political Affiliations. Knoke (1976).
Political affiliations of 6026 individuals and their fathers.
Table 4.P.10(a): 201.
Problem 4.10: 201.

Federal District Court Opinions. Carp and Rowland (1983).
Numbers of liberal decisions by Federal District Courts 1969–1977.
Table 5.2.7: 227.
Example 5.2.5: 227–228.

Author Index

Subject Index

V

Springer Texts in Statistics

(continued)

Peters	Counting for Something: Statistical Principles and Personalities
Santner and Duffy	The Statistical Analysis of Discrete Data